Lecture Notes in Computer Science 8546

Commenced Publication in 1973
Founding and Former Series Editors:
Gerhard Goos, Juris Hartmanis, and Jan van Leeuwen

T0212831

Qianping Gu   Pavol Hell   Boting Yang (Eds.)

Algorithmic Aspects in Information and Management

10th International Conference, AAIM 2014
Vancouver, BC, Canada, July 8-11, 2014
Proceedings

 Springer

Volume Editors

Qianping Gu
Simon Fraser University, School of Computing Science
Burnaby, BC, V5A 1S6, Canada
E-mail: qgu@cs.sfu.ca

Pavol Hell
Simon Fraser University, School of Computing Science
Burnaby, BC, V5A 1S6, Canada
E-mail: pavol@cs.sfu.ca

Boting Yang
University of Regina, Department of Computer Science
Regina, SK, S4S 0A2, Canada
E-mail: boting@cs.uregina.ca

ISSN 0302-9743   e-ISSN 1611-3349
ISBN 978-3-319-07955-4   e-ISBN 978-3-319-07956-1
DOI 10.1007/978-3-319-07956-1
Springer Cham Heidelberg New York Dordrecht London

Library of Congress Control Number: 2014940379

LNCS Sublibrary: SL 1 – Theoretical Computer Science and General Issues

Typesetting: Camera-ready by author, data conversion by Scientific Publishing Services, Chennai, India

Printed on acid-free paper

Springer is part of Springer Science+Business Media (www.springer.com)

Preface

The papers in this volume were presented at the 10th International Conference on Algorithmic Aspects of Information and Management (AAIM 2014), held during July 8–11, 2014, at Harbour Centre, Simon Fraser University, Vancouver, Canada. It was the first time for the AAIM conference series to be held in Canada. The topics covered most areas in discrete algorithms and their applications.

Submissions to the conference were handled electronically. A total of 45 papers were submitted, of which 30 were accepted. The papers were evaluated by an international Program Committee overseen by the Program Committee co-chairs: Qianping Gu, Pavol Hell, and Boting Yang. The Program Committee consists of Hee-Kap Ahn, Binay Bhattacharya, Anthony Bonato, Zhi-zhong Chen, Leizhen Cai, Francis Chin, Chuangyin Dang, Xiaotie Deng, Ding-Zhu Du, Michael Fellows, Bin Fu, Gena Hahn, Kazuo Iwama, David Kirkpatrick, Guohui Lin, Tian Liu, Tom McCormick, Daniel Paulusma, Lorna Stewart, Xuehou Tan, Dimitrios Thilikos, Takeshi Tokuyama, Lusheng Wang, Peter Widmayer, Jinhui Xu, Yinfeng Xu, Guochuan Zhang, Kaizhong Zhang, Xiao Zhou, and Binhai Zhu. It is expected that most of the accepted papers will appear in a more complete form in scientific journals.

The submitted papers were from 16 countries/regions: Brazil, Canada, China, France, Germany, Hong Kong, India, Japan, Korea, Mexico, The Netherlands, Switzerland, Taiwan, Tunisia, UK, and USA. Each paper was evaluated by at least three Program Committee members, assisted in some cases by subreviewers. In addition to the 30 selected papers, the conference also included two invited talks, one by Ming Li on "Approximating Semantics," and the other by Christos H. Papadimitriou on "Computational Insights and the Theory of Evolution."

We thank everyone who made the meeting a success, the invited speakers, the authors, the Program Committee members and external reviewers (listed in the proceedings). Finally, we thank Simon Fraser University for their support and the local organizers and colleagues for their assistance.

April 2014

Qianping Gu
Pavol Hell
Boting Yang

Organization

Program Committee Co-chairs

Qianping Gu — Simon Fraser University, Canada
Pavol Hell — Simon Fraser University, Canada
Boting Yang — University of Regina, Canada

Program Committee

Hee-Kap Ahn	Pohang University of Science and Technology, Korea
Binay Bhattacharya	Simon Fraser University, Canada
Anthony Bonato	Ryerson University, Canada
Zhi-zhong Chen	Tokyo Denki University, Japan
Leizhen Cai	Chinese University of Hong Kong, SAR China
Francis Chin	Hong Kong University, SAR China
Chuangyin Dang	City University of Hong Kong, SAR China
Xiaotie Deng	Shanghai Jiaotong University, China
Ding-Zhu Du	University of Texas at Dallas, USA
Michael Fellows	Charles Darwin University, Australia
Bin Fu	University of Texas-Pan American, USA
Gena Hahn	University of Montreal, Canada
Kazuo Iwama	Kyoto University, Japan
David Kirkpatrick	University of British Columbia, Canada
Guohui Lin	University of Alberta, Canada
Tian Liu	Peking University, China
Tom McCormick	University of British Columbia, Canada
Daniel Paulusma	Durham University, UK
Lorna Stewart	University of Alberta, Canada
Xuehou Tan	Tokai University, Japan
Dimitrios Thilikos	National University of Athens, Greece
Takeshi Tokuyama	Tohoku University, Japan
Lusheng Wang	City University of Hong Kong, SAR China
Peter Widmayer	ETH, Switzerland
Jinhui Xu	University at Buffalo, the State University of New York, USA
Yinfeng Xu	Xi'an Jiao Tong University, China
Guochuan Zhang	Zhejiang University, China
Kaizhong Zhang	University of Western Ontario, Canada
Xiao Zhou	Tohoku University, Japan
Binhai Zhu	Montana State University, USA

Organizing Committee

Qianping Gu   Simon Fraser University, Canada
Boting Yang   University of Regina, Canada

External Reviewers

Danyang Chen   Spyridon Maniatis
Lin Chen   Julian Mestre
Hu Ding   Andrzej Pelc
Pengbo Feng   Akiyoshi Shioura
Archontia Giannopoulou   Somnath Sikdar
Chengwei Guo   Weiping Sun
Tsunehiko Kameda   Akira Suzuki
Sandi Klavzar   Weitian Tong
Stavros Kolliopoulos   Kei Uchizawa
Wei Li   Xiangyu Wang
Zhewei Liang   Jinshan Zhang
Yi Liu   Yong Zhang

Invited Talks

Computational Insights
and the Theory of Evolution

Chrisos H. Papadimitrious

Computer Science Division
University of California at Berkeley
Berkeley, CA 94720, USA
chrisos@cs.berkeley.edu

Covertly computational ideas have influenced the Theory of Evolution from the very start. This talk is about recent work on Evolution that was inspired and informed by computational insights. Considerations about the performance of genetic algorithms led to a novel theory of the role of sex in Evolution based on the concept of mixability, while the equations describing the evolution of a species can be reinterpreted as a repeated game between genes played through the multiplicative updates algorithm. Finally, a theorem on Boolean functions helps us understand better Waddington's genetic assimilation as well as mechanisms for the emergence of novelty in Life.

Approximating Semantics

Ming Li

David R. Cheriton School of Computer Science
University of Waterloo
Waterloo, ON N2L3G1, Canada
mli@uwaterloo.ca

Latent search engines and question-answering (QA) engines fundamentally depend on our intuitive notion of semantics and semantic distance. However, such a semantic distance is likely undefinable, certainly un-computable, and often blindly approximated. Can we develop a theoretical framework for this area?

I will describe a theory, using the well-defined information distance, to approximate the elusive semantic distance such that it is mathematically proven that our approximation is "better than" any computable approximation of the intuitive concept of semantic distance. Although information distance itself is obviously also not computable, it does allow a natural approximation by compression. We will then describe a natural language encoding system to implement our theory followed by experiments on a QA system.

Table of Contents

Contributed Papers

Local Event Boundary Detection with Unreliable Sensors: Analysis of the Majority Vote Scheme[*]

Peter Brass[1], Hyeon-Suk Na[2], and Chan-Su Shin[3]

[1] Dept. of Computer Science, City College, New York, USA
`peter@cs.ccny.cuny.edu`
[2] School of Computing, Soongsil University, Seoul, Korea
`hsnaa@ssu.ac.kr`
[3] Dept. of Digital Information Engineering, Hankuk University of Foreign Studies,
Yongin, Korea
`cssin@hufs.ac.kr`

Abstract. In this paper we study the identification of an event region X within a larger region Y, in which the sensors are distributed by a Poisson process of density λ to detect this event region, i.e., its boundary. The model of sensor is a 0-1 sensor that decides whether it lies in X or not, and which might be incorrect with probability p. It also collects information on the 0-1 values of the neighbors within some distance r and revises its decision by the majority vote of these neighbors. In the most general setting, we analyze this simple majority vote scheme and derive some upper and lower bounds on the expected number of misclassified sensors. These bounds depend on several sensing parameters of p, r, and some geometric parameters of the event region X. By making some assumptions on the shape of X, we prove a significantly improved upper bound on the expected number of misclassified sensors; especially for convex regions with sufficiently round boundary.

1 Introduction

Suppose we have distributed many sensors in a region, each of which detects if it rains at that point. We want to obtain a summary: in which sub-region is it raining? Just listing all the positions at which a raindrop has been detected is not a helpful answer, first because a long list of positions is not the answer a user would want on the question "Where does it rain?," but also because each individual answer is subject to random errors such as measurement errors.

The abstract model underlying this question is as follows: we have a region Y, in which there is a set S of sensors. There is an unknown region $X \subseteq Y$ in which the event happens. We want to detect the event region X, most importantly, its boundary, where it is far from the boundary of Y and has a 'nice' topology, not being highly irregular, random or fractal. The sensors $s \in S$ are 0-1 sensors who decide whether $s \in X$ or $s \notin X$, making an error with probability p in this measurement, called the measurement error.

---

[*] Work by C.-S. Shin was supported by Hankuk University of Foreign Studies Research Fund.

1.1   Simple Majority Voting Scheme

Our aim is to reduce the error rate in detecting the boundary of X by allowing each sensor to compare its initial measurement result with those of its neighbors. To reduce the error rate by local communication, we assume that each sensor knows the values measured by all neighboring sensors within distance r. The most straightforward method to use the neighbors' sensing information is to follow the majority.

This majority vote scheme is as follows: if a sensor has k neighbors and knows its own and those k other measurements, then in its revised decision it just follows the majority of the measurements of its k neighbors, with itself as tie-breaking if necessary. This scheme was already proposed by Chintalapudi and Govindan [6] and further in [12]. It does not use the position information of the neighboring sensors. It was stated in [12] that this scheme gives a good correction of measurement errors for sensor error p up to 0.2. However, we think this observation needs further qualification, since the situation really depends on several sensing parameters such as the measurement error p and the neighborhood radius r, and some geometric parameters of the event region X such as the convexity, the perimeter and the boundary curvature.

In this paper, we analyze the majority scheme further, and make explicit and precise the dependency on these parameters. To analyze this, we first need an assumption on the distribution of sensors in the region of interest Y. We adapt the most important model; the sensors are randomly distributed by a Poisson process of density λ, which is independent from the measurement error p that the sensors make. To our best knowledge, this gives the first bounds on the expected number of incorrectly classified sensors in the majority vote scheme with all these parameters, λ, p, r, and the shape of X.

It should be noticed that our sensors are point sensors; there is no sensing range, but a yes/no decision about the situation at the sensor. Many other papers have dealt with continuous-valued sensors, but then the dependence of a sensor's decision on the neighboring values is much less clear. Also, for many practical applications a yes/no decision is ultimately the desired answer: 'does it rain?', 'is there a forest fire?', etc. The distance r within which we compare the sensor values is not related to the communication distance of the network nodes; it is a choice made depending on our a-priori knowledge on the size and shape of X, that is, choosing the right radius r is an important aspect. Finally, our majority vote scheme does not need absolute positions of the sensors, but it is not efficient in detecting thin and long event regions; to identify such event regions we would need sensors with known positions with the help of GPS units and by more complicated decision algorithms.

1.2   Previous Work

Local event boundary detection problem has been studied in several previous papers.

Chintalapudi and Govindan [6] were the first to analyze the local event boundary detection problem. They proposed three different types of algorithms; among them a simple neighborhood counting scheme that does not use the position information of the neighboring sensors, and a scheme that finds the optimum line separating in the neighborhood the event-sensors from the no-event-sensors. They found by simulation that the separating-line scheme performs best, but provided no analysis of that scheme, and assumed for the other schemes that the event boundary is a straight line.

Krishnamachari and Iyengar [12] discussed a model with a similar counting scheme, not using the neighbor's position information; in their simulation, however, the sensors are always distributed in a square grid. Wu et al. [22] discussed continuous-valued sensors, looking for threshold events, and proposed several methods based on comparing a sensor's value with the median of a set of neighbor's values to identify faulty or boundary sensors. Similar was the discussion by Jin and Nittel [9], who used the mean instead of the median. Ding and Cheng [7] fit a mixture of multivariate gaussian distributions to the observed sensor values and decided on the base of that fitted model which sensors are boundary sensors.

Wang et al. [21] considered a model that can be interpreted as only no-event sensors being available, e.g., because the event like the fire in the forest destroyed all sensors in its region; they reconstruct a boundary of the regione containing sensors, based on neighbor connectivity information without the neighbor positions. Nowak and Mitra [19] use a non-local communication model based on a hierarchical partitioning scheme to identify event boundaries.

A different line of related works contains the fusion of different information sources for the same event; e.g., combining the output of multiple classifiers in a pattern-recognition problem. This has been studied in [14,10,2,5,13,18], but that problem abstracts from the geometric structure which is the core of our considerations. Yet another related line of works contains the opinion formation in social networks [1,16,20,11,17,23].

1.3   Our Results

For a set A in the plane, we denote its area, perimeter, boundary and number of components, by $\text{area}(A), \text{peri}(A), \text{bd}(A)$ and $\text{components}(A)$, respectively.

We assume that the set S of sensors is generated by a Poisson process of density λ on our region of interest Y. Within Y, an event happens in the region X. Each sensor $s \in S$ makes a 0-1 event detection (or measurement), whether $s \notin X$ or $s \in X$, which might be incorrect with probability p. The sensor errors are independent from each other, and from the Poisson process placing the sensors. Each sensor knows the measurement results of all other sensors within radius r and revises his own measurement based on that information.

The comparison with neighboring sensors gives information only if the sensor has neighboring sensors. The expected number of neighbors in this model is $\lambda \pi r^2$ and thus the probability that a sensor has no neighbors is $e^{-\lambda \pi r^2}$, which should be much smaller than the measurement error p. The expected number of sensors in Y is $\lambda \, \text{area}(Y)$, so without correction by neighborhood comparison, the expected number of incorrect sensors, i.e., misclassified sensors, is $\lambda \, \text{area}(Y)p$. Theorem 1 and Theorem 2 show that the expected number of misclassified sensors is improved significantly by the majority vote scheme.

Let Z_r be the set of points within distance r to the boundary of X. Any sensor in this dubious region Z_r has potentially neighbors inside and outside X, in other words, we possibly have both of correct 0- and 1-answers within the same neighborhood, which would lead such sensors to make the wrong decision after the majority vote. Thus the analysis on the expected number of misclassified sensors in Z_r is a key in the majority vote scheme.

In Section 2, we analyze the majority vote scheme in a most general setting and derive the following bounds on the expected number of misclassified sensors in $Y \setminus Z_r$ and on the expected number of misclassified sensors in Z_r.

Theorem 1. *For $p \leq 1/2$, the expected number of sensors in $Y \setminus Z_r$ that are misclassified by the simple majority rule in the neighborhood of radius r is at most*

$$2\lambda\sqrt{p(1-p)}e^{-(1-2\sqrt{p(1-p)})\lambda\pi r^2} \, \text{area}(Y \setminus Z_r)$$

and at least

$$\frac{\sqrt{p(1-p)}}{4\pi r^2}\left(e^{-(1-2\sqrt{p(1-p)})\lambda\pi r^2} - e^{-\lambda\pi r^2}\right) \text{area}(Y \setminus Z_r).$$

Theorem 2. *For $p \leq 1/2$, the expected number of sensors in Z_r that are misclassified by the simple majority rule in the neighborhood of radius r is at most*

$$2\lambda r \, \text{peri}(X) + \lambda\pi r^2 \, \text{components}(X).$$

There exists some event region X such that the expected number of misclassified sensors in Z_r is at least $\Omega(\lambda r \, \text{peri}(X))$.

The ratio of the upper bound to the lower bound in Theorem 1 grows linearly with the expected number of neighbors $\lambda\pi r^2$. However, since $2\sqrt{p(1-p)} \leq 1$ for any $p \geq 0$, they both decrease exponentially with the expected number of neighbors $\lambda\pi r^2$, so the expected number of misclassified sensors outside Z_r decreases exponentially with the expected number of neighbors $\lambda\pi r^2$. Theorem 2 tells us that the expected number of misclassified sensors in Z_r grows with the parameters λ and r, and the perimeter of X. Thus a region X with long boundary or with many components would be the worst in the majority vote scheme. Moreover such worst examples exist. As a result, Theorems 1 and 2 illustrate the trade-off between the error outside Z_r, which decreases exponentially with λ and r, and the error inside Z_r, which increases with λ and r.

Sensors very near to the boundary of X can be unavoidably misclassified according to Theorem 2. If X is thin so that $X \subset Z_r$, then there are no sensors sufficiently deep inside X whose neighbors are mainly inside X, so the region will not be recognized by the majority vote scheme. We thus need to make some (seemingly strong) assumptions on the shape of X such that $X \not\subset Z_r$ is guaranteed. In this paper, we consider X as a convex event region with a bounded curvature, i.e., with sufficiently rounded boundary. For such X, in Section 3, we prove a significantly improved upper bound on the expected number of misclassified sensors in Z_r, which is a main result in this paper.

Theorem 3. *Let $p \leq 1/2$. If the event region X is convex and the radius of curvature at each point on the boundary is at least r, then the expected number of sensors in Z_r that are misclassified by the simple majority rule in the neighborhood of radius r is less than*

$$\frac{\pi\sqrt{\lambda}}{\sqrt{2}(1-2p)} \operatorname{peri}(X) + 3\lambda\pi r^2 \ln \frac{\operatorname{peri}(X)}{r}.$$

Finally we perform some simulation for convex and round event regions and check the effect of the various parameters in the majority vote scheme such as p, r, λ, and the perimeter of X, and present a refinement method to improve the performance particularly for the tricky cases, i.e., for small r and large p.

The detail of the simulation and the proof of some lemmas can be found in the full version [4].

2   Analysis for General Event Regions

In this section we analyze the simple majority rule and prove Theorem 1 and Theorem 2. Throughout the paper, we will use the following lemma.

Lemma 1. *For $p \leq 1/2$, the probability $B(n)$ of at least $\lceil \frac{n}{2} \rceil$ successes among n independent Bernoulli trials of success probability p is*

$$\frac{\sqrt{p(1-p)}}{2n} \left(2\sqrt{p(1-p)}\right)^n \leq B(n) \leq \left(2\sqrt{p(1-p)}\right)^n.$$

2.1   Proof of Theorem 1

Recall that Z_r is the set of points of Y within distance r to $\operatorname{bd}(X)$, the boundary of X. The expected number of the misclassified sensors in $Y \setminus Z_r$ by the majority vote is $\lambda \operatorname{area}(Y \setminus Z_r)$ times the probability of a sensor s in $Y \setminus Z_r$ being misclassified by the majority rule in the neighborhood of radius r.

Suppose that s has k neighbors. Since $s \in Y \setminus Z_r$, its k neighbors lie all inside X or all outside X. The probability of $s \in Y \setminus Z_r$ being misclassified is the one that at least half of measurements of the neighbors should be erroneous. When k is odd, at least $\lceil \frac{k}{2} \rceil$ errors among k measurements must happen. But when k is even, the measurement of s can be served as a tie breaker, thus at least $\lceil \frac{k+1}{2} \rceil$ errors must happen among $k+1$ measurements including a measurement of s.

Let $B(k)$ be the probability that at least $\lceil \frac{k}{2} \rceil$ successes among k trials with success probability $p \leq 1/2$. For odd k, it holds from binomial distribution that $B(k) = \frac{1}{2(1-p)}B(k+1) \leq B(k+1)$ because $\frac{1}{2(1-p)} \leq 1$ for $p \leq 1/2$. The probability of $s \in Y \setminus Z_r$ being misclassified is simplified as follows:

$$\sum_{k=0}^{\infty} \Pr(s \text{ has } k \text{ neighbors}) \Pr(s \text{ makes a wrong decision by majority rule})$$

$$= \sum_{\text{odd } k} \Pr(s \text{ has } k \text{ neighbors})B(k) + \sum_{\text{even } k} \Pr(s \text{ has } k \text{ neighbors})B(k+1)$$

$$= \sum_{\text{odd } k} \Pr(s \text{ has } k \text{ neighbors})\frac{1}{2(1-p)}B(k+1) + \sum_{\text{even } k} \Pr(s \text{ has } k \text{ neighbors})B(k+1)$$

$$\leq \sum_{k=0}^{\infty} \Pr(s \text{ has } k \text{ neighbors})B(k+1).$$

The first probability is $\frac{1}{k!}(\lambda \pi r^2)^k e^{-\lambda \pi r^2}$ by the definition of the Poisson process. The second probability $B(k+1)$ is at most $\left(2\sqrt{p(1-p)}\right)^{k+1}$ by Lemma 1. Thus we get the upper bound of the probability of $s \in Y \setminus Z_r$ being misclassified as follows:

$$\Pr(s \in Y \setminus Z_r \text{ is misclassified}) \leq \sum_{k=0}^{\infty} \frac{1}{k!}(\lambda \pi r^2)^k e^{-\lambda \pi r^2} \left(2\sqrt{p(1-p)}\right)^{k+1}$$

$$\leq 2\sqrt{p(1-p)}e^{-\lambda \pi r^2} \sum_{k=0}^{\infty} \frac{1}{k!}\left(\lambda \pi r^2 \cdot 2\sqrt{p(1-p)}\right)^k$$

$$\leq 2\sqrt{p(1-p)}e^{-(1-2\sqrt{p(1-p)})\lambda \pi r^2}.$$

Multiplying this with $\lambda \operatorname{area}(Y \setminus Z_r)$ gives the upper bound of the theorem.

For the lower bound, we can prove it similarly; refer to [4].

2.2 Proof of Theorem 2

For the upper bound of the theorem, we simply assume that any sensor in Z_r always makes the wrong decision. The expected number of sensors in Z_r is $\lambda \operatorname{area}(Z_r)$, and we have the geometric bound $\operatorname{area}(Z_r) \leq 2r \operatorname{peri}(X) + \pi r^2 \cdot \operatorname{components}(X)$ for any general set X. Thus the expected number of misclassified sensors in Z_r by the majority vote rule is at most $\lambda \left(2r \operatorname{peri}(X) + \pi r^2 \cdot \operatorname{components}(X)\right)$.

We now explain that this bound is asymptotically the best we can obtain for the expected number of misclassified sensors in Z_r for any X with $\operatorname{components}(X) = O(\operatorname{peri}(X)/r)$. Indeed, if X is a thin and long rectangle of height $r/2$ and of width $4r$ as shown in Figure 1(a), then all sensors in X will be in Z_r, i.e., $X \subset Z_r$. The perimeter of X is $9r$. Consider any sensor s in X at distance at least r from the both vertical edges of X. Let A be a disk of radius r around s. Since the height of X is $r/2$, A consists of three parts as in Figure 1(a); two circle segments of $A \setminus X$ and the middle part, $A \cap X$, between the circle segments. The expected numbers of sensors in $A \setminus X$ and $A \cap X$ whose initial measurement is "not in X" are $(1-p)\lambda \, \text{area}(A \setminus X)$ and $p\lambda \, \text{area}(A \cap X)$, respectively. Thus s has at least $(1-p)\lambda \, \text{area}(A \setminus X) + p\lambda \, \text{area}(A \cap X)$ neighbors in A whose initial measurement is "not in X". We can prove that this is at least half of the number of sensors in A, i.e., $\geq \frac{1}{2}\lambda\pi r^2$ for any $p \leq \frac{1}{2}$ by simple calculation. This results in making a wrong decision of s by the majority vote scheme. The expected number of such misclassified sensors in X is $\lambda(2r \times \frac{r}{2}) = \lambda r^2$, which is at least $\frac{1}{9}\lambda r(9r) = \frac{1}{9}\lambda r \, \text{peri}(X) = \Omega(\lambda r \, \text{peri}(X))$.

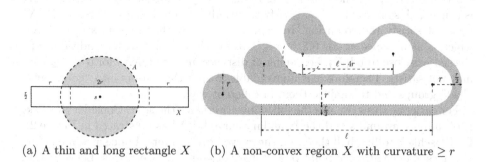

(a) A thin and long rectangle X (b) A non-convex region X with curvature $\geq r$

Fig. 1. Two lower bound examples

We can also find such a worst example even when X is not convex but has a round boundary satisfying some curvature constraint: the radius of curvature is at least r everywhere on the boundary. Figure 1(b) illustrates an event region X of the curvature radius r, but not convex. All sensors in roughly $\lfloor \ell/4r \rfloor$ thin rectangular strips have a majority of neighbors outside X, thus they will make wrong decisions. Assume that ℓ is a multiple of $4r$, so $\ell \geq 4r$. Total area of thin strips is at least $(r/2)(\ell + (\ell - 4r) + \ldots + 4r + 0) \geq \ell^2/16$. The boundary of X consists of circular arcs at its both ends and linear segments of thin strips. Since the length of any circular arc is at most $3\pi r$, the total length of circular arcs is at most $6\pi r\ell/(4r)$. Then $\text{peri}(X) \leq 6\pi r\ell/(4r) + 2(\ell + (\ell - 4r) + \ldots + 4r) \leq (\ell^2/r)((3\pi/2 + 1)(r/\ell) + 1/4) \leq 1.68\ell^2/r$ since $r/\ell \leq 1/4$. Thus the expected number of misclassified sensors in Z_r of this non-convex region X with bounded curvature r is at least $\lambda \, \text{area}(\text{thin strips}) \geq \frac{1}{16}\lambda\ell^2 \geq \frac{1}{16}\lambda\left(\frac{r}{1.68}\text{peri}(X)\right) \geq \frac{1}{27}\lambda r \, \text{peri}(X) = \Omega(\lambda r \, \text{peri}(X))$. We now complete the proof of Theorem 2.

3   Analysis for Convex Event Regions with Round Boundary

We now prove our main result, Theorem 3 that if X is a convex region with a round boundary of the curvature radius r, then the expected number of misclassified sensors in Z_r significantly decreases. As a result, the convexity and the curvature constraint both are important for a better bound.

Let s be a sensor in Z_r whose nearest point to $bd(X)$ is s' in distance δr for some constant $\delta > 0$. Let A be the disc of radius r around s. Let α be the fraction of A on the same side of $bd(X)$ as s, i.e., $\alpha := \text{area}(A \cap X)/\text{area}(A)$ if $s \in X$, $\alpha := \text{area}(A \setminus X)/\text{area}(A)$ if $s \notin X$. Then we can prove the following:

Lemma 2. *If $p \leq 1/2$ and $\alpha \geq 1/2$, then the probability of s being misclassified is at most*

$$e^{-\frac{1}{2}\lambda \, \text{area}(A)(1-2p)^2(2\alpha-1)^2}.$$

This lemma provides us a good bound on the expected number of misclassified sensors, but only for those satisfying its necessary condition, $\alpha \geq 1/2$. Since X is convex, all sensors in $Z_r \setminus X$ satisfy the condition, but some sensors in $Z_r \cap X$ may not satisfy the condition. Indeed, we can show that sensors $s \in Z_r \cap X$, which is at distance δr for $0.2 < \delta \leq 1$ from $bd(X)$, satisfy the condition, but the sensors close to $bd(X)$, i.e., being at distance δr from $bd(X)$ for $0 < \delta \leq 0.2$, may not satisfy. Therefore it is unavoidable to split the sensors into "good" and "bad" groups and to analyze them in different ways.

For easier analysis, we use the following criteria for the split. Consider the disk A of radius r around s and the boundary curve $bd(X)$ as illustrated in Figure 2. By Blaschke's rolling ball theorem, the curve $bd(X)$ enters A once at some point p, leaves A once at some point p' and it never enters A afterwards. Let q be the diametrically opposite point of p in A. Then there are two possibilities when s is in Z_r: q lies in the same side of $bd(X)$ as s, or in the opposite side of $bd(X)$ to s. For the two cases when s lies in $Z_r \cap X$, see Figure 2. We call $s \in Z_r$ *good* if q lies in the side of $bd(X)$ as s, and otherwise *bad*. By the convexity of X, it is clear that all sensors in $Z_r \setminus X$ are good.

3.1   Upper Bound on the Misclassified Good Sensors

For good sensors in Z_r, we get the following bound on the probability of being misclassified.

Lemma 3. *Let s be a good sensor in Z_r whose distance to the boundary of X is δr. Then the probability of s being misclassified is at most*

$$e^{-\frac{2}{\pi^2}\lambda \, \text{area}(A)(1-2p)^2\delta^2}.$$

Proof. As in Figure 2(a), we first consider the case that $s \in Z_r \cap X$. Let A be a disk of radius r around s on the inner parallel curve at distance δr from $bd(X)$, and let s' be the closest point on $bd(X)$ from s. Using the curvature constraint and the fact that $q \in X$, the half of A bounded by line pq is contained in X. In addition, on the other side of pq, the triangle $\triangle pqt$ of height $h \geq \delta r$, where t is the intersection of the bisector of pq with $bd(X)$, is also contained in X. Thus $\text{area}(A \cap X) \geq \frac{1}{2}\text{area}(A) + \text{area}(\triangle pqt) = \text{area}(A)(\frac{1}{2} + \frac{\delta}{\pi})$. This gives us a lower bound on $\alpha(\delta)$, where the area of the portion of A lying inside X is expressed as $\alpha(\delta) \, \text{area}(A)$, that is, $\alpha(\delta) = \text{area}(A \cap X)/\text{area}(A)$. Then $\alpha(\delta) \geq \frac{1}{2} + \frac{\delta}{\pi}$. Similarly, for good sensors s lying in $Z_r \setminus X$, the portion of A lying outside X satisfies that $\alpha(\delta) \geq \frac{1}{2} + \frac{\delta}{\pi}$. Plugging the lower bound $\frac{1}{2} + \frac{\delta}{\pi}$ into α of Lemma 2, we get the result.

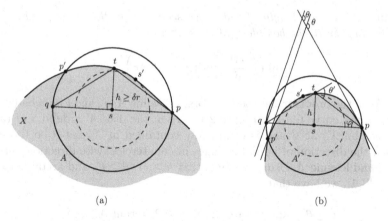

Fig. 2. Classification of sensors $s \in Z_r \cap X$. (a) s is good since $q \in X$. (b) s is bad since $q \notin X$.

We integrate this probability over all inner and outer parallel curves and get an upper bound on the expected number of misclassified good sensors in Z_r:

Expected number of misclassified good sensors in $Z_r \cap X$

+ Expected number of misclassified good sensors in $Z_r \setminus X$

$$\leq \lambda r \int_0^1 e^{-\frac{2}{\pi^2}\lambda \, \text{area}(A)(1-2p)^2\delta^2}(\text{peri}(X) - 2\pi\delta r)d\delta$$

$$+\lambda r \int_0^1 e^{-\frac{2}{\pi^2}\lambda \, \text{area}(A)(1-2p)^2\delta^2}(\text{peri}(X) + 2\pi\delta r)d\delta$$

$$= 2\lambda r \, \text{peri}(X) \int_0^1 e^{-\frac{2}{\pi^2}\lambda \, \text{area}(A)(1-2p)^2\delta^2} \, d\delta$$

$$< 2\lambda r \, \text{peri}(X)\frac{\sqrt{\pi}}{2\sqrt{\frac{2}{\pi^2}\lambda \, \text{area}(A)(1-2p)^2}} < \frac{\pi\sqrt{\lambda}}{\sqrt{2}(1-2p)}\text{peri}(X). \qquad (1)$$

For the upper bound on the integrals, we used that $\int_0^1 e^{-cx^2}dx = \frac{\sqrt{\pi}}{2\sqrt{c}}\text{erf}(\sqrt{c}) < \frac{\sqrt{\pi}}{2\sqrt{c}}$, where $\text{erf}(x)$ is an error function appeared in integrating the Gaussian function with $\text{erf}(x) < 1$ for any $x < \infty$.

3.2 Upper Bound on the Misclassified Bad Sensors

Now we derive a bound on the expected number of misclassified bad sensors in Z_r. Note that all bad sensors of Z_r appear only in $Z_r \cap X$, and see Figure 2(b) for illustration. For bad points(sensors), we do not know any upper bound but 1 on the probability of being misclassified. However, we can get an upper bound on the total length of disjoint bad curve segments, which consist of bad points only, on C_δ, an inner parallel curve at distance δr from $\text{bd}(X)$ for some $0 < \delta \leq 1$.

Lemma 4. *The total length of bad curve segments on the inner parallel curve C_δ at distance δr to the boundary of X is at most*

$$\min\left(\frac{3\pi r}{\delta}, \mathrm{peri}(X) - 2\pi\delta r\right).$$

Proof. As in the proof for the good sensors, we define A, s', and t for a bad sensor s on the inner curve C_δ. See Figure 2(b). Then the disk A' around s of radius δr touches $\mathrm{bd}(X)$ at s' and is completely contained in X. Let θ be the angle by which the direction of $\mathrm{bd}(X)$ changes in counterclockwise direction between entering and leaving A. Using the facts that $h \geq \delta r$, $\theta \geq \theta'$, and $\arctan(x) \geq \frac{\pi}{4}x$ over $x \in [0,1]$, we have that

$$\theta \geq \theta' = 2\varphi = 2\arctan(h/r) \geq 2\arctan(\delta) \geq \frac{\pi}{2}\delta.$$

Since the total direction change of $\mathrm{bd}(X)$ traversing a simple curve once around is at most 2π, we cannot have more than $\frac{4}{\delta}$ such curve segments on C_δ whose interiors are disjoint, each with a direction change of at least $\frac{\pi}{2}\delta$.

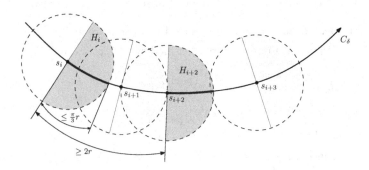

Fig. 3. Left half disks around each picked bad points

We now pick bad points on C_δ at distance of at least r along C_δ as traversing it in counterclockwise direction as follows. If the points on C_δ are all bad, then its length becomes the perimeter of C_δ, i.e., at most $\mathrm{peri}(X) - 2\pi\delta r$. Otherwise, there must be at least one good point on C_δ. We traverse C_δ from the good point in counterclockwise direction. We will meet the first bad point, then we pick this bad point and call it s_0. We next pick the bad point s_1 on C_δ at distance of at least r from s_0 along the curve. Continuing this picking process, we can pick k bad points $s_0, s_1, \ldots, s_{k-1}$ where the distance from s_{k-1} to s_0 might be less than r. As in Figure 3(a), we denote by H_i a right half of the disk of radius r around each picked bad point s_i with respect to the traversing direction. Then it is clear that the union of such right half disks covers all the bad points on C_δ because any bad point s between s_i and s_{i+1}[1] is contained in H_i.

---

[1] The addition on the indices is a modular addition with k.

Without loss of generality, we assume that k is odd. Let us now consider the intersections of C_δ with the right half disks $H_0, H_2, \ldots, H_{k-1}$ around every even picked bad points $s_0, s_2, \ldots, s_{k-1}$. These intersections $H_i \cap C_\delta$ for even i result in curve segments (or arc intervals) of C_δ whose the left endpoint is s_i. We claim that these segments except from the first and last ones are disjoint; $H_{k-1} \cap C_\delta$ can overlap with $H_0 \cap C_\delta$. As in Figure 3, we consider two consecutive intersections, $H_i \cap C_\delta$ and $H_{i+2} \cap C_\delta$ for even $i < k-1$. It suffices to show that the right endpoint of $H_i \cap C_\delta$ lies in the left of the left endpoint of $H_{i+2} \cap C_\delta$ on the curve. The arc length between s_i and s_{i+2} is at least $2r$ by picking rule, and the length of $H_i \cap C_\delta$ is at most $\frac{\pi}{3}r$ by curvature constraint. Thus the distance between the right endpoint of $H_i \cap C_\delta$ and the left endpoint of $H_{i+2} \cap C_\delta$ is at least $2r - \frac{\pi}{3}r > 0$, so the claim is proved.

The number of disjoint bad curve segments on C_δ is already proved to be no more than $\frac{4}{5}$, so the sum of their length (excluding the length of $H_{k-1} \cap C_\delta$) is at most $\frac{4}{5} \cdot \frac{\pi r}{3} = \frac{4\pi r}{3\delta}$. For the last curve segment $H_{k-1} \cap C_\delta$, we simply add its length $\frac{\pi r}{3}$, which gives the length of $\frac{(4+\delta)\pi r}{3\delta} \leq \frac{5\pi r}{3\delta}$ for $\delta \leq 1$. Considering the curve segments generated by every odd picked bad points, the sum of their length is at most $\frac{4\pi r}{3\delta}$. Note here that the first odd segment does not overlap with the last odd one. Thus the total length of bad curve segments on C_δ is at most $\frac{5\pi r}{3\delta} + \frac{4\pi r}{3\delta} \leq \frac{3\pi r}{\delta}$. Furthermore, the total length should be no more than the length of C_δ, $\mathrm{peri}(X) - 2\pi\delta r$, which completes the lemma.

Integrating this over all inner parallel curves, we get an upper bound on the expected number of all misclassified bad sensors in Z_r as follows:

Expected number of misclassified bad sensors in Z_r

$$\leq \lambda r \int_0^1 \min\left(\frac{3\pi r}{\delta}, \mathrm{peri}(X) - 2\pi\delta r\right) d\delta \leq \lambda r \int_0^1 \min\left(\frac{3\pi r}{\delta}, \mathrm{peri}(X)\right) d\delta$$

$$= \lambda r \left(\int_0^{\frac{3\pi r}{\mathrm{peri}(X)}} \mathrm{peri}(X) d\delta + \int_{\frac{3\pi r}{\mathrm{peri}(X)}}^1 \frac{3\pi r}{\delta} d\delta\right)$$

$$= 3\lambda\pi r^2 \left(1 + \ln\frac{\mathrm{peri}(X)}{3\pi r}\right) \leq 3\lambda\pi r^2 \ln\frac{\mathrm{peri}(X)}{r}. \tag{2}$$

Now we put both (1) and (2) together to obtain the upper bound on the expected number of misclassified points in Z_r, completing the proof of Theorem 3:

$$\frac{\pi\sqrt{\lambda}}{\sqrt{2}(1-2p)}\mathrm{peri}(X) + 3\lambda\pi r^2 \ln\frac{\mathrm{peri}(X)}{r}.$$

References

1. Agur, Z., Frankel, A.S., Klein, S.T.: The Number of Fixed Points of the Majority Rule. Discrete Mathematics 70, 295–302 (1988)
2. Alkoot, F., Kittler, J.: Experimental Evaluation of Expert Fusion Strategies. Pattern Recognition Letters 20, 1361–1369 (1999)

3. Blaschke, W.: Kreis und Kugel. Zweite Auflage, Walter de Gruyter AG, Berlin (1956)
4. Brass, P., Na, H.-S., Shin, C.-S.: Local Event Boundary Detection with Unreliable Sensors: Analysis of the Majority Vote Scheme. CoRR abs/1311.3149 (2013)
5. Chen, D., Cheng, X.: An Asymptotic Analysis of Some Expert Fusion Methods. Pattern Recognition Letters 22, 901–904 (2001)
6. Chintalapudi, K.K., Govindan, R.: Localized Edge Detection in Sensor Fields. Ad Hoc Networks 1, 273–291 (2003)
7. Ding, M., Cheng, X.: Robust Event Boundary Detection in Sensor Networks–A Mixture Model Approach. In: Proc. IEEE INFOCOM 2009, pp. 2991–2995 (2009)
8. Hagerup, T., Rüb, C.: A Guided Tour of Chernoff Bounds. Information Processing Letters 33(6), 305–308 (1990)
9. Jin, G., Nittel, S.: NED: An Efficient Noise-Tolerant Event and Event Boundary Detection Algorithm in Wireless Sensor Networks. In: Proc. 7th Int. IEEE Conference on Mobile Data Management, vol. 153 (2006)
10. Kittler, J., Hatef, M., Duin, R., Matas, J.: On Combining Classifiers. IEEE Transactions on Pattern Analysis and Machine Intelligence 20, 226–239 (1998)
11. Královič, R.: On Majority Voting Games in Trees. In: Pacholski, L., Ružička, P. (eds.) SOFSEM 2001. LNCS, vol. 2234, pp. 282–291. Springer, Heidelberg (2001)
12. Krishnamachari, B., Iyengar, S.: Distributed Bayesian Algorithms for Fault-Tolerant Event Region Detection in Wireless Sensor Networks. IEEE Transactions on Computers 53(3), 241–250 (2004)
13. Kuncheva, L.I.: A Theoretical Analysis of Six Classifier Fusion Strategies. IEEE Transactions on Pattern Analysis and Machine Intelligence 24, 281–286 (2002)
14. Lam, L., Suen, C.Y.: Application of Majority Voting to Pattern Recognition: An Analysis of its Behavior and Performance. IEEE Transactions on Systems, Man, and Cybernetics 27, 553–568 (1997)
15. Matousek, J., Vondrak, J.: Lecture Notes: The Probabilistic Method, pp. 1–71 (2008)
16. Mustafa, N.A., Pekeč, A.: Majority Consensus and the Local Majority Rule. In: Orejas, F., Spirakis, P.G., van Leeuwen, J. (eds.) ICALP 2001. LNCS, vol. 2076, pp. 530–542. Springer, Heidelberg (2001)
17. Mustafa, N.A., Pekeč, A.: Listen to your Neighbors: How (not) to Reach a Consensus. SIAM Journal of Discrete Mathematics 17, 634–660 (2004)
18. Narasimhamurthy, M.: A Framework for the Analysis of Majority Voting. In: Bigun, J., Gustavsson, T. (eds.) SCIA 2003. LNCS, vol. 2749, pp. 268–274. Springer, Heidelberg (2003)
19. Nowak, R.D., Mitra, U.: Boundary Estimation in Sensor Networks: Theory and Methods. In: Zhao, F., Guibas, L.J. (eds.) IPSN 2003. LNCS, vol. 2634, pp. 80–95. Springer, Heidelberg (2003)
20. Peleg, D.: Local Majorities, Coalitions and Monopolies in Graphs: A Review. Theoretical Computer Science 282, 231–257 (2002)
21. Wang, Y., Gao, J., Mitchell, J.S.B.: Boundary Recognition in Sensor Networks by Topological Methods. In: Proc. 12th International Conference on Mobile Computing and Networking, pp. 122–133 (2006)
22. Wu, W., Cheng, X., Ding, M., Xing, K., Liu, F., Deng, P.: Localized Outlying and Boundary Data Detection in Sensor Networks. IEEE Transactions on Knowledge and Data Engineering 19(8), 1145–1157 (2007)
23. Zollman, K.J.S.: Social Structure and the Effects of Conformity. Synthese 172, 317–340 (2010)

On the Exact Block Cover Problem

Haitao Jiang[1], Bing Su[2], Mingyu Xiao[3], Yinfeng Xu[4],
Farong Zhong[5], and Binhai Zhu[6]

[1] School of Computer Science and Technology, Shandong University, Jinan, China
htjiang@sdu.edu.cn
[2] School of Economics and Management,
Xi'an Technological University, Xi'an, Shaanxi, China
subing684@sohu.com
[3] School of Computer Science and Engineering,
University of Electronic Science and Technology of China, Chengdu, Sichuan, China
myxiao@gmail.com
[4] Business School, Sichuan University, Chengdu, Sichuan, China
yfxu@scu.edu.cn
[5] College of Math, Physics and Information Technology,
Zhejiang Normal University, Jinhua, China
zfr@zjnu.cn
[6] Department of Computer Science, Montana State University, Bozeman, MT 59717, USA
bhz@cs.montana.edu

Abstract. Minimum Common String Partition (MCSP) has drawn a lot of attention due to its application in genome rearrangement. The best approximation algorithm has a factor $O(\log n \log^* n)$ and it was shown most recently that it is FPT (but with a very high running time). In this paper, we consider the decision version of the one-sided MCSP problem (formally called the *exact block cover* problem); namely, when one sequence is already partitioned into k blocks, how to decide whether the other sequence can be partitioned accordingly. While this decision problem is obviously in FPT, we show interesting results in this paper: (1) If each letter is allowed to appear at most twice (or three times), then the problem is polynomially solvable, (2) There is an FPT algorithm which runs in $O^*(2^k)$ time, improving the trivial bound of $O^*(k!)$, and (3) If $|\Sigma| = c$, c being a constant at least 2, then the problem is NP-complete.

1 Introduction

Computing the similarity of strings is an important problem in sequence analysis, computational biology, etc. In this paper, we revisit the Minimum Common String Partition (MCSP) problem which originates from genome rearrangement problems and provides a natural measure for string similarity.

We define a *partition* P of a string X as a sequence $P = (P_1, P_2, \ldots, P_m)$ of strings whose concatenation is equal to X, that is, $P_1 \cdot P_2 \cdots P_m = X$. The strings P_i are called the blocks of P. Given a partition P of a string X and a partition Q of a string Y, the pair $\pi = (P, Q)$ is called a common partition of X and Y if Q is a permutation of P, i.e., there exists a permutation σ on $[m]$ such that $P_i = Q_{\sigma_i}$. The *minimum common string partition problem* is to find a common partition of X, Y with

Q. Gu, P. Hell, and B. Yang (Eds.): AAIM 2014, LNCS 8546, pp. 13–22, 2014.

the minimum number of blocks. In the Minimum Common String Partition (MCSP) problem, we are given two strings X and Y of length n over an alphabet Σ. Let each symbol appear the same number of times in X and Y. Throughout this paper, we assume that X and Y always satisfy this condition. Clearly, this is a necessary and sufficient condition for X and Y to have a common string partition. For example, two strings $X = abebcdb$ and $Y = abdbebc$ have a common partition $((ab, eb, c, db), (ab, db, eb, c))$. There are several versions of MCSP. The restricted version where each letter occurs at most d times in each input string, is called d-MCSP. Another important version where the input strings are over an alphabet with size bounded by c, is called $MCSP^c$. Most of the known results, notably approximation algorithms, only hold for d-MCSP and $MCSP^c$ [11,14,16,3,4,13,15]. The general MCSP problem admits an approximation algorithm of factor $O(\log n \log^* n)$ [5], which is still the current best.

On the framework of parameterized complexity, Damaschke first solved MCSP by an FPT algorithm referring to parameters k (size of the optimum solution), r (the repetition number) and d (the distance radio depending on the shortest block in the optimum solution) [6]. Recently, d-MCSP is shown to be in FPT [12]. In [9], exact and polynomial time algorithms (with certain conditions) are also considered for MCSP.

Most recently, it was also shown that MCSP is FPT, but the high running time indicates that it is probably impractical for most of the real datasets [1,2]. On the other hand, the One-sided MCSP problem, i.e., assuming a partition of X being given decide whether Y admits a common partition, admits a simple FPT algorithm: Let X be partitioned into k blocks, Y has a common partition iff there is a permutation of the k blocks in X which results in a sequence identical to Y. The running time of this algorithm is $O(k!n) = O^*(k!)$. From now on, we call this the Exact Block Cover problem.

The above observation triggers the start of this research: what is the complexity of the Exact Block Cover problem? Similar to MCSP, the restricted version where each letter occurs at most d times in each input string, is called d-EBC. When the input strings are over an alphabet with size bounded by c, the problem is called EBC^c.

Our Contribution

We prove that the problem EBC^c is NP-complete when $c = 2$. For 2-EBC, we present a simple decision algorithm which runs in $O(n^2)$ time. For 3-EBC, we present a more involved decision algorithm which also runs in $O(n^2)$ time. Finally we present an FPT algorithm for EBC in $O^*(2^k)$ time, improving the trivial $O^*(k!)$ bound.

This paper is organized as follows. In Section 2 we present some basic definitions and notations. In Section 3 we show the hardness result for EBC^2. In Section 4, we show that 2-EBC is polynomially solvable. In Section 5, we show that 3-EBC is also polynomially solvable. In Section 6, we present an $O^*(2^k)$ time FPT algorithm for EBC. In Section 7, we conclude the paper.

2   Preliminaries

As aforementioned, we recall the formal definition of EBC.

Exact Block Cover

Input: two strings X, Y of length n over an alphabet Σ where each letter appears the same number of times in X and Y, a partition of X, $P = (P_1, P_2, ..., P_m)$.

Question: Can Y have a partition Q such that Q is a permutation of P?

We define d-EBC as the restricted version where each letter occurs at most d times in each input string and EBC^c as the version where the input strings are over an alphabet with size bounded by c.

An FPT (Fixed-Parameter Tractable) algorithm for a decision problem Π with parameter (say, solution value) k is an algorithm which solves the problem in $O(f(k)n^c) = O^*(f(k))$ time, where f is any function only on k, n is the input size and c is some fixed constant not related to k. For convenience we also say that Π is in FPT. More details on FPT algorithms can be found in [7,8].

As described in the introduction, both EBC and MCSP are FPT though the running time for MCSP is $O^*(k^{k^2})$. The latter indicates that such an FPT algorithm is probably impractical for most of the real datasets.

3   Hardness for EBC^c

In this section, we prove that EBC^c is strongly NP-complete when $c = 2$ by a reduction from 3-PARTITION [10]. Firstly, we quote the formal definition of 3-PARTITION here.

3-PARTITION

Input: Positive integers n and B, and positive integers set $A = \{a_1, a_2, \ldots, a_{3n}\}$, with $B/4 < a_i < B/2$ and $\sum_{a_i \in A} a_i = nB$.

Question: Can A be partitioned into n disjoint sets $S_1, S_2, \ldots, S_n$ such that, for $1 \leq i \leq n$, $\sum_{a_j \in S_i} a_j = B$.

The problem 3-PARTITION is strongly NP-hard: that is, there is a polynomial $p(n)$ such that it is still NP-hard when all the a_i are at most $p(n)$. Our reduction is polynomially bounded for instances of this type.

Given an instance of 3-PARTITION with weights $a_1, a_2, \ldots, a_{3n}$, we construct two strings X, Y for an instance of EBC^2 as follows

$$X = 0^{a_1} 1 0^{a_2} 1 \cdots 0^{a_{3n}} 1$$

$$Y = (0^B 111)^n.$$

Moreover, assume that X is already partitioned as $0^{a_i}, i = 1, \ldots, 3n$ and $3n$ blocks of 1's.

Theorem 1. EBC^2 is NP-complete.

Proof. EBC^2 is obviously in NP. For the NP-hardess part, we reduce 3-PARTITION to EBC^2. We prove that the 3-PARTITION problem has an precise partition if and only if X, Y have a common partition with $6n$ blocks.

On the necessary side, assume that $S_1, S_2, \ldots, S_n$ satisfy $\sum_{a_j \in S_i} a_j = B$ for all i. For any $S_i = \{a_p, a_q, a_r\}$, we obtain three blocks form X: $0^{a_p}, 0^{a_q}, 0^{a_r}$ and one block from Y: 0^B. Obviously, the block 0^B from Y can be divided into three blocks corresponding to the blocks from X. Note that in Y we also have $3n$ blocks of 1's. Consequently, we obtain a common partition of X, Y with $6n$ blocks.

On the sufficient side, if there is a common partition of X, Y with $6n$ blocks, then $3n$ of them are in the form of 0^{a_i} and $3n$ of them are 1's (due to the given blocks in X). Then, as $B/4 < a_i < B/2$, each block 0^B can be partitioned exactly three times. This implies that 3-PARTITION has a solution.   □

Note that the above proof implies that the EBC problem remains NP-complete if $|\Sigma| = c$, where c is a constant at least two. The reason is that new letters can be added into X and Y as blocks. Next, we consider the d-EBC problem, which, symmetric to EBC^c, has a large alphabet.

4   2-EBC Is Polynomially Solvable

In this section, we show that the *2-EBC* problem is polynomially solvable. As in the previous section, we assume that X has already been partitioned into m blocks: X_1, X_2, ..., X_m. Moreover, if a letter is only covered once by the blocks of X (e.g., X_i) then we have to use X_i to cover that letter. So, from now on we assume that each letter in *2-EBC* appears twice in X and twice in Y.

Firstly, we preprocessing the blocks by tackling those blocks that only appears once in Y. If a block, say X_i, appears only once in Y, obviously, we can match X_i to the unique location in Y.

Now, we assume that all the remaining blocks appears exactly twice in Y. We construct a block graph G_X where the vertices are the remaining blocks in X. There is an edge between X_i, X_j if X_i is a prefix, suffix or substring of X_j, or there is a suffix X' of X_i such that X' is a prefix of X_j with $X_i - X' \neq \phi$ and $X_j - X' \neq \phi$. We call these two kinds of edges *type-1* and *type-2* respectively.

For example, let $\boxed{abcd}$, $\boxed{cded}$, $\boxed{xy}$, $\boxed{xyzw}$ be 4 of the m blocks in X. Then, the edge between $\boxed{abcd}$, $\boxed{cded}$ is type-2 and the edge between $\boxed{xy}$, $\boxed{xyzw}$ is type-1.

Clearly we have the following lemma.

Lemma 1. *2-EBC has a solution if and only if the corresponding graph G_X contains no cycle.*

Proof. Consider a type-1 edge (X_i, X_j) in G_X, W.L.O.G, assume that X_i is a prefix, suffix or substring of X_j, then X_i is of degree one in G_x, since each letter appears exactly twice and the letters of X_i appear once in X_i and once in X_j. Therefore, the type-1 edges will not be contained in any cycle.

For a block X_i, there are at most two blocks X_j and X_k such that (X_i, X_j) and (X_i, X_k) are type-2 edges. That is because the identical substring X' as a suffix of X_i and a prefix of X_j, or vice versa, appears once in X_i and once in X_j. Therefore, each block connects at most two type-2 edges.

($\Rightarrow$) Assume to the contrary that there is a cycle $C = [X_1, X_2, \ldots, X_k]$ in G_X, where $(X_i, X_i + 1)(1 \leq i \leq k)$ and (X_k, X_1) are all type-2 edges. Since Y is a linear sequence, any sub-permutation of Y has two ends. If the blocks could be matched to Y, they must be matched in a linear order. Assume that $\{X_1, X_2, \ldots, X_k\}$ are matched to Y in the order of $[X_{p(1)}, X_{p(2)}, \ldots, X_{p(k)}]$, then $X_{(p(1)-1)mod\ k}$ can not appear twice, since its suffix is on the immediate left of $X_{p(1)}$ and its prefix is on the immediate right of $X_{(p(1)-2)mod\ k}$ in Y, the two locations are separated. Therefore, $X_{(p(1)-1)mod\ k}$ should have been matched to Y before G_X is constructed, a contradiction.

($\Leftarrow$) Assume that G_X contains no cycle. Then G_X is a forest, each connected component of which is a tree. From the above argument, we know that there are at most one path of type-2 edges in each tree, and each type-1 edges connects a leaf of the tree. Since a tree is a bipartite graph, we can firstly arrange the blocks on the two sides of the bipartite graph alternatively along the type-2-edge-path, then arrange the leaves accordingly. Since each block appears twice in Y, it could be formed by the concatenation of all its neighbors. Therefore, the concatenation of blocks on each side of the bipartite graph is a sequence in Y. Moreover, the two sequences are disjoint since each block is separated with its neighbors in Y. □

Here is an example. Let $Y = abcdefgabcdefg$, the blocks of X are { ab , bc , cdef , fg , de , ga }. The block ga appears only once in Y, so we have to match it to the substring ga. We then obtain two sequences $abcdef$ and $bcdefg$. The (bipartite) graph G_X is composed of 5 nodes ab , bc , cdef , fg , and de , and there are 4 edges in G_X: (ab , bc), (bc , cdef), (cdef , fg), which are type-2, and (cdef , de), which is type-1. G_X is a tree in this case. The blocks on the two sides of the type-2-edge-path in G_X can be arranged as ab cdef and bc fg (where de can be easily placed).

We present the general algorithm as follows:

Algorithm *EBC-2(B,S)*

Input: *Two strings X,Y such that each symbol appears the same number of times, and, at most twice in each string; and a partition B of X. Initially $S = \{Y\}$.*

Output: *A partition Q of S which is common to B (if exists)*

1   If $B = S = \emptyset$, return YES.

2   Compute the occurrence of each block of B in a string in S.

3   If the number of occurrence for some $B_i \in B$ is 0, return NO.

4   If the number of occurrence for some $B_i \in B$ is 1

  4.1   then update $B \leftarrow B - \{B_i\}$

  4.2   Put B_i in Q, together with its location on Y.

  4.3   Let $Y = S_1 \cdot B_i \cdot S_2$. Update $S \leftarrow \{S_1, S_2\}$.

  4.4   Call *EBC-2(B,S)*.

5   If all blocks in B appear twice in S, then build the block graph G_B for all blocks in B.

  5.1   For each connected component H' of G_B, convert it into a bipartite graph H.

  5.1.1   Check whether S contains two disjoint substrings of the same contents corresponding to each side of the vertices of H. If the answer is negative, return NO. Otherwise, delete these two substrings and store them in Q (with location information), delete the blocks corresponding to the vertices of H, update S and continue with another connected component of G_B.

  5.2   If $B = S = \emptyset$, then return YES and retrieve the partition Q.

Theorem 2. *There is an $O(n^2)$ time algorithm which decides 2-EBC.*

Proof. The correctness of Algorithm EBC-2 can be seen from Lemma 1. As for the running time of Algorithm EBC-2, the dominating part is on Step 5, where we could have $O(n)$ blocks and we need to check whether there is an edge between any pair of two blocks. Step 2 can be done in $o(n^2)$ time with the suffix tree, but it will not change the overall running time — so we can even use a naive $O(n^2)$ time algorithm. The remaining steps all take linear time.   □

In the next section, we show that 3-EBC is also polynomially decidable. The technique is more involved.

5   3-EBC Is Polynomially Solvable

In this section, we show that 3-EBC is also polynomial. Let $I = (P, Y)$ be an instance of 3-EBC, where $P = \{P_1, P_2, \ldots, P_m\}$. Let Σ' be a new alphabet where $\Sigma \cap \Sigma' = \emptyset$. A *duo* is a substring of length two. A *specific duo* is an appearance of a duo in P or Y. The following three lemmas are trivial due to the definition of exact block cover.

Lemma 2. *For each block P_i, if it appears $TB_1(P_i)$ times in $P = \{P_1, P_2, \ldots, P_m\}$ and $TB_2(P_i)$ in Y, then $TB_1(P_i) \le TB_2(P_i)$; otherwise there is no partition of Y which is a permutation of P. If $TB_1(P_i) = TB_2(P_i)$, then in a partition Q of Y, all the P_i's must be blocks.*

Lemma 3. *For each duo D_i, if it appears $TD_1(D_i)$ times in $P = \{P_1, P_2, \ldots, P_m\}$ and $TD_2(D_i)$ in Y, then $TD_1(D_i) \le TD_2(D_i)$; otherwise, there is no partition of Y which is a permutation of P.*

Lemma 4. *For each duo D_i, if $TD_1(D_i) = TD_2(D_i)$, by replacing D_i with a new letter from Σ', we obtain another EBC instance $I' = (P' = \{P_1', P_2', \ldots, P_m'\}, Y')$ such that I is a Yes-instance if and only if I' is a Yes-instance.*

Next, we show how to reduce the total size of the instance I.

Lemma 5. *For a letter a, if there exist two duos ax and ay ($x \ne y$) in P, then at least one of these two duos could be replaced by a new letter from Σ' to have a reduced equivalent instance I'.*

Proof. Since a appears at most three times in Y, there are at most three duos of the form $a?$. So, we have $TD_2(ax) + TD_2(ay) \le 3$. Then the lemma holds because $1 \le TD_1(ax) \le TD_2(ax)$, $1 \le TD_1(ay) \le TD_2(ay)$ and, either $TD_1(ax) = TD_2(ax)$ or $TD_1(ay) = TD_2(ay)$.   □

Corollary 1. *For a letter b, if there exist two duos ub and vb ($u \ne v$) in P, then at least one of these two duos could be replaced by a new letter from Σ' to have a reduced equivalent instance I'.*

Thus, in the following part of this section, we assume that for each block P_i, $TB_1(P_i) < TB_2(P_i)$, and for each duo D_i, $TD_1(D_i) < TD_2(D_i)$. Since each letter appears at most three times in P, as well as in Y, then there are three cases for $(TB_1(P_i), TB_2(P_i))$ and $(TD_1(D_i), TD_2(D_i))$, which are (2,3),(1,3),(1,2). We will handle each case respectively.

From Lemma 5, if $(TD_1(xy), TD_2(xy)) = (2,3)$, there must be two blocks of the form αx and $y\beta$ in P (where α, β represents a substring of a block); if $(TD_1(xy), TD_2(xy)) =(1,3)$ or $(TD_1(xy), TD_2(xy)) = (1,2)$, there must be two blocks of the form $\alpha_1 x$ and $\alpha_2 x$ in P, as well as two blocks of the form $y\beta_1$ and $y\beta_2$;

Lemma 6. *For the duo xy, if $(TD_1(xy), TD_2(xy)) = (2,3)$, by constructing a new block $\alpha xy\beta$, we can obtain another EBC instance $I' = (P' = \{P_1', P_2', \ldots, P_{m-1}'\}, Y')$ such that I is a Yes-instance if and only if I' is a Yes-instance.*

Proof. ($\Rightarrow$) If I is a Yes-instance, then αx and $y\beta$ must match to two adjacent blocks in Y, equally, the block $\alpha xy\beta$ match to some block containing duo xy in Y. So, we can delete the block $\alpha xy\beta$.

   ($\Leftarrow$) If I' is a Yes-instance, let $\alpha xy\beta$ match to a block Q_i in Y, it is trivial that αx and $y\beta$ can both match to disjoint parts of Q_i.   □

Lemma 7. *For the duo xy, if $(TD_1(xy), TD_2(xy)) = (1,3)$, by constructing four new block $\alpha_1 xy$, $\alpha_2 xy$, β_1, β_2, we can obtain another EBC instance $I' = (P' = \{P_1', P_2', \ldots, P_m'\}, Y')$ such that I is a Yes-instance if and only if I' is a Yes-instance.*

Proof. ($\Rightarrow$) If I is a Yes-instance, then the blocks $\alpha_1 x$, $\alpha_2 x$, $y\beta_1$, $y\beta_2$ must form two pairs. Each of $\alpha_i x$, $y\beta_j$ must match to two adjacent blocks in Y, equally, the block $\alpha_i xy$ and β_j match to some block containing duo xy in Y.

   ($\Leftarrow$) If I' is a Yes-instance, let $\alpha_i xy$ and β_j match to two adjacent blocks Q_s and Q_t in Y, it is trivial that αx and $y\beta$ can match to $Q_s - y$ and $y + Q_t$.   □

Note that after the above modifications of the instance I, if $TD_1(xy) = TD_2(xy) = 3$ then xy can be replaced by a new letter from Σ' due to Lemma 4.

It remains to handle the case where all duos appear once in P and twice in Y. Similar to the block graph in Section 4, we now construct a block graph G_X, where the vertices are those blocks containing at least two letters. There is an edge between X_i, X_j if the last letter of X_i is equal to the first letter of X_j. Since there does not exist common duos between blocks, each block can act the role of X_i at most once and the role of X_j at most once, and the maximum degree of vertices in G_X is two. So, G_X is composed of cycles, paths and isolated vertices. (Note that G_X is not necessarily bipartite.)

Lemma 8. *If for each duo X_i, $(TD_1(X_i), TD_2(X_i)) = (1,2)$, then it is a Yes-instance.*

Proof. We prove this lemma by finding disjoint blocks of the X_is from Y.

1. For those blocks which are isolated vertices in G_X, it is trivial that we can find them from Y.
2. For those blocks appearing in some path $x_1 y_1 \alpha_1 x_2, x_2 y_2 \alpha_2 x_3, \ldots, x_k y_k \alpha_k x_{k+1}$ of G_X, we can find them from Y in the order along the path. Note that each block also appears twice in Y, otherwise it should have been handled by Lemma 2. Since splitting $x_i y_i \alpha_i x_{i+1}$ from Y can destroy at most one specific duo $x_{i+1} y_{i+1}$, we can detect the block $x_{i+1} y_{i+1} \alpha_{i+1} x_{i+2}$ and so on.

3. For those blocks appearing in some cycle $x_1y_1\alpha_1x_2, x_2y_2\alpha_2x_3, \ldots, x_ky_k\alpha_kx_1$ of G_X, we assert that there exists a block $x_jy_j\alpha_jx_{(j+1)mod\ k}$, such that splitting it from Y will not destroy the duo $x_{(j+1)mod\ k}y_{(j+1)mod\ k}$. Assume to the contrary that splitting $x_jy_j\alpha_jx_{(j+1)mod\ k}$ from Y will destroy the duo $x_{(j+1)modk}y_{(j+1)modk}$ for all $1 \le j \le k$, then $x_1y_1\alpha_1x_2y_2\alpha_2x_3 \cdots x_ky_k\alpha_kx_1$ forms a cyclic sequence, a contradiction. Therefore, let $x_iy_i\alpha_ix_{i+1}$ be such a block, then after splitting it from Y, we can detect other blocks in a backward order along the cycle.

This completes the proof of the lemma.   □

Theorem 3. *3-EBC is polynomially solvable.*

Proof. Given a 3-EBC instance, we firstly handle it by Lemma 2, then by Lemma 4, followed by Lemma 6, Lemma 7, and Lemma 4, and finally by Lemma 8. The time complexity is $O(n^2)$.   □

6   A Better FPT Algorithm for EBC

As we discussed in the introduction section, let k be the solution value for EBC, then the problem admits an FPT algorithm with a running time of $O^*(k!)$. This running time might still be too high for application purposes. Here we present a better FPT algorithm which runs in $O^*(2^k)$ time.

Recall that in the EBC problem we are given two strings X and Y of length n over an alphabet Σ and a partition $(X_1, X_2, \cdots, X_k)$ of X. The question is whether Y is a string jointed by a permutation of $\{X_i\}_{i=1}^k$. We consider the parameterized problem with the parameter being k. Here we give a dynamic programming algorithm that runs in $O^*(2^k)$ time.

Let $\mathcal{Y}$ be the set of all nonempty substrings of Y. It is easy to see that

$$|\mathcal{Y}| = \binom{n+1}{2} < n^2.$$

Let S be a nonempty subset of $P = \{X_i\}_{i=1}^l$ and x_j be an element in S. We use $EBC(S, x_j)$ to store strings satisfying the following properties:
1. Y' is a string jointed by a permutation of S and x_j is the last element in the permutation; and
2. $Y' \in \mathcal{Y}$ (i.e., Y' is a substring of Y).
When no string satisfies the condition, we simply let $EBC(S, x_j)$ be an empty set.

Since $EBC(S, x_j)$ is a subset of $\mathcal{Y}$, we know that for each $EBC(S, x_j)$,

$$|EBC(S, x_j)| \le |\mathcal{Y}| < n^2.$$

It is clear that EBC is an yes instance if and only if there exists an x_{j_0} such that $EBC(S = P, x_{j_0})$ is not an empty set. Our dynamic programming algorithm will compute $EBC(S, x_j)$ for every pair (S, x_j). It is trivial to compute $EBC(S, x_j)$ when S contains only one element x_j. For the cases $|S| > 1$, we use the following method to compute $EBC(S, x_j)$ from $EBC(S', x_j')$ with $|S'| < |S|$: Let $S_{-j} = S \setminus \{x_j\}$.

For each element $x'_j \in S_{-j}$, for each string $S^* \in EBC(S_{-j}, x'_j)$, if $S^* x_j$ (the string jointed by S^* and x_j) is an element in $\mathcal{Y}$, then add $S^* x_j$ into $EBC(S, x_j)$.

It is easy to observe the correctness of the above step. To compute $EBC(S, x_j)$, we need to compute at most $|EBC(S_{-j}, x'_j)||S_{-j}| \leq n^3$ strings and check each of them is a substring of Y or not. So we use most n^4 basic computation steps.

The above method to compute $EBC(S, x_j)$ can be transformed in a dynamic programming algorithm by solving $EBC(S, x_j)$ increasing the size of S. The number of different $EBC(S, x_j)$ is $\sum_{i=1}^{k} \binom{k}{i} i < 2^k k$. Therefore, we have the following theorem.

Theorem 4. *The Exact Block Cover problem can be solved in $O(k2^k n^4)$ time.*

7 Closing Remarks

In this paper, we consider the one-sided version of the famous MCSP (Minimum Common String Partition) problem — Exact Block Cover (EBC). EBC obviously has a simple FPT algorithm with a running time of $O^*(k!)$. We show that (1) EBC admits a better FPT algorithm with a running time of $O^*(2^k)$; (2) EBC^2, i.e., when the alphabet is binary, is NP-complete and (3) 2-EBC (resp. 3-EBC), i.e., when each letter appears at most twice (resp. three times) in both of the input strings, is decidable with an $O(n^2)$ running time. An immediate question is whether 4-EBC (or in general, d-EBC, where d is a constant at least 4) is decidable in polynomial time.

Acknowledgments. This research is partially supported by NSF of China under grant 60928006, 71071123, 61221069, 61202014 and 61370071, and by the Open Fund of Top Key Discipline of Computer Software and Theory in Zhejiang Provincial Colleges at Zhejiang Normal University.

References

1. Bulteau, L., Fertin, G., Komusiewicz, C., Rusu, I.: A fixed-parameter algorithm for minimum common string partition with few duplications. In: Darling, A., Stoye, J. (eds.) WABI 2013. LNCS, vol. 8126, pp. 244–258. Springer, Heidelberg (2013)
2. Bulteau, L., Komusiewicz, C.: Minimum common string partition parameterized by partition size is fixed-parameter tractable. ArXiv: CoRR abs/1305.0649 (2013)
3. Chen, X., Zheng, J., Fu, Z., Nan, P., Zhong, Y., Lonardi, S., Jiang, T.: Computing the assignment of orthologous genes via genome rearrangement. In: Proc. APBC 2005, pp. 363–378 (2005)
4. Chrobak, M., Kolman, P., Sgall, J.: The greedy algorithm for the minimum common string partition problem. In: Jansen, K., Khanna, S., Rolim, J.D.P., Ron, D. (eds.) APPROX and RANDOM 2004. LNCS, vol. 3122, pp. 84–95. Springer, Heidelberg (2004)
5. Cormode, G., Muthukrishnan, S.: The string edit distance matching problem with moves. In: Proceedings of the 13th ACM-SIAM Symp. on Discrete Algorithms (SODA 2002), pp. 667–676 (2002)
6. Damaschke, P.: Minimum Common String Partition Parameterized. In: Crandall, K.A., Lagergren, J. (eds.) WABI 2008. LNCS (LNBI), vol. 5251, pp. 87–98. Springer, Heidelberg (2008)

7. Downey, R., Fellows, M.: Parameterized Complexity. Springer (1999)
8. Flum, J., Grohe, M.: Parameterized Complexity Theory. Springer (2006)
9. Fu, B., Jiang, H., Yang, B., Zhu, B.: Exponential and polynomial time algorithms for the minimum common string partition problem. In: Wang, W., Zhu, X., Du, D.-Z. (eds.) COCOA 2011. LNCS, vol. 6831, pp. 299–310. Springer, Heidelberg (2011)
10. Garey, M.R., Johnson, D.S.: Computers and Intractability: A Guide to the Theory of NP-Completeness. W.H. Freeman (1979)
11. Goldstein, A., Kolman, P., Zheng, J.: Minimum common string partition problem: Hardness and approximations. In: Fleischer, R., Trippen, G. (eds.) ISAAC 2004. LNCS, vol. 3341, pp. 484–495. Springer, Heidelberg (2004); Also in: The Electronic Journal of Combinatorics 12, paper R50 (2005)
12. Jiang, H., Zhu, B., Zhu, D., Zhu, H.: Minimum common string partition revisited. J. of Combinatorial Optimization 23(4), 519–527 (2012)
13. Kaplan, H., Shafrir, N.: The greedy algorithm for edit distance with moves. Inf. Process. Lett. 97(1), 23–27 (2006)
14. Kolman, P., Waleń, T.: Reversal Distance for Strings with Duplicates: Linear Time Approximation Using Hitting Set. In: Erlebach, T., Kaklamanis, C. (eds.) WAOA 2006. LNCS, vol. 4368, pp. 279–289. Springer, Heidelberg (2007)
15. Kolman, P.: Approximating reversal distance for strings with bounded number of duplicates. In: Jedrzejowicz, J., Szepietowski, A. (eds.) MFCS 2005. LNCS, vol. 3618, pp. 580–590. Springer, Heidelberg (2005)
16. Kolman, P., Walen, T.: Approximating reversal distance for strings with bounded number of duplicates. Discrete Applied Mathematics 155(3), 327–336 (2007)

Minimax Regret k-sink Location Problem in Dynamic Path Networks*

Guanqun Ni[1,**], Yinfeng Xu[1,2], and Yucheng Dong[1]

[1] Business School of Sichuan University, 24 South Section 1, Yihuan Road, Chengdu, 610065, P.R. China
[2] State Key Lab for Manufacturing Systems Engineering, Xi'an 710049, China
{gqni,yfxu,ycdong}@scu.edu.cn

Abstract. Recently, Cheng et al. [1] proposed the minimax regret 1-sink location problem in dynamic path networks and presented an $O(n \log^2 n)$ time algorithm for the proposed problem, where n is the number of vertices. In this paper, we study the general problem, i.e., minimax regret k-sink location problem in the dynamic path networks. Based on the algorithm for the 1-sink location problem, we design an $O(n^2 (\log n)^{1+\log k} C_n^{k-1})$ time algorithm for the general problem, where C_n^{k-1} is the number of combination choosing $k-1$ from n.

Keywords: minimax regret, k-sink location, dynamic flow, path networks, evacuation problem.

1 Introduction

Within five years, there are two big earthquakes happening in Sichuan province China, i.e., Wenchuan Big Earthquake and Lushan Big Earthquake. Today, there are so many big disasters occurring worldwide, such as the 2010 Haiti Earthquake, the 2011 Tohoku-Pacific Ocean Earthquake and Tsunamis and so on. When a big disaster happens, in order to counter potential emergency effectively, it is need to evacuate all evacuees (*disaster victims*) of every settlement (*disaster point*) in the disaster region as fast as possible to an evacuation building (*rescue point*) waiting for further assist. We have to consider where evacuation buildings are assigned and how to partition a large area into small regions so that one evacuation building is designated in each region. There are several considerable criterions for this evacuation problem. To minimize the time required to complete the evacuation, [5] has studied the tree network problem where a nonnegative weight that represents the number of evacuees at each vertex is known. For this certain problem, [5] proposed an $O(n \log^2 n)$ time algorithm to find an optimal location of one *sink* (the location of an evacuation building).

---

* This work was supported by China Postdoctoral Science Foundation under Grant 2013M530404, NSF of China under Grants 71371129, 71071123 and 61221063 and Program for Changjiang Scholars and Innovative Research Team in University under Grant IRT1173.
** Corresponding author.

Q. Gu, P. Hell, and B. Yang (Eds.): AAIM 2014, LNCS 8546, pp. 23–31, 2014.

However, it's important to note that the number of evacuees at each vertex (the vertex weight) is changing over time and uncertain when the potential emergency occurs. For example, in an office area in a big city there are many people during the daytime on weekdays while there are much less people on weekends or during the night time. Sometimes, it is also impossible or inappropriate to assume any specific probability distribution on the unknown information. Therefore, we consider a minimax regret criterion assuming that the weight of any vertex can take any value within a corresponding pre-specified interval and use the *minimax regret approach* to deal with this uncertain problem. Supposing a particular realization (assignment of a weight to each vertex) is called a scenario, our objective is to choose one evacuation building point (1-sink) to minimize the maximum *regret* for any possible scenario of weight. [1] considered the simplest case, i.e. minimax regret 1-sink location problem, for which the disaster region consists of a single road and the number of evacuation buildings is just equal to 1. For this uncertain problem, [1] proposed an $O(n \log^2 n)$ algorithm. Not long after, [2] and [6] independently improved the algorithm to $O(n \log n)$. Recently, [3] proposed an $O(n^2 \log^2 n)$ algorithm for the corresponding problem on the tree networks. And [4] extended 1-sink location problem on the path networks to 2-sink problem and proposed an $O(n^3 \log n)$ algorithm. In this paper, we extend the problem on the path networks to more general case where there are k sinks to be assigned on the road line, and propose an $O(n^2 (\log n)^{1+\log k} C_n^{k-1})$ algorithm based on [1].

2 Preliminaries

In this section, we formulate the general problem with k sinks based on the formulation of 1-sink problem. Thus, we first state the the simplest minimax regret 1-sink location problem in dynamic path networks formulated by [1] as follows.

Let $P = (V, E)$ be a path where $V = \{v_0, v_1, ..., v_n\}$ and $E = \{e_1, e_2, ..., e_n\}$ such that v_{i-1} and v_i are endpoints of e_i for $1 \leq i \leq n$. Let $\mathcal{N} = (P, l, W, c, \tau)$ be a dynamic flow network with the underlying undirected graph being a path P, where l is a function that associates each edge $e_j \in V$ with the positive length $l(e_j)$, W is also a function that associates each vertex $v_i \in V$ with an interval of the weight (the number of the evacuees) $W(v_i) = [\underline{w}_i, \overline{w}_i]$ with $0 \leq \underline{w}_i \leq \overline{w}_i$, c is a constant representing the capacity of each edge: the least upper bound for the number of the evacuees passing a point in an edge per unit time, and τ is also a constant representing the time required for traversing the unit distance of each evacuee.

Let $\mathcal{S} = \prod_{1 \leq i \leq n} [\underline{w}_i, \overline{w}_i]$ denote the Cartesian product of all $W(v_i)$ for $1 \leq i \leq n$. When a scenario $s \in \mathcal{S}$ is given, we use the notation $w_i(s)$ to denote the weight of each vertex $v_i \in V$ under the scenario s. Suppose that a path P is embedded on a real line and each vertex $v_i \in V$ is associated with the line coordinate x_i such that $x_i = x_0 + \sum_{1 \leq j \leq i} l(e_j)$ for $1 \leq i \leq n$. For a point $x \in P$, we also use a notation x to denote the line coordinate of the point, and the *left side* of x (resp.

the *right side* of x) to denote the part of P consisting of all points $t \in P$ such that $t < x$ (resp. $t > x$). Suppose that a sink (evacuation building) is located at a point $x \in P$. Let $\Theta_L(x, s)$ (resp. $\Theta_R(x, s)$) denote the minimum time required for all evacuees on the left side (resp. the right side) of x to complete evacuation to x under a scenario $s \in \mathcal{S}$. For the ease of exposition, we assume that $c = 1$ (the case of $c > 1$ can be treated in essentially the same manner). Note that we assume that the capacity of the entrance of an evacuation building is infinite, and thus, if we place a sink in a vertex v_i, all evacuees of v_i can finish their evacuation in no time. Then, $\Theta_L(x, s)$ and $\Theta_R(x, s)$ are expressed as follows:

$$\Theta_L(x, s) = \max_{0 \le i \le n-1} \{(x - x_i)\tau + \sum_{0 \le j \le i} w_j(s) | x_0 \le x_i < x\}, \tag{1}$$

$$\Theta_R(x, s) = \max_{1 \le i \le n} \{(x_i - x)\tau + \sum_{i \le j \le n} w_j(s) | x < x_i \le x_n\}. \tag{2}$$

Now, under $s \in \mathcal{S}$, the minimum time required for the evacuation to $x \in P$ of all evacuees is defined by

$$\Theta(x, s) = \max\{\Theta_L(x, s), \Theta_R(x, s)\}. \tag{3}$$

Under any $s \in \mathcal{S}$, let $x_{opt}(s)$ be the optimal sink location, then the *regret* for x under s is defined as

$$R(x, s) = \Theta(x, s) - \Theta(x_{opt}(s), s). \tag{4}$$

Moreover, we also define the *maximum regret* of x as

$$R_{max}(x) = \max\{R(x, s) | s \in \mathcal{S}\}. \tag{5}$$

If $R_{max}(x) = R(x, s^*)$ for a scenario s^*, then we call s^* the *worst case scenario* for x. The goal is to find a point $x^* \in P$, called the *minimax regret sink*, which minimizes $R_{max}(x)$ over $x \in P$, i.e., the objective is

$$\text{minimize}\{R_{max}(x) | x \in P\}. \tag{6}$$

For this minimization problem (6), we have presented an $O(n \log^2 n)$ time algorithm [1]. For the general case, minimax regret k-sink location problem, we have to assign $k \ge 1$ sinks on the path and the objective is also to minimize the time regret under the worst case scenario.

Note that it is necessary to partition the path first for the general problem. We define $PT_k = \{V_1, V_2, ..., V_k\}$ a k-partition for all the vertices on the path. Where $V_1 = \{v_0, ..., v_{k_1}\}$, $V_2 = \{v_{k_1+1}, ..., v_{k_2}\}, ..., V_k = \{v_{k_{k-1}+1}, ..., v_{k_k}\}$, $v_{k_0+1} = v_0$, $v_{k_k} = v_n$ and $k_i < k_j$ for $\forall i < j$. For any $V_i \in PT_k$, we have to locate one sink (evacuation building) at a point $v \in V_i$ and all the evacuees in the region V_i will evacuate to v.

Given a k-partition PT_k, under $s \in \mathcal{S}$, we can compute the optimal one sink location for all V_is in $O(n \log n)$ time [1]. Let $\theta(V_i, s)$ denote the minimum time to complete the evacuation of all evacuees in V_i under $s \in \mathcal{S}$. Then for the k-partition PT_k, under $s \in \mathcal{S}$, the minimum time required to complete the evacuation of all evacuees on the path is defined by

$$\Theta(PT_k, s) = \max\{\theta(V_i, s)|V_i \in PT_k\}. \tag{7}$$

Under any $s \in \mathcal{S}$, let $PT_{opt}(s)$ be the optimal k-partition, then the *regret* for any k-partition PT_k under s is defined as

$$R(PT_k, s) = \Theta(PT_k, s) - \Theta(PT_{opt}(s), s). \tag{8}$$

Moreover, we also define the *maximum regret* of PT_k as

$$R_{max}(PT_k) = \max\{R(PT_k, s)|s \in \mathcal{S}\}. \tag{9}$$

If $R_{max}(PT_k) = R(PT_k, s^*)$ for a scenario s^*, then we call s^* the *worst case scenario* for PT_k. The goal is to find a k-partition $PT_k^* \in \mathcal{PT}$ which minimizes $R_{max}(PT_k)$ over $PT_k \in \mathcal{PT}$, where $\mathcal{PT}$ is the set of all k-partitions, i.e., the objective is

$$\text{minimize}\{R_{max}(PT_k)|PT_k \in \mathcal{PT}\}. \tag{10}$$

Obviously, during an optimal evacuation process there is sure no traffic flow conflict which means that all the evacuees between one point v and one sink x should evacuate to x as long as the evacuees at v evacuate to x. Additionally, we assume all the evacuees at one point have to evacuate to the same sink in order to prevent traffic disturbance happening.

3   Properties and Algorithms for Certain k-sink Problem

For the proposed minimax regret k-sink location problem, we first study the certain problem where the weight scenario s is given and then present some properties and design algorithms in this section.

Given a scenario s, the certain k-sink location problem is to assign k sinks (evacuation buildings) at the path minimizing the time to complete the evacuation, i.e., the objective is

$$\text{minimize}\{\Theta(PT_k, s)|PT_k \in \mathcal{PT}\}. \tag{11}$$

For a k-partition PT_k, let v be the boundary point of any V_b and V_{b+1} where $v \in V_b$ and $v \notin V_{b+1}$. We also call v a *b-partition-boundary point*. Suppose the *left-side* b-partition and the *right-side* $(k - b)$-partition are both optimal, i.e., the minimum completing times of both sides of v are respectively equal to

$$\Theta_L^v(PT_b, s) = \min\{\Theta(PT_b, s)|PT_b \in \mathcal{PT}_b^{L_v}\}, \tag{12}$$

$$\Theta_R^v(PT_{k-b}, s) = \min\{\Theta(PT_{k-b}, s)|PT_{k-b} \in \mathcal{PT}_{k-b}^{R_v}\}. \tag{13}$$

where $\mathcal{PT}_b^{L_v}$ (resp. $\mathcal{PT}_{k-b}^{R_v}$) is the set of all b-partitions for the left side of v (resp. all $(k-b)$-partitions for the right side of v).

Now, under $s \in \mathcal{S}$, the minimum time required to complete the evacuation of a k-partition with v as the b-partition-boundary point is defined by

$$\Theta_b^v(PT_k, s) = \max\{\Theta_L^v(PT_b, s), \Theta_R^v(PT_{k-b}, s)\}. \tag{14}$$

Under any $s \in \mathcal{S}$, the minimum time required to complete the evacuation of an optimal k-partition is defined as

$$\Theta_{opt}(PT_k, s) = \min\{\Theta_b^v(PT_k, s) | v \in V\}. \tag{15}$$

That is, if let $v_{optb}(s)$ be the optimal b-partition-boundary point, then

$$v_{optb}(s) = \arg\min_{v \in V}\{\Theta_b^v(PT_k, s)\}. \tag{16}$$

Then, similar to the Propositions 1 and 2 in [1] for the certain 1-sink location problem, we correspondingly have the following two propositions for the certain k-sink location problem.

Proposition 1. *Under a scenario $s \in \mathcal{S}$, $v_{optb}(s)$ is unique.*

Proposition 2. *Under a scenario $s \in \mathcal{S}$,*
(i) for any vertex l at the left of $v_{optb}(s)$, $\Theta_L^l(PT_b, s) < \Theta_R^l(PT_{k-b}, s)$ holds;
(ii) for any vertex r at the right of $v_{optb}(s)$, $\Theta_L^r(PT_b, s) > \Theta_R^r(PT_{k-b}, s)$ holds.

Based on Propositions 1 and 2, we first design the following algorithm BSA_2 for the simplest certain case with $k = 2$.

Binary Search Algorithm BSA_2: Given any scenario $s \in \mathcal{S}$, do

Step 1. Initially, partition the vertex set V into two conjoined subsets, V_l and V_r, and let $V_l = \Phi$ and $V_r = V$. Define $F = V$ initially.

Step 2. For both V_l and V_r, solve the certain 1-sink location problem based on [1]. Compute their evacuation times defined as τ_l and τ_r, respectively. Then the time required to complete the evacuation of 2-sink problem is equal to $max\{\tau_l, \tau_r\}$. If $\tau_l < \tau_r$, then move the vertices of *left half* of set F to V_l where the number of *half* of set F equals $\left\lceil \frac{|F|}{2} \right\rceil - 1$ and $\lceil x \rceil$ means the smallest integer larger than x. Update the sets both V_l and V_r, and turn to **Step 3**. If $\tau_l > \tau_r$, then move the vertices of *right half* of set F to V_r. Update the sets both V_l and V_r, and turn to **Step 3**. If it is the case $\tau_l = \tau_r$ before $|F| = 0$, then the two optimal 1-sink locations are also the optimal locations for the 2-sink problem and then **stop**. If it is always the case $\tau_l \neq \tau_r$ till $|F| = 0$, then turn to **Step 4**.

Step 3. For the updated V_l and V_r, compute and update their corresponding evacuation times τ_l and τ_r, respectively. If $\tau_l < \tau_r$, then update the set F and let $F = F \cap V_r$ and turn to **Step 2**. If $\tau_l > \tau_r$, then update the set F and let $F = F \cap V_l$ and turn to **Step 2**.

Step 4. Compute and compare the evacuation times of 2-sink location problem during the last two iterations. Choose the smaller one as the optimal evacuation time and the corresponding sink positions are also the optimal sink locations for the certain 2-sink location problem. And then **stop**.

Theorem 1. *The sink location of BSA_2 is the optimal solution to the certain 2-sink location problem. And the time complexity of BSA_2 is $O(n \log^2 n)$.*

Proof. First, for the correctness, each iteration of Step 2 improves the sink performance, i.e., reduces the time to complete the evacuation based on Propositions 1 and 2. Which guarantees the correctness.

Next, we prove the time complexity. For Steps 2, 3 and 4 in each iteration, we all solve two small-scale certain 1-sink location problems and thus the time complexity of Steps 2, 3 and 4 is $O(n \log n)$ [1]. For Step 1, it spends $O(1)$ time. Moreover, BSA_2 has at most $\log n$ iterations. Thus, The time complexity of BSA_2 is $O(n \log^2 n)$.   □

For the algorithm BSA_2, if we solve the certain 1-sink location problem for V_l and solve the certain 2-sink location problem for V_r, then we can easily get the optimal sink locations for the certain 3-sink location problem in $O(n \log^3 n)$ time. Therefore, we have the following corollary for the general certain k-sink problem.

Corollary 1. *For the general certain k-sink location problem, **Binary Search Algorithm** can solve the optimal sink locations in $O(n \log^k n)$ time.*

However, we can design an alternative **Binary Search Algorithm** BSA_k for the general problem in $O(n(\log n)^{1+\log k})$ time as follows.

Binary Search Algorithm BSA_k: Given any scenario $s \in S$, suppose that we already have an algorithm solving the certain $\left\lceil \frac{|k|}{2} \right\rceil$-sink problem and the certain $\left\lfloor \frac{|k|}{2} \right\rfloor$-sink problem. Where $\lceil x \rceil$ means the smallest integer larger than x and $\lfloor x \rfloor$ means the largest integer smaller than x. And do

Step 1. Initially, partition the vertex set V into two conjoined subsets, V_l and V_r, and let $V_l = \Phi$ and $V_r = V$. Define $F = V$ initially.

Step 2. For V_l and V_r, solve the certain $\left\lceil \frac{|k|}{2} \right\rceil$-sink location problem and the certain $\left\lfloor \frac{|k|}{2} \right\rfloor$-sink location problem, respectively. Compute their evacuation times defined as τ_l and τ_r, respectively. Then the time required to complete the evacuation of k-sink location problem is equal to $max\{\tau_l, \tau_r\}$. If $\tau_l < \tau_r$, then move the vertices of *left half* of set F to V_l where the number of *half* of set F equals $\left\lceil \frac{|F|}{2} \right\rceil - 1$. Update the sets both V_l and V_r, and turn to **Step 3.** If $\tau_l > \tau_r$, then move the vertices of *right half* of set F to V_r. Update the sets both V_l and V_r, and turn to **Step 3.** If it is the case $\tau_l = \tau_r$ before $|F| = 0$, then the corresponding sink locations are the optimal locations for the k-sink problem and then **stop.** If it is always the case $\tau_l \neq \tau_r$ till $|F| = 0$, then turn to **Step 4.**

Step 3. For the updated V_l and V_r, compute and update their corresponding evacuation times τ_l and τ_r, respectively. If $\tau_l < \tau_r$, then update the set F and let $F = F \cap V_r$ and turn to **Step 2.** If $\tau_l > \tau_r$, then update the set F and let $F = F \cap V_l$ and turn to **Step 2.**

Step 4. Compute and compare the evacuation times of k-sink location problem during the last two iterations. Choose the smaller one as the optimal evacuation time and the corresponding sink positions are also the optimal sink locations for the certain k-sink location problem. And then **stop**.

Theorem 2. *The sink location of BSA_k is the optimal solution to the certain k-sink location problem. And the time complexity of BSA_k is $O(n(\log n)^{1+\log k})$.*

Proof. First, the correctness is obviously based on Propositions 1 and 2 and the proof of Theorem 1.

Next, we prove the time complexity. The time complexity of BSA_k depends on two parts, the time complexity solving the certain $\left\lceil \frac{|k|}{2} \right\rceil$-sink location problem and the iterations of algorithm BSA_k, where $\lceil x \rceil$ means the smallest integer larger than x. It is obviously that the iterations of algorithm BSA_k are no more than $\log n$. Suppose the time complexity solving the certain $\left\lceil \frac{|k|}{2} \right\rceil$-sink problem is equal to $T(\left\lceil \frac{|k|}{2} \right\rceil)$, then the time complexity of BSA_k is defined as

$$T(\lceil |k| \rceil) = T\left(\left\lceil \frac{|k|}{2} \right\rceil\right)\log n = T\left(\left\lceil \frac{|k|}{4} \right\rceil\right)\log^2 n = \ldots = T(2)(\log n)^{\log k - 1.} \quad (17)$$

Because the time complexity of BSA_2 solving the certain 2-sink location location problem is $T(2) = O(n \log^2 n)$, the time complexity of BSA_k is equal to $T(\lceil |k| \rceil) = O(n(\log n)^{1+\log k})$. □

4   Properties and Algorithms for Minimax Regret k-sink Problem

Because there are C_n^{k-1} (C_n^{k-1} is the number of combination choosing $k-1$ from n) possible k-partitions in all for the problem with n vertices and k sinks, it is initially to compare all these partitions and then choose the one with minimum maximal regret as the optimal partition.

Given a k-partition $PT_k = \{V_1, V_2, ..., V_k\}$, a scenario $s \in \mathcal{S}$ is said to be i-*left-dominant* (resp. i-*right-dominant*) if for one and at most one set $V_i = \{v_{k_{i-1}+1}, ..., v_{k_i}\}$ with $v_x \in V_i$, $w_j(s) = \overline{w}_j$ for $k_{i-1} + 1 \leq j < x$, $w_j(s) = \underline{w}_j$ for $x \leq j \leq k_i$ and $w_j(s) = \underline{w}_j$ for $\forall v_j \notin V_i$ hold (resp. $w_j(s) = \underline{w}_j$ for $k_{i-1} + 1 \leq j < x$, $w_j(s) = \overline{w}_j$ for $x \leq j \leq k_i$ and $w_j(s) = \overline{w}_j$ for $\forall v_j \notin V_i$ hold). Given any V_i, let $\mathcal{S}_L^i$ (resp. $\mathcal{S}_R^i$) denote the set of all i-left-dominant (resp. i-right-dominant) scenarios. $\mathcal{S}_L^i$ consists of the following scenarios:

$$s_L^{ij} = (\overline{w}_{k_{i-1}+1}, \ldots, \overline{w}_j, \underline{w}_{j+1}, \ldots, \underline{w}_{k_i}) \quad \text{for } j = k_{i-1} + 1, \ldots, k_i - 1, \quad (18)$$

$$s_L^{ik_i} = (\overline{w}_{k_{i-1}+1}, \overline{w}_{k_{i-1}+2}, \ldots, \overline{w}_{k_i}), \quad (19)$$

and $\mathcal{S}_R^i$ consists of the following scenarios:

$$s_R^{ij} = (\underline{w}_{k_{i-1}+1}, \ldots, \underline{w}_j, \overline{w}_{j+1}, \ldots, \overline{w}_{k_i}) \quad \text{for } j = k_{i-1} + 1, \ldots, k_i - 1, \quad (20)$$

$$s_R^{ik_i} = (\underline{w}_{k_{i-1}+1}, \underline{w}_{k_{i-1}+2}, \ldots, \underline{w}_{k_i}), \quad (21)$$

Let $\mathcal{S}_L = \bigcup_{1 \le i \le k} \mathcal{S}_L^i$ and $\mathcal{S}_R = \bigcup_{1 \le i \le k} \mathcal{S}_R^i$, then both $\mathcal{S}_L$ and $\mathcal{S}_R$ consist of $n+1$ scenarios. That is, $\mathcal{S}_L$ and $\mathcal{S}_R$ consist of $2(n+1)$ scenarios in all. The following is a key theorem.

Theorem 3. *For any k-partition $PT_k \in \mathcal{PT}$, there is a worst case scenario for PT_k which belongs to $\mathcal{S}_L \cup \mathcal{S}_R$.*

Proof. The proof of Theorem 3 is straightforward from Theorem 1 of [1]

Given a k-partition PT_k, assume that the worst case scenario s' does not belong to $\mathcal{S}_L \cup \mathcal{S}_R$, and the time to complete the evacuation under s' results from V_i, i.e., equal to $\theta(V_i, s')$. Let $\Theta(PT_{opt}(s'), s')$ be the evacuation time of one optimal k-partition under s'. Then the *regret* for PT_k under s' is defined as

$$R(PT_k, s') = \theta(V_i, s') - \Theta(PT_{opt}(s'), s'). \tag{22}$$

If we decrease the weights of all vertices excluded in V_i to their corresponding lower bound, then the value of $\theta(V_i, s')$, i.e. the evacuation time for partition PT_k is surely unchanged. However, the evacuation time for one optimal k-partition under the renewed scenario likely decreases and impossibly increases at least. Further, if the weight structure in V_i is not consistent with either i-left-dominant scenario or i-right-dominant scenario, then we can design an alternative i-left-dominant scenario or i-right-dominant scenario to largen (at least to maintain) the regret directly based on Theorem 1 of [1]

In one word, given a k-partition PT_k, we can always renew the *worst case scenario* s' to an alternative scenario $s^* \in \mathcal{S}_L \cup \mathcal{S}_R$ resulting in a larger regret or an unchanged regret at least. □

Based on Theorem 3 and algorithm BSA_k, we design an algorithm $MMRA_k$ solving minimax regret k-sink problem as follows.

MiniMax-Regret-Algorithm $MMRA_k$:

Step 1. For any k-partition PT_k, (C_n^{k-1} k-partitions in all) do

Step 1.1. Under any worst case scenario $s \in \mathcal{S}_L \cup \mathcal{S}_R$, compute the evacuation time of PT_k, $\Theta(PT_k, s)$.

Step 1.2. Solve the certain k-sink location problem under the same scenario s based on algorithm BSA_k. And compute the optimal evacuation time, $\Theta(PT_{opt}(s), s)$.

Step 1.3. Define the regret of PT_k under s as $R(PT_k, s) = \Theta(PT_k, s) - \Theta(PT_{opt}(s), s)$. Repeat Step 1.1 and Step 1.2 for any $s \in \mathcal{S}_L \cup \mathcal{S}_R$. Compare the regrets among all the $2(n+1)$ worst case scenarios and define the largest one as the maximum regret of PT_k.

Step 2. Compare the maximum regrets among all the C_n^{k-1} k-partitions and define the smallest one as the *minimax regret* and choose the corresponding sink locations as the optimal solution to the minimax regret k-sink problem.

Obviously, $MMRA_k$ solves the minimax regret k-sink location problem based on Theorem 3 and the fact that there are C_n^{k-1} k-partitions in all. For any $s \in \mathcal{S}_L \cup \mathcal{S}_R$, Step 1.1 solves k certain 1-sink location problem and spends $O(n \log n)$ time [1] and Step 1.2 spends $O(n(\log n)^{1+\log k})$ time to solve the certain k-sink location problem based on Theorem 2. Because Step 1.3 has $2(n+1)$ iterations, for any k-partition PT_k, Step 1 spends $O(n^2(\log n)^{1+\log k})$ time to compute its maximum regret. Step 2 needs to compute C_n^{k-1} maximum regrets. Thus, we have the following theorem.

Theorem 4. *The sink location of $MMRA_k$ is the optimal solution to the minimax regret k-sink location problem. And the time complexity of $MMRA_k$ is $O(n^2 C_n^{k-1}(\log n)^{1+\log k})$.*

5 Future Directions

In this paper, we extend the minimax 1-sink location problem in dynamic path networks proposed by [1] to the general case, i.e., the minimax regret k-sink location problem. We find several observations and facts based on which we present an $O(n^2 C_n^{k-1}(\log n)^{1+\log k})$ time algorithm to solve the general problem. However, the presented algorithm is very initial. And thus, one interesting direction is to improve the presented algorithm. Another direction is to extend the problem to more general graphs, like trees.

References

1. Cheng, S.-W., Higashikawa, Y., Katoh, N., Ni, G., Su, B., Xu, Y.: Minimax regret 1-sink location problems in dynamic path networks. In: Chan, T.-H.H., Lau, L.C., Trevisan, L. (eds.) TAMC 2013. LNCS, vol. 7876, pp. 121–132. Springer, Heidelberg (2013)
2. Higashikawa, Y., Augustine,J., Cheng, S.W., Golin, G.J., Katoh, N., Ni, G., Su, B., Xu, Y.: Minimax regret 1-sink location problems in dynamic path networks. Theoretical Computer Science (to appear, 2014)
3. Higashikawa, Y., Golin, M.J., Katoh, N.: Minimax regret sink location problems in dynamic tree networks with uniform capacity. In: Pal, S.P., Sadakane, K. (eds.) WALCOM 2014. LNCS, vol. 8344, pp. 125–137. Springer, Heidelberg (2014)
4. Li, H., Xu, Y., Ni, G.: Minimax regret 2-sink location problem in dynamic path networks. Journal of Combinatorial Optimization (to appear, 2014)
5. Mamada, S., Uno, T., Makino, K., Fujishige, S.: An $O(n \log^2 n)$ algorithm for the optimal sink location problem in dynamic tree networks. Discrete Applied Mathematics 154, 2387–2401 (2006)
6. Wang, H.: Minimax regret 1-facility location on uncertain path networks. In: Cai, L., Cheng, S.-W., Lam, T.-W. (eds.) ISAAC 2013. LNCS, vol. 8283, pp. 733–743. Springer, Heidelberg (2013)

Competitive Algorithms
for Unbounded One-Way Trading

Francis Y.L. Chin[1,*], Bin Fu[2], Minghui Jiang[3],
Hing-Fung Ting[1,**], and Yong Zhang[1,4,***]

[1] Department of Computer Science, The University of Hong Kong, Hong Kong
{chin,hfting,yzhang}@cs.hku.hk
[2] Department of Computer Science,
University of Texas-Pan American, Edinburg, TX 78539, USA
bfu@utpa.edu
[3] Department of Computer Science,
Utah State University, Logan, UT 84322, USA
mjiang@cc.usu.edu
[4] Shenzhen Institutes of Advanced Technology, Chinese Academy of Sciences, China

Abstract. In the one-way trading problem, a seller has some product to be sold to a sequence σ of buyers $u_1, u_2, \ldots, u_\sigma$ arriving online and he needs to decide, for each u_i, the amount of product to be sold to u_i at the then-prevailing market price p_i. The objective is to maximize the seller's revenue. We note that most previous algorithms for the problem need to impose some artificial upper bound M and lower bound m on the market prices, and the seller needs to know either the values of M and m, or their ratio M/m, at the outset. Moreover, the performance guarantees provided by these algorithms depend only on M and m, and are often too loose; for example, given a one-way trading algorithm with competitive ratio $\Theta(\log(M/m))$, its actual performance can be significantly better when the actual highest to actual lowest price ratio is significantly smaller than M/m.

This paper gives a one-way trading algorithm that does not impose any bounds on market prices and whose performance guarantee depends directly on the input. In particular, we give a class of one-way trading algorithms such that for any positive integer h and any positive number ϵ, we have an algorithm $A_{h,\epsilon}$ that has competitive ratio $O(\log r^* (\log^{(2)} r^*) \ldots (\log^{(h-1)} r^*)(\log^{(h)} r^*)^{1+\epsilon})$ if the value of $r^* = p^*/p_1$, the ratio of the highest market price $p^* = \max_i p_i$ and the first price p_1, is large and satisfy $\log^{(h)} r^* > 1$, where $\log^{(i)} x$ denotes the application of the logarithm function i times to x; otherwise, $A_{h,\epsilon}$ has a constant competitive ratio Γ_h. We also show that our algorithms are near optimal by showing that given any positive integer h and any one-way trading algorithm A, we can construct a sequence of buyers σ with $\log^{(h)} r^* > 1$ such that the ratio between the optimal revenue and the revenue obtained by A is at least $\Omega(\log r^* (\log^{(2)} r^*) \ldots (\log^{(h-1)} r^*)(\log^{(h)} r^*))$.

[*] Research supported by HK RGC grant HKU-711709E and Shenzhen basic research project (NO.JCYJ20120618143038947).
[**] Research supported by HK RGC grant HKU-716412E.
[***] Corresponding author. Research supported by NSFC 11171086 and Natural Science Foundation of Hebei Province A2013201218.

Q. Gu, P. Hell, and B. Yang (Eds.): AAIM 2014, LNCS 8546, pp. 32–43, 2014.

1 Introduction

The *one-way trading problem*, which was introduced by El-Yaniv *et al.*[10, 11] and Borodin *et al.*[6], involves selling a fixed amount of a product to a sequence of buyers, with the objective of maximizing the seller's revenue. A major difference between this problem and other general revenue maximization problems commonly studied in economics and computer science is that for the general problems, the seller has some control of the prices; he can determine the amount and the price of product to be sold to each buyer. However, for the one-way trading problem, a seller has no control of the prices, and when a buyer arrives, he can only determine the amount of the product to be sold at the then-prevailing market price. There are many applications that can be modeled as a one-way trading problem. One example is money-exchange, in which a seller has some initial asset, say US dollars, and he wants to sell them at the price of some target asset, say yen. In fact, the one-way trading problem is formulated as a money exchange problem in [10]. The exchange rate fluctuates everyday. To maximize the amount of yen gained, the seller needs to decide, for each day, the right amount of US dollars to be changed at the exchange rate used on that day. Other applications such as stock selling in a stock market and electricity selling in a power grid can also be modeled naturally as one-way trading problem.

It is easy to solve the offline version of the problem; if the seller knows all the future prices, he can simply wait for the highest price and then sell all his product at that price. However, our problem is online in nature, and without knowledge of future prices, a player cannot be sure whether the current price is the highest. More formally, in our one-way trading problem, there is a seller who has L units of product to be sold, and there is a sequence of buyers $u_1, u_2, \ldots, u_\sigma$ arriving. When a buyer u_i arrives, the then-prevailing unit price p_i is revealed and the seller needs to decide the amount x_i of product to be sold to u_i at price p_i, and the objective is to maximize $\sum_i p_i x_i$ subject to $\sum_i x_i \leq L$. The main features of the problem that make it difficult and interesting include: (1) the seller has no control of the prices, which fluctuates with time, and (2) he does not have any knowledge about the future prices, i.e., when u_i arrives, he does not know any price p_j where $j > i$, and (3) he needs to decide the amount of product to be sold to a buyer u_i as soon as u_i arrives.

Previous results
After introducing the one-way trading problem, El-Yaniv *et al.* gave in [11] an algorithm for the problem that works under the assumption that there are a lower bound m and an upper bound M on the market prices such that $p_i \in [m, M]$ for all p_i, and that these bounds m and M are known to the algorithm. They proved that their algorithm has competitive ratio $O(\log(M/m))$, and showed that it is optimal by deriving a matching lower bound. They also studied the case when only the ratio M/m is known, and gave an optimal algorithm for this case. Without knowledge of M/m in advance, an algorithm with

competitive ratio $O(\log(M/m)\log^{1+\epsilon}(\log(M/m)))$ was given in [11]. [1] More recently, Fujiwara *et al.*[12] have studied the one-way trading problem under the assumption that the input prices follow some given probability distribution. In [9], Chen *et al.* introduced the *planning game problem*, which is similar to the one-way trading problem, and they gave an algorithm for their problem which imposes some different constraint on the prices: instead of assuming that $p_i \in [m, M]$ for some price range $[m, M]$, their algorithm assumes that the difference between any two consecutive prices p_i and p_{i+1} is not too large, or more precisely, they assumed that for any i, $p_i/\beta \le p_{i+1} \le \alpha p_i$ for some fixed $\alpha, \beta > 1$. They showed that if there are n buyers, their algorithm has competitive ratio $\frac{n\alpha\beta-(n-1)(\alpha+\beta)+(n-2)}{\alpha\beta-1}$.

In [11], El-Yaniv *et al.* also studied another problem similar to the one-way trading problem, namely the *1-max-search* problem, in which there is a sequence of prices coming online, and when a price arrives, we have to decide immediately whether we accept the price or not. The objective is to accept the highest price. By assuming that all prices fall in the range $[m, M]$ and these bounds m and M are known, they gave an algorithm for this problem with competitive ratio $O(\sqrt{M/m})$, i.e., the ratio of the highest price and the price accepted by the algorithm is $O(\sqrt{M/m})$. In [14], Lorenz *et al.* generalized the 1-max-search problem to the *k-max-search* problem, in which the objective is to accept the k highest prices. By requiring that the bounds m and M are known, they gave an optimal algorithm for the problem, which has competitive ratio $\sqrt[k+1]{k^k(M/m)}$.

For recent related research on revenue maximization that allows price setting, we mention the *auction problem* [4, 13] and the *pricing problem* [1–3, 5, 7, 8]. For the auction problem, there are bidders competing for the products by sending their bids to the auctioneer, and the auctioneer chooses some bidders, and determines the price and amount of products to be sold to each chosen bidder. For the pricing problem, we have studied an interesting version in [16] in which the seller has m units of products to sell and each buyer has a valuation (i.e., price at which he is willing to buy) represented by a function $v(x)$, which gives the valuation per unit if x units are purchased. When the highest valuation v^* is known, we gave an algorithm with competitive ratio $O(\log v^*)$. Moreover, this algorithm was shown to be asymptotically optimal by giving a matching lower bound. We also studied in [17] an extension of this problem, in which there are multiple types of products and each user is interested in a particular bundle of products.

Our Contribution

In this paper, we consider the unbounded one-way trading algorithm that does not need to impose any constraint on the market prices, and we derive a bound on its competitive ratio that depends directly on the input, or more precisely, depends on $r^* = p^*/p_1$, the ratio of the highest price $p^* = \max_i p_i$ and the

---

[1] Remark 4 in [11] said that it is possible to achieve an upper bound of $O(\log(M/m)(\log(\log^{(k)}(M/m)))^{1+\epsilon})$, however, this bound contradicts with the general lower bound of the competitive ratio in this paper. Thus, the claimed upper bound in [11] does not hold for general case.

first price p_1 (in fact, our algorithm will treat p_1 as the lowest price and ignore any prices lower than p_1). Furthermore, the algorithm does not make any assumption on the number of prices p_i in the input sequence and an adversary can terminate the sequence at any time by sending buyers with extremely low prices. In fact, we propose a generic one-way trading algorithm whose behavior depends on some given function $f(x)$, which can be any function satisfying the following conditions: (i) It is non-increasing, and (ii) $\int_1^\infty f(t)dt$ is bounded. Roughly speaking, $f(x)$ helps us determine the amount of products the seller should sell at price x. We show that by using $f(x)$ in our generic algorithm, we have a one-way trading algorithm with competitive ratio $O(\frac{1}{r^* f(r^*)})$. Thus, to get a small competitive ratio, it suffices to find a $f(x)$ that satisfies (i) and (ii), and $f(x)$ is as large as possible. We observe that the following class of functions satisfies our requirements:

$$\frac{1}{x \log x (\log^{(2)} x) \ldots (\log^{(h-1)} x)(\log^{(h)} x)^{1+\epsilon}},$$

where h is any positive integer and ϵ is any positive real number, and where $\log^{(k)} x$ denotes the function $\log \log \ldots \log x$, which applies the logarithm function k times to x. Based on these functions, (a different function for each different value of h and ϵ) our generic algorithm gives us a class of one-way trading algorithms such that for any fixed positive integer h and positive number ϵ, we have an algorithm $A_{h,\epsilon}$ such that when $\log^{(h)} r^* > 1$, $A_{h,\epsilon}$ has competitive ratio $O((\log r^*) \ldots (\log^{(h-1)} r^*)$ $(\log^{(h)} r^*)^{1+\epsilon})$; otherwise, its competitive ratio is bounded by some constant Γ_h depending only on h. We also show that the bounds are almost tight by employing the divergence of the same class of function when $\epsilon = 0$ to design an adversary such that, given any online algorithm A for the problem, the adversary gives a sequence of buyers σ such that the ratio between the revenue obtained by an optimal offline algorithm on σ and that obtained by A is $\Omega((\log r^*) \ldots (\log^{(h-1)} r^*)(\log^{(h)} r^*))$ for any positive integer h. Moreover, we show that our results still hold if the amount of products sold to each buyer is constrained to be at most a maximum amount specified by the buyer.

2   Upper Bound

Since products could be sold fractionally, we may assume, without loss of generality, that the seller has one unit of product to sell. The offline version is easy to solve: the whole product is assigned to the buyer with the highest market price. However, for the online version, we have no information about the future prices, including the bound of the highest market price. If the whole amount of product has been sold by the time a buyer with very high market price arrives, the performance will be poor. Thus, we must keep or reserve some amount in case there is a future buyer with a higher market price. On the other hand, if we reserve too much for the possible buyer with higher market price and assign very little to the buyers who have come already, the performance will be also poor since the possible buyer with higher price may not come. Thus, to have a good performance, the amount sold and the amount remaining should be balanced nicely.

For the purposes of illustrating the main ideas of our algorithm only, consider the case when all prices are non-negative integers; in general, our algorithm is not restricted to integer prices. We make the following observations.

- Our algorithm should only sell products when the price is strictly higher than the maximum price that we have seen so far. For example, suppose the input sequence of prices is 1, 4, 2, 3, 6, 5, 12. We can ignore the prices 2, 3, 5 and do not sell any at these prices because the optimal offline algorithm will ignore these prices anyway; if our solution is competitive for the input 1, 4, 6, 12, it will also be competitive for the input 1, 4, 2, 3, 6, 5, 12. Therefore, we can focus on handling price sequences that are strictly increasing.
- If we have a good solution for a sequence of strictly increasing and consecutive prices, i.e., for the price sequence $1, 2, 3, ..., p^*$, then we can easily modify it to get a good solution for any price sequences that are strictly increasingly with the highest price p^*. For example, suppose that for the prices 1, 2, 3, 4, our algorithm sells an amount δ_1, δ_2, δ_3, δ_4 of products at prices 1, 2, 3, 4, respectively and thus obtains a revenue of $R = \delta_1 + 2\delta_2 + 3\delta_3 + 4\delta_4$. Then, for the strictly increasing price sequence 1, 3, 4, we can sell an amount of δ_1 at price 1, $\delta_2 + \delta_3$ at price 3, and δ_4 at price 4. Then, the revenue we obtain is $\delta_1 + 3(\delta_2 + \delta_3) + 4\delta_4 \geq R$.

Therefore, our algorithm can focus on strictly increasing and consecutive price sequences. For these sequences, we only need to determine the amount δ_i of products to be sold at price p_i. Since there is only one unit of product, we must have $\sum_{i=1}^{+\infty} \delta_i \leq 1$. Another property that is desirable is that the δ_is should be decreasing, i.e., $\delta_1 > \delta_2 > \delta_3 > ...$[2]; the leading δ_is should be large so that we can sell enough products even if the market crashes very early, i.e., the adversary declares immediately that there are no more buyers, or buyers with extremely low market prices. Then, for any input price sequence with highest value p^*, our algorithm will have revenue at least $\delta_1 + 2\delta_2 + ... + p^*\delta^* \geq (p^*)^2\delta^*/2$, and since no algorithm (including the offline optimal algorithm) can have revenue higher than p^*, the competitive ratio of this algorithm is $O(1/(p^*\delta^*))$ (Lemma 1).

Now we give the algorithm. The algorithm assigns amounts based on a non-increasing function $f(x)$, which computes the value of δ_i such that $\int_0^{+\infty} f(x)dx = 1$, $\int_0^1 f(x)dx = \delta_1$, $\int_1^2 f(x)dx = \delta_2$, ..., $\int_{i-1}^i f(x)dx = \delta_i$.

Let $(p_1, p_2, ..., p^*)$ be the sequence of strictly increasing transacted prices, i.e. prices at which the seller sells some (non-zero) amount to the buyer. For ease of analysis, we can normalize this sequence to be $(r_1, r_2, ..., r^*) = (1, p_2/p_1, ..., p^*/p_1)$ where the first price r_1 is 1 and the normalized maximum r^* is the ratio of the highest transacted price p^* to the lowest transacted price p_1. Any buyers with market price less than p_1 will be ignored. For the sake of simplicity, we shall

---

[2] The decreasing of δ_i can be argued easily. WLOG, assume that $p_1 < p_2$ and $\delta_1 \leq \delta_2$, we can show that the competitive ratio can be decreased by moving a small amount from δ_2 to δ_1. This process can continue until $\delta_1 > \delta_2$.

denote r_i as the normalized price of the i-th buyer and r^* as the highest normalized price. The online selling strategy is described below as Algorithm 1. Note that Algorithm 1 can handle non-integer prices.

Algorithm 1. Online Selling

    Initially, let $cr^* \leftarrow 0$. {}cr^* is the current highest normalized price.
    repeat
        when a buyer with normalized market price r comes
        if $r > cr^*$ **then**
            Assign $\int_{cr^*}^{r} f(x)dx$ products to this buyer.
            $cr^* \leftarrow r$
        end if
    until no buyer comes

Lemma 1. *Suppose r^* is the highest normalized market price, if $f(.)$ is a non-increasing function, the competitive ratio is at most $O(\frac{1}{r^* \cdot f(r^*)})$.*

Proof. The revenue received from Algorithm 1 is

$$r_1 \int_0^{r_1} f(r)dr + r_2 \int_{r_1}^{r_2} f(r)dr + \ldots + r^* \int_{r^{*-}}^{r^*} f(r)dr \geq \int_0^{r^*} r \cdot f(r)dr$$

where r^{*-} is the second highest normalized market price in the sequence.

    Since $f(r)$ is non-increasing, the revenue received from Algorithm 1 is at least

$$\int_0^{r^*} r \cdot f(r)dr \geq f(r^*) \int_0^{r^*} r\, dr = f(r^*) \cdot \frac{(r^*)^2}{2}.$$

Note that the maximum revenue is r^* given that the seller has only one unit to sell, and therefore, the competitive ratio is at most

$$\frac{r^*}{\int_0^{r^*} r \cdot f(r)dr} = O(\frac{1}{r^* \cdot f(r^*)}).$$

$\square$

In order to get a good performance, we need to find a non-increasing function $f(x)$ such that $\int_0^\infty f(x)dx$ converges to 1, or more simply, $\int_0^\infty f(x)dx = c$ for some constant c (as we can normalize it to 1 later), and for any $x > 1$, $f(x)$ is as large as possible. After assuming the first market price is 1, we may just analyze the property of $\int_1^\infty f(x)dx$. It is well known that $\int_1^\infty \frac{1}{x}dx$ diverges and thus $f(x) = 1/x$ is too large. Similarly as $\int_1^\infty \frac{1}{x^{1+\epsilon}}dx$ converges for any $\epsilon > 0$, $f(x) = 1/(x \cdot x^\epsilon)$ is too small. This suggests that $f(x) = 1/(x\xi(x))$ where $\xi(x)$ is an increasing function and $\xi(x) = o(x^\epsilon)$ for any $\epsilon > 0$. A good candidate for $\xi(x)$ is a poly-log function of x. This motivates us to focus on the class of functions $f(x) = 1/(x \log x \log^{(2)} x \ldots (\log^{(i)} x)^{1+\epsilon})$ where $\epsilon > 0$ and $\log^{(i)} x$ denotes the application of the logarithm function i times to x, where $i \geq 0$. Now we define the class of functions formally.

Definition 1. *Assume real number $\epsilon \geq 0$, integer $i \geq 0$, $b_0 = 1$, and $b_{i+1} = e^{b_i}$, define function $q_{i,\epsilon}(x)$ for $x \geq b_i$ as follows.*

$$q_{i,\epsilon}(x) = \begin{cases} x^{1+\epsilon} & \text{if } i = 0 \\ x \cdot q_{i-1,\epsilon}(\ln x) & \text{if } i > 0 \end{cases}$$

Thus, $q_{1,\epsilon}(x) = x \cdot (\ln x)^{1+\epsilon}$, $q_{2,\epsilon}(x) = x \cdot (\ln x) \cdot (\ln^{(2)} x)^{1+\epsilon}$, and $q_{i,\epsilon}(x) = x \cdot (\ln x) \cdot (\ln^{(2)} x) \cdot \ldots \cdot (\ln^{(i)} x)^{1+\epsilon}$. The following lemma gives the condition when $\int_{b_i}^{+\infty} \frac{1}{q_{i,\epsilon}(x)} dx$ converges.

Lemma 2. *For each integer $i \geq 0$, $\int_{b_i}^{+\infty} \frac{1}{q_{i,\epsilon}(x)} dx$ converges if and only if $\epsilon > 0$, in particular, $\int_{b_i}^{+\infty} \frac{1}{q_{i,0}(x)} dx$ diverges.*

Proof. By induction on i. When $i = 0$, $b_0 = 1$, it is easy to see that $\int_{b_0}^{+\infty} \frac{1}{q_{0,\epsilon}(x)} dx$ $= \int_1^{+\infty} \frac{1}{x^{1+\epsilon}} dx$ converges if and only if $\epsilon > 0$. Assume that the hypothesis is true for $i - 1$. As $b_i = e^{b_{i-1}}$, we have $\int_{b_i}^{+\infty} \frac{1}{q_{i,\epsilon}(x)} dx = \int_{e^{b_{i-1}}}^{+\infty} \frac{1}{x \cdot q_{i-1,\epsilon}(\ln x)} dx = \int_{b_{i-1}}^{+\infty} \frac{1}{q_{i-1,\epsilon}(y)} dy$, where $y = \ln x$. Thus, $\int_{b_i}^{+\infty} \frac{1}{q_{i,\epsilon}(x)} dx$ converges if and only if $\epsilon > 0$. $\qquad\square$

The following theorem shows the competitive ratio of Algorithm 1 by constructing $f(x)$ from $q_{i,\epsilon}(x)$, i.e., proving that the area under $f(x)$ when $x > 0$ is bounded and $f(x)$ is non-increasing and defined for all $x > 0$.

Theorem 1. *Suppose r^* is the highest normalized market price, there exists an online algorithm $A_{h,\epsilon}$ for the unbounded one-way trading problem with competitive ratio $O(1)$ if $r^* < b_h$ and $O(q_{h-1,\epsilon}(\log r^*))$ if $r^* \geq b_h$ for any fixed positive integer h and any real number $\epsilon > 0$.*

Proof. For any fixed positive integer h, b_h is a constant such that $\ln^{(h)} b_h = 1$. From Lemma 2, for any real number $\epsilon > 0$, suppose $\int_{b_h}^{+\infty} \frac{1}{q_{h,\epsilon}(x)} dx$ converges to a constant, say c. As $\ln^{(h)}(x) \geq 1$ when $x \geq b_h$, we define function $f_{h,\epsilon}(x)$ as follows.

$$f_{h,\epsilon}(x) = \begin{cases} \frac{1}{b_h + c \cdot q_{h,\epsilon}(b_h)} & \text{if } 0 < x < b_h \\ \frac{q_{h,\epsilon}(b_h)}{b_h + c \cdot q_{h,\epsilon}(b_h)} \cdot \frac{1}{q_{h,\epsilon}(x)} & \text{if } x \geq b_h \end{cases}$$

It can be verified that $\int_0^{+\infty} f_{h,\epsilon}(x) dx = 1$ and $f_{h,\epsilon}(x)$ is non-increasing (since $f_{h,\epsilon}(x) = f_{h,\epsilon}(b_h)$ is a constant when $0 < x < b_h$ and $f_{h,\epsilon}(x)$ is decreasing when $x \geq b_h$), i.e., $f_{h,\epsilon}(x)$, which depends on h and ϵ, satisfies the requirement of Algorithm 1, which gives $A_{h,\epsilon}$. By Lemma 1, we can analyze the competitive ratio w.r.t. the highest market price r^*.

- If $r^* < b_h$, the competitive ratio is $O(\frac{b_h + c \cdot q_{h,\epsilon}(b_h)}{r^*})$, which is $O(1)$.
- If $r^* \geq b_h$, the competitive ratio is $O(\frac{1}{r^* \cdot f_{h,\epsilon}(r^*)})$, which is $O(q_{h-1,\epsilon}(\log r^*))$,
  i.e., $O(\log r^* \log^{(2)} r^* ... (\log^{(h)} r^*)^{1+\epsilon})$.

$\square$

Now we consider the case where each buyer has a maximum amount of products he wants to buy at the market price. This variant can be regarded as an extension of the previous part. Algorithm 1 assigns products only based on the buyer's market price with no regard for how much the buyer is able to buy, i.e., the buyer's quota. Modify Algorithm 1 by taking into consideration the buyer's quota, we have the following conclusion. (For details, please see appendix.)

Theorem 2. *For the unbounded one-way trading problem, if each buyer has a maximum amount of products he wants to buy at the market price, there is an online selling strategy with competitive ratio $O(1)$ if $r^* < b_h$ and $O(q_{h-1,\epsilon}(\log r^*))$ if $r^* \geq b_h$ for any fixed integer h and any real number $\epsilon > 0$, where r^* is the highest normalized market price.*

3 Lower Bound

In this part, we present a lower bound for the competitive ratio of the unbounded one-way trading problem. We will show that the lower bound and the upper bound given in Section 2.1 are almost tight; in another words, Algorithm 1 is near optimal.

To derive a lower bound on the competitive ratio, we give an adversary that determines the sequence of prices $p_1, p_2, p_3 \ldots$, and whenever the seller has sold some products, the adversary checks the total revenue the seller has accumulated so far, and if it is not competitive, the adversary declares immediately that there are no more buyers, or buyers with extremely low market price, i.e., the market "crashes". The prices p_i's grow exponentially, i.e., $p_i = \Theta(e^i)$. The adversary also determines for each i a bound Δ_i, which is the minimum amount of product sold during the first i prices in order to prevent the market crashes. In other words, if the amount of product sold at price $p_1, p_2, \ldots, p_k$ are $s_1, s_2, \ldots, s_k$, respectively, and $s_1 \geq \Delta_1, s_1 + s_2 \geq \Delta_2, \ldots, \sum_{k=1}^{j-1} s_k \geq \Delta_{j-1}$, and $\sum_{k=1}^{j} s_k < \Delta_j$, the market crashes immediately at price p_j. Note that in such case, the seller has sold at most $\Delta_j - \Delta_{j-1}$ unit of product at p_j, and since p_j is much larger than all previous prices, we would be able to show that the total revenue obtained by the seller will be dominated by the last transaction and is $O((\Delta_j - \Delta_{j-1})p_j)$. On the other hand, an offline algorithm can sell the whole unit of product at p_j and gets the maximum revenue p_j. Thus the competitive ratio of the algorithm is $\Omega(\frac{1}{\Delta_j - \Delta_{j-1}})$ if the adversary "crashes" the market after p_j. The challenge for getting a large lower bound is to decide the Δ_is such that (i) they are unbounded (i.e., $\Delta_i \to \infty$ when $i \to \infty$) so that the seller will fail eventually to meet the requirement on the minimum amount of product sold, and (ii) $\Delta_i - \Delta_{i-1}$ is as small as possible. The bound $\Delta_i = \frac{1}{e+1} + \frac{1}{e+2} + \ldots + \frac{1}{e+i}$ can be considered as a good candidate, which will lead us to a lower bound of $\Omega(i)$ or $\Omega(\log p_i)$ when the highest price $p_i = O(e^i)$. Below, we describe some other Δ_i's that will lead us to a substantially larger bound.

From Lemma 2, we know that $q_{h,0}(x)$ is a good candidate such that there is a $b_h > 0$ causing $\int_{b_h}^{+\infty} \frac{1}{q_{h,0}(x)} dx$ to diverge, where $\ln^{(h)} b_h = 1$. The adversary in Algorithm 2 uses $\sum_{k=1}^{j} \frac{1}{q_{h,0}(b_h+k-1)}$ as a candidate for Δ_j as mentioned before and s_j is the amount of products assigned to buyer u_j. Since $1/q_{h,0}(x)$ is monotone decreasing and $\int_{b_h}^{+\infty} \frac{1}{q_{h,0}(x)} dx$ diverges for any fixed integer $h > 0$, the sum $\sum_{k=1}^{\infty} \frac{1}{q_{h,0}(b_h+k-1)}$ diverges. Therefore, Algorithm 2 must be terminated on some buyer since the seller has only one unit of product.

Algorithm 2. Adversary for online selling

Assume that the seller has one unit of product to sell.
Let $j \leftarrow 0$.
repeat
  Let $j \leftarrow j + 1$.
  Send buyer u_j with market price e^{b_h+j-1} to the seller.
  The seller sells s_j product to buyer u_j.
until $\sum_{k=1}^{j} s_k \leq \sum_{k=1}^{j} \frac{1}{q_{h,0}(b_h+k-1)}$

Assume the adversary stops sending buyers after the arrival of buyer u_j. From Algorithm 2, the total revenue received is $\sum_{k=1}^{j} s_k \cdot e^{b_h+k-1}$, while the maximum offline revenue is e^{b_h+j-1}. The following lemma estimates the total revenue received from Algorithm 2.

Lemma 3. $\sum_{k=1}^{j} s_k \cdot e^{b_h+k-1} = O(\frac{e^{b_h+j-1}}{q_{h,0}(b_h+j-1)})$

Proof. From the adversary's strategy, at any step $j' < j$,

$$\sum_{k=1}^{j'} s_k > \sum_{k=1}^{j'} \frac{1}{q_{h,0}(b_h + k - 1)},$$

and in the last step j,

$$\sum_{k=1}^{j} s_k \leq \sum_{k=0}^{j} \frac{1}{q_{h,0}(b_h + k - 1)}.$$

Therefore,

$$\sum_{k=1}^{j} s_k \cdot e^{b_h+k-1} \leq \sum_{k=1}^{j} \frac{e^{b_h+k-1}}{q_{h,0}(b_h + k - 1)}.$$

In Lemma 4, we show that

$$\frac{e^{b_h+k}}{q_{h,0}(b_h + k)} \cdot \frac{q_{h,0}(b_h + k - 1)}{e^{b_h+k-1}} = \frac{e \cdot q_{h,0}(b_h + k - 1)}{q_{h,0}(b_h + k)} \geq c$$

for some constant $c > 1$ and any $k \geq 1$. Thus,

$$\sum_{k=1}^{j} \frac{e^{b_h+k-1}}{q_{h,0}(b_h+k-1)} \leq \frac{e^{b_h+j-1}}{q_{h,0}(b_h+j-1)} \cdot \frac{1}{1-1/c} = O\left(\frac{e^{b_h+j-1}}{q_{h,0}(b_h+j-1)}\right).$$

$\square$

Lemma 4. *For any integer $h \geq 1$ and $k \geq 1$, there exist a constant $c > 1$, such that*

$$\frac{e \cdot q_{h,0}(b_h+k-1)}{q_{h,0}(b_h+k)} \geq c$$

Proof. Based on the logarithmic characteristic of $q_{h,0}(x)$, i.e., the increasing rate is decreasing with the increase of x, $\frac{q_{h,0}(b_h+k-1)}{q_{h,0}(b_h+k)}$ achieves the lowest value when $k = 1$. Thus, it is sufficient to prove the following inequality for any integer $h \geq 1$.

$$\frac{e \cdot q_{h,0}(b_h)}{q_{h,0}(b_h+1)} \geq c > 1 \tag{1}$$

We prove Inequality (1) by induction on h.

Basis step: $h = 1$. As $b_h = e$,

$$\frac{e \cdot q_{1,0}(b_1)}{q_{1,0}(b_1+1)} = \frac{e \cdot b_1 \cdot \ln b_1}{(b_1+1) \cdot \ln(b_1+1)} \approx 1.513 > 1$$

Induction step: Assume Inequality (1) is true for h,

$$\frac{e \cdot q_{h,0}(b_h)}{q_{h,0}(b_h+1)} = \frac{e \cdot b_h \cdot \ln b_h \cdot \ldots \cdot \ln^{(h)} b_h}{(b_h+1) \cdot \ln(b_h+1) \cdot \ldots \cdot \ln^{(h)}(b_h+1)}$$

$$= \frac{e \cdot b_h}{b_h+1} \prod_{h'=1}^{h} \frac{\ln^{(h')} b_h}{\ln^{(h')}(b_h+1)} \geq c > 1 \tag{2}$$

Then for $h + 1$,

$$\frac{e \cdot q_{h+1,0}(b_{h+1})}{q_{h+1,0}(b_{h+1}+1)} = \frac{e \cdot b_{h+1} \cdot \ln b_{h+1} \cdot \ldots \cdot \ln^{(h+1)} b_{h+1}}{(b_{h+1}+1) \cdot \ln(b_{h+1}+1) \cdot \ldots \cdot \ln^{(h+1)}(b_{h+1}+1)}$$

$$= \frac{e \cdot b_{h+1}}{b_{h+1}+1} \prod_{h'=1}^{h+1} \frac{\ln^{(h')} b_{h+1}}{\ln^{(h')}(b_{h+1}+1)} \tag{3}$$

We shall prove that $\frac{e \cdot b_{h+1}}{b_{h+1}+1} \prod_{h'=1}^{h+1} \frac{\ln^{(h')} b_{h+1}}{\ln^{(h')}(b_{h+1}+1)}$ as given in Equation (3) is larger than

$\frac{e \cdot b_h}{b_h+1} \prod_{h'=1}^{h} \frac{\ln^{(h')} b_h}{\ln^{(h')}(b_h+1)}$ as given in Equation (2) term by term. Since $\ln b_{h+1} = b_h$, we have $\ln(b_{h+1}+1) < b_h + 1$. Thus, for any $2 \le h' \le h+1$,

$$\frac{\ln^{(h')} b_{h+1}}{\ln^{(h')}(b_{h+1}+1)} > \frac{\ln^{(h'-1)} b_h}{\ln^{(h'-1)}(b_h+1)}$$

As for the first few terms, because $\ln b_{h+1} = b_h$ and $\ln(b_{h+1}+1) = b_h + \delta$, where $\delta \ll 1$, we have

$$\frac{b_{h+1} \cdot \ln b_{h+1}}{(b_{h+1}+1) \cdot \ln(b_{h+1}+1)} \ge \frac{b_h}{b_h + 1}$$

Thus, we have shown that

$$\frac{e \cdot q_{h,0}(b_h + k)}{q_{h,0}(b_h + k + 1)} \ge c > 1.$$

□

Based on the above analysis, Theorem 3 gives the lower bound on the competitive ratio of the unbounded one-way trading problem.

Theorem 3. *The competitive ratio of the unbounded one-way trading problem is at least* $\Omega(q_{h,0}(\log r^*)) = \Omega(\log r^* \cdot \log^{(2)} r^* \cdot \ldots \cdot \log^{(h+1)} r^*)$ *where* r^* *is the highest normalized market price and* $h > 0$ *is any fixed integer.*

Proof. Assume that Algorithm 2 terminates on some buyer u_j. As mentioned before, the revenue received from Algorithm 2 is $\sum_{k=1}^{j} s_k \cdot e^{b_h+k-1} = O(\frac{e^{b_h+j-1}}{q_{h,0}(b_h+j-1)})$ (Lemma 3), and the maximum offline revenue is e^{b_h+j-1} by assigning the whole product to buyer u_j with the market price e^{b_i+j-1}. As $p_j = e^{b_h+j-1}$, the performance ratio is at least $\Omega(q_{h,0}(b_h + j - 1)) = \Omega(\log p_j \log^{(2)} p_j \ldots \log^{(h+1)} p_j)) = \Omega(\log r^* \log^{(2)} r^* \ldots \log^{(h+1)} r^*))$ since b_h can be regarded as a constant and $r^* = p_j/p_1 = e^{j-1}$. □

4 Conclusion

There are many real applications where the market price fluctuates and cannot be controlled by the seller. It is a problem of practical interest to find a good revenue-maximizing (or profit-maximizing) selling strategy for the seller in such a situation. This paper has made an attempt towards this direction. However, the strategy prescribed in this paper may not be too practical in the sense that, for example, products are not sold when the market price decreases. The reality is that, in practice, the seller may have a fixed time-frame to sell and cannot wait forever for the buyer with the highest price to arrive, and price movements from one moment to the next may not be drastic or arbitrary. Additional assumptions and/or constraints to the unbounded one-way trading problem to reflect such practical realities will be studied in our next attempt and hopefully could lead to more practical selling strategies.

References

1. Babaioff, M., Dughmi, S., Kleinberg, R., Slivkins, A.: Dynamic Pricing with Limited Supply. In: Proceedings of the 13th ACM Conference on Electronic Commerce, pp. 74–91 (2012)
2. Badanidiyuru, A., Kleinberg, R., Singer, Y.: Learning on a budget: posted price mechanisms for online procurement. In: Proc. of the 13th ACM Conference on Electronic Commerce, pp. 128–145 (2012)
3. Balcan, M.-F., Blum, A., Mansour, Y.: Item pricing for revenue maximization. In: Proceedings of the 9th ACM Conference on Electronic Commerce, pp. 50–59 (2008)
4. Blum, A., Hartline, J.D.: Near-optimal online auctions. In: Proceedings of the 16th Annual ACM-SIAM Symposium on Discrete Algorithms, pp. 1156–1163 (2005)
5. Blum, A., Gupta, A., Mansour, Y., Sharma, A.: Welfare and Profit Maximization with Production Costs. In: Proceedings of 52th Annual IEEE Symposium on Foundations of Computer Science, pp. 77–86 (2011)
6. Borodin, A., El-Yaniv, R.: Online Computation and Competitive Analysis. Cambridge University Press (1998)
7. Chakraborty, T., Even-Dar, E., Guha, S., Mansour, Y., Muthukrishnan, S.: Approximation schemes for sequential posted pricing in multi-unit auctions. In: Saberi, A. (ed.) WINE 2010. LNCS, vol. 6484, pp. 158–169. Springer, Heidelberg (2010)
8. Chakraborty, T., Huang, Z., Khanna, S.: Dynamic and non-uniform pricing strategies for revenue maximization. In: Proceedings of 50th Annual IEEE Symposium on Foundations of Computer Science, pp. 495–504 (2009)
9. Chen, G.-H., Kao, M.-Y., Lyuu, Y.-D., Wong, H.-K.: Optimal buy-and-hold strategies for financial markets with bounded daily returns. SIAM J. Compt. 31(2), 447–459 (2001), A preliminary version appeared in STOC 1999, pp. 119–128
10. El-Yaniv, R., Fiat, A., Karp, R.M., Turpin, G.: Competitive analysis of financial games. In: Proceedings of 50th Annual IEEE Symposium on Foundations of Computer Science, pp. 372–333 (1992)
11. El-Yaniv, R., Fiat, A., Karp, R.M., Turpin, G.: Optimal search and one-way trading online algorithms. Algorithmica 30(1), 101–139 (2001)
12. Fujiwara, H., Iwama, K., Sekiguchi, Y.: Average-case competitive analyses for one-way trading. Journal of Combinatorial Optimization 21(1), 83–107 (2011)
13. Koutsoupias, E., Pierrakos, G.: On the Competitive Ratio of Online Sampling Auctions. In: Saberi, A. (ed.) WINE 2010. LNCS, vol. 6484, pp. 327–338. Springer, Heidelberg (2010)
14. Lorenz, J., Panagiotou, K., Steger, A.: Optimal algorithms for k-search with application in option pricing. Algorithmica 55, 311–328 (2009); A preliminary version appeared in Arge, L., Hoffmann, M., Welzl, E. (eds.) ESA 2007. LNCS, vol. 4698, pp. 275–286. Springer, Heidelberg (2007)
15. Myerson, R.B.: Optimal auction design. Mathematics of Operations Research 6, 58–73 (1981)
16. Zhang, Y., Chin, F.Y.L., Ting, H.-F.: Competitive Algorithms for Online Pricing. In: Fu, B., Du, D.-Z. (eds.) COCOON 2011. LNCS, vol. 6842, pp. 391–401. Springer, Heidelberg (2011)
17. Zhang, Y., Chin, F.Y.L., Ting, H.-F.: Online pricing for bundles of multiple items. Journal of Global Optimization 58(2), 377–387

The Complexity of Degree Anonymization by Vertex Addition

Robert Bredereck*, Vincent Froese**, Sepp Hartung, André Nichterlein,
Rolf Niedermeier, and Nimrod Talmon***

Institut für Softwaretechnik und Theoretische Informatik, TU Berlin, Germany
{robert.bredereck,vincent.froese,sepp.hartung,andre.nichterlein,
rolf.niedermeier}@tu-berlin.de, nimrodtalmon77@gmail.com

Abstract. Motivated by applications in privacy-preserving data publishing, we study the problem of making an undirected graph k-anonymous by adding few vertices (together with incident edges). That is, after adding these "dummy vertices", for every vertex degree d in the resulting graph, there shall be at least k vertices with degree d. We explore three variants of vertex addition (justified by real-world considerations) and study their (parameterized) computational complexity. We derive mostly (worst-case) intractability results, even for very restricted cases (including trees or bounded-degree graphs) but also obtain a few encouraging fixed-parameter tractability results.

1 Introduction

This work is concerned with making an undirected graph k-*anonymous*, that is, transforming it (at "low cost") into a graph where every vertex degree occurs either zero or at least k times. This graph modification scenario is motivated by data privacy requests in social networks; it focuses on degree-based attacks on identity disclosure of network nodes. Liu and Terzi [12] (also see Clarkson et al. [5] for an extended version) pioneered degree-based identity anonymization in graphs, which recently developed into a very active research field [1, 2, 3, 4, 10, 14, 18] with theoretical as well as practical work. So far, the most common models have relied on edge modifications (allowing either only edge addition or both edge addition and deletion) [2, 5, 10, 14, 12, 18]. We are aware of one theoretical work [1] that considers vertex deletion as modification operation; there mostly computational hardness results have been achieved. Chester et al. [3] started to investigate vertex addition; here we follow this line of research.

There is good reason why vertex addition may be preferred to other graph modification operations when aiming at k-anonymity. The central point here is the "utility" of the anonymized graph. For instance, in the edge addition scenario

---

* Supported by the DFG, project PAWS (NI 369/10).

** Supported by the DFG, project DAMM (NI 369/13).

*** Supported by DFG Research Training Group "Methods for Discrete Structures" (GRK 1408).

Q. Gu, P. Hell, and B. Yang (Eds.): AAIM 2014, LNCS 8546, pp. 44–55, 2014.

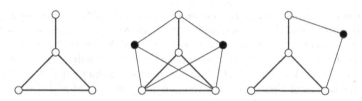

Fig. 1. Example ($k = 2$): The input graph on the left is not yet 2-anonymous. The graph in the middle shows a solution for the vertex cloning variant. The two added vertices (black) are clones of the middle vertex. Note that it is not possible to 2-anonymize the graph by adding only one clone. The graph on the right shows a solution for the general variant where the new vertex can be connected arbitrarily.

inserting a new edge destroys distance properties between vertices and indeed may introduce undesirable and misleading "fake relations". Adding new vertices and connecting them to some of the vertices of the original graph could avoid this problem and gives at least a better chance to preserve essential graph properties such as connectivity, shortest paths, or diameter. Chester et al. [3] provide a more thorough discussion of the benefits of vertex addition. The basic decision version of the problem we study is as follows.

DEGREE ANONYMIZATION (VA)
Input: A simple undirected graph $G = (V, E)$ and $k, t \in \mathbb{N}$.
Question: Is there a k-anonymous graph $G' = (V \cup V', E \cup E')$ such that $|V'| \leq t$ and $E' \subseteq \{\{u, v\} \subseteq V \cup V' \mid u \in V' \vee v \in V'\}$, where a graph is said to be k-anonymous if and only if every vertex degree in it appears either zero or at least k times ?

It is important to note that Chester et al. [3] studied a slightly different model, with decisive consequences for computational complexity: Their model gets as input a graph $G = (V, E)$, and integers t and k, and also a vertex subset $X \subseteq V$, and the task is to k-anonymize the degree sequence (that is, the vertex degrees sorted in ascending order) of $X \cup V'$ *and* the degree sequence of X. On the contrary, we consider the simpler model where $X = V$, and we require to k-anonymize only the degree sequence of $X \cup V'$ ($= V \cup V'$). Consider the following example highlighting this difference: Let $G = (V, E)$ be an eight-vertex graph containing one star with five leaves plus an edge (that is, $K_{1,5} \cup P_2$). Let $k = 2$ and $X = V$. Since G contains seven vertices of degree 1 and one vertex of degree 5, the solution of Chester et al. [3] will give four as the minimum number of vertices needed to 2-anonymize the degree sequence. In our model, however, this instance can be solved by adding only one new vertex, and connecting it to exactly five old vertices of degree 1 (e.g., transforming the $K_{1,5}$ into a $K_{2,5}$). This happens because the new vertex and the old vertex of degree 5 will form *together* a 2-anonymized "block". However, we believe that our results extend to the model of Chester et al. [3].

Table 1. Overview of our results: Each column represents a different problem variant, where VC (respectively Π, VA) stands for DEGREE ANONYMIZATION (VC) (respectively Π-PRESERVING DEGREE ANONYMIZATION (VA), DEGREE ANONYMIZATION (VA)). The first row refers to standard complexity analysis, while the remaining rows show results with respect to several parameters. Here, Δ denotes the maximum degree of the input graph, k is the degree of anonymity, s is the maximum number of added edges, and t is the maximum number of added vertices.

parameter	VC	Π	VA
-	NP-h. [Th. 1]	NP-h. [Th. 3]	weakly NP-h. [Th. 5]
Δ	NP-h., $\Delta = 3$ [Th. 1]	open	open
k	NP-h.[a], $k = 2$ [Th. 2]	NP-h.[a], $k = 2$ [Th. 3]	open
s	open	W[1]-h.[b] [Th. 4]	FPT [Th. 9]
t	W[2]-h. [Th. 2]	W[2]-h. [Th. 3]	XP[c] [Th. 6]
(Δ, k)	open	open	FPT [Th. 8]
(Δ, t)	open	open	FPT [Th. 7]
(t, k)	W[2]-h. [Th. 2]	W[2]-h. [Th. 3]	XP[c] [Th. 6]

[a] Even on trees.
[b] Only for Π = Distances.
[c] Open whether in FPT.

Our contributions. Partially answering an open question of Chester et al. [3], we show that DEGREE ANONYMIZATION (VA) is weakly NP-hard for a compact encoding of the input. Based on this, we provide several (fixed-parameter) tractability results, exploiting parameterizations by the maximum vertex degree of the input graph, the number of added vertices, and the number of (implicitly) added new edges. Moreover, we also study variants of DEGREE ANONYMIZATION (VA) where we only allow "cloned" vertices to be added (that is, identical copies of existing vertices with exactly the same neighborhood; this problem variant is denoted by DEGREE ANONYMIZATION (VC)) or where we explicitly demand the preservation of some desirable features such as distance properties (this problem variant is denoted by Π-PRESERVING DEGREE ANONYMIZATION (VA)). For these practically interesting variants we prove computational hardness already for very restricted cases (for instance even on trees). Table 1 surveys most of our results.

Due to the lack of space, most proofs are deferred to a full version.

2   Preliminaries and Problem Definitions

Preliminaries. We consider simple undirected graphs $G = (V, E)$. We denote by $\deg(v)$ the degree of a vertex $v \in V$ and by $\Delta := \max_{v \in V} \deg(v)$ the maximum degree of G. For an integer $0 \le i \le \Delta$, we define $B_i := \{v \in V \mid \deg(v) = i\}$, the *block of degree* i. We say that B_i is *empty* (*full*) if $B_i = \emptyset$ ($|B_i| \ge k$). For a full block B_i, we say that it has $z := |B_i - k|$ many *spare* vertices. We call a block B_i

good if it is empty or full, otherwise we call it *bad* (that is, $0 < |B_i| < k$). The *block sequence* $B(G) := \{(i, |B_i|) \mid B_i \neq \emptyset\}$ of G contains the degrees and sizes of each non-empty block. We call a block sequence *realizable* if it is the block sequence of a graph. For any graph G and for any pair of vertices u and v, we define $\text{dist}_G(u, v)$ to be the length of the shortest path between u and v in G (and $\text{dist}_G(u, v) = \infty$ if there is no path connecting u and v in G). For $n \in \mathbb{N}$, we define $[n] := \{1, 2, \ldots, n\}$.

Problem Definitions. DEGREE ANONYMIZATION (VA) allows to add vertices and edges incident to the new vertices. For a given solution of some yes-instance, we denote the actual number of new vertices by t' (obviously, $0 \leq t' \leq t$) and the total number of newly inserted edges by s.

Π-PRESERVING DEGREE ANONYMIZATION (VA) adds some constraints on the new edges. The idea is to preserve some desirable properties of the input graph. A general definition reads as follows.

Π-PRESERVING DEGREE ANONYMIZATION (VA)
Input: An undirected graph $G = (V, E)$ and $k, t \in \mathbb{N}$.
Question: Is there a k-anonymous graph $G' = (V \cup V', E \cup E')$ such that $|V'| \leq t$, $E' \subseteq \{\{u, v\} \subseteq V \cup V' \mid u \in V' \vee v \in V'\}$, and Π is preserved?

We now discuss what "Π is preserved" means for three properties we consider here. First, we say that the *connectedness* remains unchanged if any pair of disconnected vertices in G remains disconnected in G'. As introducing vertices and edges cannot disconnect vertices, this property can be formalized as $\forall\, u, v \in V\colon \text{dist}_G(u, v) = \infty \iff \text{dist}_{G'}(u, v) = \infty$. Second, we say that the *distances* remain unchanged if, for any pair of vertices in G, their distance is the same in G and G', formally, $\forall\, u, v \in V\colon \text{dist}_G(u, v) = \text{dist}_{G'}(u, v)$. Third, we say that the *diameter* remains unchanged if the diameter of G and G' is the same, formally, $\max_{u,v \in V} \text{dist}_G(u, v) = \max_{u,v \in V \cup V'} \text{dist}_{G'}(u, v)$. Note that the diameter property also considers paths between newly added vertices, whereas this is not the case for the first two properties. The reason for this is that the diameter is naturally defined as a single number, whereas the other properties store information for each pair of vertices.

A further restricted variant of DEGREE ANONYMIZATION (VA) is to use vertex cloning for modifying the graph. Here, cloning a vertex v means to introduce a new vertex v' and make v' adjacent to all neighbors of v. Formally, we arrive at the following problem:

DEGREE ANONYMIZATION (VC)
Input: An undirected graph $G = (V, E)$ and $k, t \in \mathbb{N}$.
Question: Can G be transformed into a k-anonymous graph by at most t vertex cloning operations?

We remark that there are different cloning variants: Consider two adjacent vertices u and v. If both u and v are cloned, then although the clone u' is adjacent to v and the clone v' is adjacent to u, the clones u' and v' may or may not be adjacent depending on the variant. If the clones are inserted simultaneously at the same time, then u' and v' are not adjacent. If the clones are inserted one after the other, then u' and v' are adjacent (no matter in what order they are inserted). Our results for DEGREE ANONYMIZATION (VC) (Theorems 1 and 2) hold for both variants.

Parameterized Complexity. An instance (I, k) of a parameterized problem consists of the actual instance I and an integer k being the *parameter* [6, 9, 16]. A parameterized problem is called *fixed-parameter tractable* (FPT) if there is an algorithm solving it in $f(k) \cdot |I|^{O(1)}$ time, whereas an algorithm with running time $O(|I|^{f(k)})$ only shows membership in the class XP (clearly, FPT $\subseteq$ XP). One can show that a parameterized problem L is (presumably) not fixed-parameter tractable with a *parameterized reduction* from a W[1]-hard or W[2]-hard problem (such as CLIQUE or SET COVER parameterized by solution size) to L. A parameterized reduction from a parameterized problem L to another parameterized problem L' is a function that, given an instance (I, k), computes in $f(k) \cdot |I|^{O(1)}$ time an instance (I', k') (with $k' \leq g(k)$) such that $(I, k) \in L \Leftrightarrow (I', k') \in L'$.

3   Constrained Degree Anonymization

Cloning seems a natural and well-motivated modification operation for social networks. Unfortunately, we face computational intractability even on very restricted input graphs with maximum degree three. The corresponding reduction is from INDEPENDENT SET.

Theorem 1. DEGREE ANONYMIZATION (VC) *is NP-hard, even on graphs with maximum degree three.*

Also from the viewpoint of fixed-parameter algorithms, we have no good news with respect to the standard parameter "solution size" t, even on trees. The corresponding reduction is from SET COVER.

Theorem 2. DEGREE ANONYMIZATION (VC) *is NP-hard and W[2]-hard with respect to the number t of clones, even if the degree k of anonymity is two and the graph is a tree.*

We can adjust the reduction from Theorem 2 to also work for Π-PRESERVING DEGREE ANONYMIZATION (VA).

Theorem 3. *For $\Pi \in \{Distances, Diameter, Connectivity\}$, Π-PRESERVING DEGREE ANONYMIZATION (VA) is NP-hard and also W[2]-hard with respect to the number t of added vertices, even if $k = 2$. For $\Pi \in \{Distances, Connectivity\}$, this is also true on trees.*

We strengthen (using a reduction from CLIQUE) parts of Theorem 3 by also showing that the problem remains intractable with respect to the typically larger parameter number s of added edges. For simplicity, we consider s as part of the input.

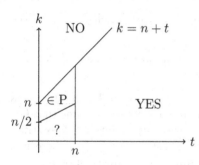

Fig. 2. Visualization of our knowledge about the complexity of DEGREE ANONYMIZATION (VA) depending on the values of k and t. The NO-cases follow from Observation 1, the YES-cases are due to Lemma 1, and the polynomial-time solvable cases follow from Lemma 2. For values inside the "?-area", the complexity is open for the graph problem (the number version is weakly NP-hard, see Theorem 5).

Theorem 4. *For $\Pi = $ Distances, Π-*PRESERVING DEGREE ANONYMIZATION *(VA) is W[1]-hard with respect to the number s of new edges.*

4 Plain Degree Anonymization

In this section, we study the general problem DEGREE ANONYMIZATION (VA), without any restrictions on how to connect the new vertices to the input graph. This freedom might raise hope to find solutions more efficiently. Indeed, settling the computational complexity of DEGREE ANONYMIZATION (VA) turns out to be tricky in that, on the one hand, we observe that several cases are fairly easy to solve, but we are not aware of any polynomial-time algorithm solving the problem in general. On the other hand, we can only prove weak NP-hardness for a number version of the problem.

In terms of fixed-parameter tractability, however, DEGREE ANONYMIZATION (VA) turns out to be more accessible. We obtain some fixed-parameter tractability results regarding, amongst others, certain (combined) parameters (for example, s, (Δ, k), and (Δ, t)), for some of which we proved the cloning and property-preserving problem variants to be W-hard.

Easy Cases. We start analyzing the complexity of DEGREE ANONYMIZATION (VA) with respect to the two input values k and t. Figure 2 provides a two-dimensional map indicating those combinations of k and t for which the problem is polynomial-time solvable or even trivial. In the following, we briefly state the corresponding results, starting with the following easy observation:

Observation 1. *Let $I = (G, k, t)$ be an instance of* DEGREE ANONYMIZATION *(VA) with G being an n-vertex graph. If $k > n + t$, then I is a no-instance.*

Next, we identify some yes-instances using the fact that—due to a result by Erdős and Kelly [7]—it is always possible to construct a regular graph if we are allowed to add enough, that is, at least n, new vertices.

Lemma 1. *Let $I = (G, k, t)$ be an instance of* DEGREE ANONYMIZATION *(VA) with G being an n-vertex graph. If $k \le n + t$ and $t \ge n$, then I is a yes-instance.*

We finish with some polynomial-time solvable instances which can be solved using f-factors [13, Chapter 10].

Lemma 2. DEGREE ANONYMIZATION (VA) *is polynomial-time solvable for* $2k > (n+t)$.

Proof. Let $I = (G = (V, E), k, t)$ be an instance of DEGREE ANONYMIZATION (VA) with $2k > (n+t)$. By Observation 1 and Lemma 1, we can assume that $k \leq n + t < 2k$ and $t < n$. Observe that in this case any solution (if existing) transforms G into a regular graph. Hence, the question is whether there is a regular graph H with at most $n+t$ vertices containing G as induced subgraph. We solve this problem by using the polynomial-time solvable f-FACTOR problem [13, Chapter 10], which is defined as follows:

f-FACTOR
Input: A graph $G = (V, E)$ and a function $f : V \to \mathbb{N}_0$.
Question: Is there an f-*factor*, that is, a subgraph $G' = (V, E')$ of G such
               that $\deg_{G'}(v) = f(v)$ for all $v \in V$?

Our algorithm is as follows. First, we guess in $O(n^2)$ time the number $t' \leq t$ of vertices that we will add and the degree d of the final regular graph H. Second, we create an f-FACTOR instance $(G' = (V', E'), f)$ as follows: We define the set $V' := V \cup X$ where X is a set of t' vertices. We start with an edgeless graph G' and then add all edges such that one endpoint is in X, formally, $E' = \{\{u, v\} \mid u \in X, v \in V'\}$. Finally, we set $f(v) := d - \deg_G(v)$ for all $v \in V$ and $f(u) = d$ for all $u \in X$. This completes the f-FACTOR instance I'. Clearly, I' is a yes-instance if and only if there exists a d-regular graph H with $n+t'$ vertices containing G as an induced subgraph. Hence, our algorithm runs in polynomial time.   □

(Weak) NP-Hardness. In Figure 2, we left open the computational complexity of DEGREE ANONYMIZATION (VA) for instances with $2k \leq n+t$. We now partially settle this question claiming that an equivalent number version of the problem is weakly NP-hard. To this end, notice that since we are not allowed to add any edges between old vertices, the actual structure of the input graph G becomes negligible and we only need to store the information of how many vertices of which degree it contains (that is, its block sequence $B(G)$):

Observation 2. *Let G and G' be two graphs with identical block sequences, that is, $B(G) = B(G')$. Then, for the DEGREE ANONYMIZATION (VA) instances $I := (G, k, t)$ and $I' := (G', k, t)$, it holds that I is a yes-instance if and only if I' is a yes-instance.*

Based on Observation 2, we can now define an equivalent number version of DEGREE ANONYMIZATION (VA).

BLOCK SEQUENCE ANONYMIZATION (VA)
Input: A realizable block sequence B and $k, t \in \mathbb{N}$.
Question: Is there a graph G with block sequence B such that (G, k, t) is a
               yes-instance of DEGREE ANONYMIZATION (VA)?

Note that BLOCK SEQUENCE ANONYMIZATION (VA) is a pure number problem. This helps us to develop a polynomial-time reduction from a weakly NP-hard version of the SUBSET SUM problem. An NP-hard problem is *weakly* NP-hard if it can be solved in polynomial-time provided that the input is encoded in unary. We conclude with the following theorem:

Theorem 5. BLOCK SEQUENCE ANONYMIZATION (VA) *is weakly NP-hard.*

Proof (sketch). The reduction is from the weakly NP-hard CHANGE MAKING problem [15]: Given integers $a_1, \ldots, a_n$, m, and b, are there nonnegative integers $x_1, \ldots, x_n$ such that $\Sigma_{i \in [n]} x_i \leq m$, and $\Sigma_{i \in [n]} x_i a_i = b$? We can assume, without loss of generality, that $\forall i, j : |a_i - a_j| \geq m^3$. If this property does not hold, then we simply multiply all numbers by m^3, that is, we set $a_i := m^3 \cdot a_i$ and $b := m^3 \cdot b$. It is easy to verify that this new instance is a yes-instance if and only if the original instance is a yes-instance.

We now create an equivalent BLOCK SEQUENCE ANONYMIZATION (VA) instance (B, k, t), with $t := m$ and $k := t(b + n + 5t + 1)$. The realizable block sequence B is the block sequence of a graph G, which is defined as follows. We introduce several gadgets, that is, subgraphs of G with distinguished vertices of specific degrees which play an important role in the correctness proof. In the following, we only specify the degrees of these *proper* vertices. To realize these gadgets, we add an appropriate number of degree-one neighbors. Our construction ensures that, when k-anonymizing G by adding t vertices, the degree-one vertices will always keep their degree. The construction works as follows.

Add a *b-gadget* consisting of $5t$ *base vertices* of degree $n + t$, b *count vertices* of degree $n + 2t - 1$, and $k - b - 5t$ *b-catch vertices* of degree $n + 2t$. For each $i \in [n]$, add one a_i-*gadget* consisting of one a_i-*vertex* of degree $a_i + n + 4t + 1$ and $k - 1$ a_i-*catch vertices* of degree $a_i + n + 5t + 1$. Finally, add a *dummy gadget* consisting of one *dummy vertex* of degree $n + 4t + 1$ and $k - 1$ *dummy catch vertices* of degree $n + 5t + 1$. This completes the construction. ◻

Tractability Results. While it remains open whether DEGREE ANONYMIZATION (VA) is NP-hard, the weak NP-hardness result for BLOCK SEQUENCE ANONYMIZATION (VA) (Theorem 5) indicates that also the graph problem may be hard to solve. Hence, a parameterized approach solving DEGREE ANONYMIZATION (VA) is reasonable. Notably we provide several (fixed-parameter) tractability results contrasting the hardness results for the constrained problem versions considered in Section 3.

A natural parameter to consider is the solution size t. Unfortunately, we do not know whether DEGREE ANONYMIZATION (VA) is fixed-parameter tractable with respect to t; we only know that DEGREE ANONYMIZATION (VA) is polynomial-time solvable when t is a constant.

Theorem 6. DEGREE ANONYMIZATION (VA) *parameterized by the maximum number t of added vertices is in XP.*

We can, however, "improve" containment in XP with respect to t to fixed-parameter tractability with respect to the combined parameter (t, Δ). Before proving the theorem, we introduce some notation and a helpful lemma.

For a set A of vertices whose addition transforms a graph $G = (V, E)$ into a k-anonymous graph, we call A an *addition set* and we write $G + A$ for the k-anonymous graph. Furthermore, the edges in $G + A$ having at least one endpoint in A (the "added" edges) are denoted by $E(A)$. Hence, $G + A = (V \cup A, E \cup E(A))$.

Clearly, for an addition set A of size t all vertices in $G + A$, except those in A, have degree at most $\Delta + t$ where Δ is the largest degree in G. It may happen that the degree of some (potentially all) vertices from A in $G + A$ is larger than $\Delta + t$. In this case, there are full blocks in $G + A$ of degree larger than $\Delta + t$ consisting only of vertices from A, implying that $t \geq k$. We call blocks corresponding to degrees greater than $\Delta + t$ *large-degree blocks*. Lemma 3 shows that we may assume that there are at most two large-degree blocks which are, in terms of their degree values, not too far away from each other. This will later allow us to guess their degrees.

Lemma 3. *Let (G, k, t) be a yes-instance of* DEGREE ANONYMIZATION (VA). *There is an addition set A of size at most t such that in $G + A$ there are no large-degree blocks, or there is only one large-degree block, or there are only two large-degree blocks whose degrees differ by exactly one.*

Theorem 7. DEGREE ANONYMIZATION (VA) *is fixed-parameter tractable with respect to the combined parameter (t, Δ).*

Proof. Our algorithm consists of three phases. First (Phase I), we guess what the solution looks like, specifically guessing the degrees of the good blocks, and the degrees of the new vertices, while respecting the guessed degrees of the good blocks. Then (Phase II), we use a bottom-up lazy method to solve the instance for the old vertices, but with respecting guessed degrees of the new vertices. Finally (Phase III), we use integer linear programming to solve the instance for the new vertices. A detailed description follows.

Phase I: we guess the subgraph induced by the new vertices (in $O(2^{t^2})$ time). We know, from Lemma 3, that the number of possible blocks in the solution is upper-bounded by $\Delta + t + 2 = O(\Delta + t)$. We guess the degrees of the large-degree blocks (in $O(n)$ time). Then, we guess, for each block, whether it is empty or full (in $O(2^{\Delta + t})$). Finally, we guess the degree of each new vertex (in $(\Delta + t)^t$ time). Phase I runs in $n \cdot O\left(2^{t^2} \cdot 2^{\Delta + t} \cdot (\Delta + t)^t\right) = n \cdot f_1(t, \Delta)$ time.

For ease of presentation, we say that we *move* a vertex up, meaning that we connect it to some new vertices, thus changing its degree and moving it to a different block of some desired degree. We can choose which new vertices to use in a round-robin way, but considering their guessed degrees (that is, each new vertex participates in the round-robin until it reaches its guessed degree).

Phase II: we perform the following bottom-up lazy method. We start from the lowest degree block, and work all the way up to the highest degree block. If the current block B_i is guessed to become empty, then we move its vertices up, to the first block above it which is guessed to become full (if there is a gap greater than t to such a block, we halt with a negative answer). Otherwise, if it is guessed to become full, then we distinguish between the following two cases: if the number of old vertices in the block plus the number of new vertices guessed to be in this block is at least k, then we do nothing, because it means that this block is already anonymized with respect to the old vertices, and continue to the next block. Otherwise, B_i has a shortage of some z_i many vertices to become full, so we find the maximum $j < i$ such that the number of old vertices in B_j plus the number of new vertices guessed to be in B_j is greater than k (specifically, equals to $k + z_j$ for some z_j spare vertices in B_j; if the gap $i - j$ is greater than t, then we halt with a negative answer, because B_i cannot be k-anonymized). We move $\min(z_i, z_j)$ spare vertices from B_j to B_i. If, after moving these spare vertices, B_i still needs some more vertices (that is, if $z_i > z_j$), then we repeat this step once more, looking for the maximum $j' < j$ such that the number of old vertices in $B_{j'}$ plus the number of new vertices guessed to be in $B_{j'}$ is greater than k, until we have enough vertices in the current block. If in the end of this phase, all of the blocks are anonymized, we continue to the next phase. The overall cost of Phase II is $O(\Delta + t)^3 = f_2(t, \Delta)$.

Our approach is lazy for two reasons. The first reason is that we use the spare vertices from the *closest* full block below the current one. The second reason is that we move the minimum number of vertices to make the blocks anonymized with respect to the old vertices, that is, we only change the bad blocks to become full, but not overfull.

Phase III: We check if we reached the exact guessed total number s of edges added. If so, then we halt with a positive answer, as this means that the new vertices reached their guessed degrees. If we reached a larger number, then we halt with a negative answer, since Phase II is lazy, it means that we cannot k-anonymize the graph using the guessed number of edges added. If we reached a smaller number, then we still have some hope of reaching s, because of the laziness of Phase II, so we try to move some more vertices, until we reach the guessed total number of edges added, while not destroying the anonymity of the blocks. To this end, denote the number of spare vertices in each full block B_i by z_i. Notice that we can move any number of up to z_i vertices from this block, to any full block above it, and no other moves are possible. Now our problem reduces to the following integer linear program:

Input: n' numbers $\{z'_1, \ldots, z'_{n'}\}$, $n' \times m'$ matrix $A = a_{ij}$, and integer Z.
Task: maximize $\sum_{i \in [n']} \sum_{j \in [m']} a_{ij} x_{ij}$ such that $\sum_{i \in [n']} \sum_{j \in [m']} a_{ij} x_{ij} \leq Z$ and $\forall j : \sum_{i \in [n']} a_{ij} \leq z_j$

Specifically, we set n' and m' to be the number of full blocks. For each full block, we set z'_i to be z_i and $a_{i,j}$ to be the gap between the jth full block and the ith full block. Fortunately, the number of variables is upper-bounded by the number of full blocks squared, (therefore, upper-bounded by $O((\Delta + t)^2)$). By a famous result of Lenstra [11], it follows that the running time is exponential only in the number of variables, therefore the cost of this phase is $\text{poly}(n) \cdot f_3(t, \Delta)$.

We now prove the correctness of the algorithm. As the algorithm only performs permitted operations (that is, adds up to t new vertices and connects up to s edges, each incident to at least one new vertex), it follows that if the input is a no-instance, then the algorithm returns a negative answer. Otherwise, if the input is a yes-instance, then at least one set of guesses from Phase I will be correct. Any solution must at least move the vertices that are moved in Phase II, and then the problem reduces to the ILP presented in Phase III.   □

The question whether fixed-parameter tractability also holds for the parameter t or Δ alone remains open. Nevertheless, we find that fixed-parameter tractability also holds for the combined parameter (Δ, k).

Theorem 8. DEGREE ANONYMIZATION (VA) *is fixed-parameter tractable with respect to the combined parameter* (Δ, k).

Contrasting the W[1]-hardness of Π-PRESERVING DEGREE ANONYMIZATION (VA) parameterized by the number s of new edges (Theorem 4), we conclude with fixed-parameter tractability for DEGREE ANONYMIZATION (VA) with respect to s. We again assume that s is given as part of the input.

Theorem 9. DEGREE ANONYMIZATION (VA) *is fixed-parameter tractable with respect to the number* s *of newly inserted edges.*

5   Conclusion

Table 1 in the introductory section overviews most of our results and leaves several specific open questions. Moreover, it is fair to say that our positive algorithmic results are basically of classification nature and require further improvement for practical relevance. Indeed, a more holistic approach in terms of a full-fledged multivariate complexity analysis [8, 17], perhaps also driven by the analysis of real-world network data characteristics, may help to derive practically useful algorithmic results. A deeper investigation of approximation algorithms (cf. [3, 4]) may be beneficial as well. Finally, typical social network properties such as measured by the clustering coefficient or the average path length are studied in experimental work [3], but the complexity of Π-PRESERVING DEGREE ANONYMIZATION (VA) with respect to these properties is unexplored so far.

References

[1] Bredereck, R., Hartung, S., Nichterlein, A., Woeginger, G.J.: The complexity of finding a large subgraph under anonymity constraints. In: Cai, L., Cheng, S.-W., Lam, T.-W. (eds.) ISAAC 2013. LNCS, vol. 8283, pp. 152–162. Springer, Heidelberg (2013)
[2] Casas-Roma, J., Herrera-Joancomartí, J., Torra, V.: An algorithm for k-degree anonymity on large networks. In: Proc. ASONAM 2013, pp. 671–675. ACM Press (2013)
[3] Chester, S., Kapron, B.M., Ramesh, G., Srivastava, G., Thomo, A., Venkatesh, S.: Why Waldo befriended the dummy? k-anonymization of social networks with pseudo-nodes. Social Netw. Analys. Mining 3(3), 381–399 (2013)

[4] Chester, S., Kapron, B.M., Srivastava, G., Venkatesh, S.: Complexity of social network anonymization. Social Netw. Analys. Mining 3(2), 151–166 (2013)

[5] Clarkson, K.L., Liu, K., Terzi, E.: Towards identity anonymization in social networks. In: Link Mining: Models, Algorithms, and Applications, pp. 359–385. Springer (2010)

[6] Downey, R.G., Fellows, M.R.: Fundamentals of Parameterized Complexity. Springer (2013)

[7] Erdős, P., Kelly, P.: The minimal regular graph containing a given graph. Amer. Math. Monthly 70, 1074–1075 (1963)

[8] Fellows, M.R., Jansen, B.M.P., Rosamond, F.A.: Towards fully multivariate algorithmics: Parameter ecology and the deconstruction of computational complexity. Eur. J. Combin. 34(3), 541–566 (2013)

[9] Flum, J., Grohe, M.: Parameterized Complexity Theory. Springer (2006)

[10] Hartung, S., Nichterlein, A., Niedermeier, R., Suchý, O.: A refined complexity analysis of degree anonymization in graphs. In: Fomin, F.V., Freivalds, R., Kwiatkowska, M., Peleg, D. (eds.) ICALP 2013, Part II. LNCS, vol. 7966, pp. 594–606. Springer, Heidelberg (2013); To appear in Information and Computation

[11] Lenstra, H.W.: Integer programming with a fixed number of variables. Math. Oper. Res. 8, 538–548 (1983)

[12] Liu, K., Terzi, E.: Towards identity anonymization on graphs. In: ACM SIGMOD Conference, SIGMOD 2008, pp. 93–106. ACM (2008)

[13] Lovász, L., Plummer, M.D.: Matching Theory. Annals of Discrete Mathematics, vol. 29. North-Holland (1986)

[14] Lu, X., Song, Y., Bressan, S.: Fast identity anonymization on graphs. In: Liddle, S.W., Schewe, K.-D., Tjoa, A.M., Zhou, X. (eds.) DEXA 2012, Part I. LNCS, vol. 7446, pp. 281–295. Springer, Heidelberg (2012)

[15] Lueker, G.S.: Two NP-complete problems in nonnegative integer programming. Technical report. Computer Science Laboratory, Princeton University (1975)

[16] Niedermeier, R.: Invitation to Fixed-Parameter Algorithms. Oxford University Press (2006)

[17] Niedermeier, R.: Reflections on multivariate algorithmics and problem parameterization. In: Proc. 27th STACS. LIPIcs, vol. 5, pp. 17–32. Schloss Dagstuhl–Leibniz-Zentrum für Informatik (2010)

[18] Zhou, B., Pei, J.: The k-anonymity and l-diversity approaches for privacy preservation in social networks against neighborhood attacks. Knowl. Inf. Syst. 28(1), 47–77 (2011)

Makespan Minimization on Multiple Machines Subject to Machine Unavailability and Total Completion Time Constraints*

Yumei Huo

Department of Computer Science,
City University of New York, College of Staten Island,
Staten Island, New York 10314, USA
yumei.huo@csi.cuny.edu

Abstract. In this paper, we study the preemptive bi-criteria scheduling problems on m parallel machines with machine unavailable intervals. The goal is to minimize the makespan subject to the constraint that the total completion time is minimized. We study the model where each machine can have multiple unavailable intervals, but at any time, there is at most one machine unavailable. We show that there is an optimal polynomial time algorithm for this model.

1 Introduction

This research concerns both bi-criteria scheduling and scheduling with limited machine availability simultaneously. As Panwalkar et al. ([17]) points out, decision makers are often faced with the problem of satisfying several different groups of people simultaneously. Thus, managers actually need to develop schedules based on multiple criteria. A lot of applications for multi-criteria scheduling have been addressed in the books ([21], [20]), surveys ([3], [1], [6])and the references therein. On the other hand, machines may not be continuously available due to breakdown, preventive maintenance or processing unfinished jobs from a previous planning horizon. Applications for scheduling subject to machine availability have been addressed in many surveys ([2], [10], [16], [19], [18]) and the references therein. In the real life, both models may co-exist in some scenarios. While manufacturers aim at optimizing multi-criteria simultaneously, resourses may not be always available. So it is natural to consider bi-criteria scheduling subject to the limited machine availability.

In this paper we study the criteria of makespan and total completion time and the goal is to optimize makespan subject to the condition that the total completion time is minimized. The makespan and total completion time are two objectives of considerable interest. Minimizing makespan can ensure a good balance of the load among the machines and minimizing the total completion time can minimize the inventory holding costs. It is quite common that the

---

* This work is supported by PSC-CUNY Research Award.

Q. Gu, P. Hell, and B. Yang (Eds.): AAIM 2014, LNCS 8546, pp. 56–65, 2014.

manufacturers wish to minimize both objectives. The motivations for bi-criteria scheduling concerned with makespan and total completion have been addressed by Gupta et al. ([5]), Leung and Young in [12], and survey papers about multi-criteria scheduling mentioned above.

We consider preemptive schedules for multiple parallel machines. A job can be preempted by another job or interrupted by machine unavailable intervals and resumed later on any available machine. Machines may not be always available. Specifically, we study the case that each machine may have multiple unavailable periods, but there is at most one machine unavailable at any time. In reality, this machine model exists in many scenarios since the preventive maintenance or periodical repair is usually done on a rotation basis instead of maintaining or repairing several machines simultaneously.

Formally speaking, there is a set of n jobs, $J = \{J_1, J_2, \cdots, J_n\}$, that need to be scheduled on m machines. Each job J_j has a processing time p_j. Without loss of generality, the processing times of jobs are assumed to be integer. Jobs are resumable, that is, after interrupted by the unavailable interval, the job can continue to be processed after the machine has become available. Let S be a feasible schedule of these jobs, the completion time of job J_j in schedule S is denoted by $C_j(S)$. If S is clear from the context, we will use C_j for short. The makespan of S is $C_{\max}(S) = \max\{C_j(S)\}$, and the total completion time of S is $\sum C_j(S)$. The goal is to schedule the set of jobs on m parallel machines so as to minimize $C_{\max}(S)$ subject to the machine unavailability constraint and the condition that $\sum C_j$ is minimized. Each machine may have multiple unavailable periods, but there is at most one machine unavailable at any time. By extending the 3-field notation $\alpha \mid \beta \mid \gamma$ introduced by Graham et al. [4], we denote our problem by $P_{m-1,1} \mid r - a, pmtn \mid C_{\max}/\sum C_j$.

1.1 Literature Review

So far there are two research papers that consider multicriteria scheduling with limited machine availability constraint [7] [8]. In [7], Huo and Zhao give optimal polynomial algorithms for three problems: (1) $P_{1,1} \mid r - a, pmtn \mid \sum C_j/C_{\max}$; (2) $P_{1,1} \mid r - a, pmtn \mid C_{\max}/\sum C_j$; and (3) $P_2 \mid r - a, pmtn \mid C_{\max}/\sum C_j$ in which both machines are unavailable during an interval $[t, t + x)$ and at most one machine is unavailable at any other time. In [8], Huo and Zhao give optimal polynomial algorithms for two problems: (1) $P_m, r_i \mid r - a, pmtn \mid \sum C_j/C_{\max} \leq T$ where each machine has a release time; (2) $P_{m-1,1} \mid r - a, pmtn \mid \sum C_j/C_{\max} \leq T$ where each machine may have multiple unavailable periods, but there is at most one machine unavailable at any time.

In the following we will review the relevant results in the area of bi-criteria scheduling and in the area of scheduling with limited machine availability, respectively. We will survey the results concerning with makespan and total completion time only. For more details about multicriteria scheduling, see [3], [1], [21], [6], [20] and the references therein. For details about scheduling with limited machine availability, see the surveys [10], [2], [16], [19] and [18].

Research on multicriteria scheduling problems on parallel machines has not been dealt with adequately in the literature. Gupta et al. ([5]) proposes an exponential algorithm to solve optimally the bi-criteria problem of minimizing the weighted sum of makespan and mean flow time on two identical parallel machines. When preemption is allowed, $P \mid pmtn \mid C_{\max}/\sum C_j$ and $P \mid pmtn \mid \sum C_j/C_{\max}$ are polynomially solved by Leung and Young in [12], and Leung and Pinedo in [13], respectively.

With limited machine availability, when there are multiple machines, if preemption is not allowed, makespan minimization and total completion time minimization problems are both NP-hard ([11],[19], [9]). When preemption is allowed and the machines have limited availability constraint, the makespan problem is shown to be solvable in P by Liu and Sanlaville ([15]); additionally if the number of available machines does not go down by 2 within a period of $p_{\max}$ (which is the largest processing time of the jobs), Leung and Pinedo([14]) solved the total completion time minimization problem using PSPT (preemptive SPT, i.e., at any time, when a machine becomes available for processing jobs, the job with the minimum remaining time gets scheduled.) rule. In [7], Huo and Zhao show that for two parallel machines with a single zero-availability interval, the problem of minimizing total completion time can be solved optimally. When each machine has a release time, that is, machine M_i is only available in interval $[r_i, \infty)$, where $r_i \geq 0$, it is easy to show that SPT (shortest processing time first) rule minimizes $\sum C_j$ while for general unavailability constraint, the problem becomes NP-hard. Approximation algorithms are developed for total completion time (see Lee and Liman (1993)) and makespan (see Lee (1991), Lin et al. (1997) and Kellerer (1998)).

1.2   New Contributions

In this paper, we study the problem $P_{m-1,1} \mid r-a, pmtn \mid C_{\max}/\sum C_j$ and we show this problem is in P by developing an optimal algorithm.

It should be noted that although the authors in [7] have solved a special case of $m = 2$ and it is natural to conjecture the problem is still in P when $m > 2$, it turns out the optimal algorithm itself and the proof of its optimality is totally different from the optimal algorithm for $m = 2$ and is much more complicated.

1.3   Organization

Our paper is organized as follows. In Section 2, we give some preliminary results. In Section 3, we study $P_{m-1,1} \mid r-a, pmtn \mid C_{\max}/\sum C_j$. In Section 4, we draw the conclusion.

2   Preliminaries

By following similar arguments from [7] and [14], one can easily show the following two lemmas:

Lemma 1. *For m machines, $m \geq 2$, such that at any time at most one machine is unavailable, PSPT generates an optimal schedule for $\sum C_j$.*

Lemma 2. *For m machines with arbitrary unavailability constraints, $m \geq 2$, there is an optimal schedule for the objective $C_{\max}/\sum C_j$, such that if $p_i < p_j$, then $C_i \leq C_j$.*

Throughout this paper, we assume jobs J_1, J_2, ..., J_n are indexed in nondecreasing order of their processing times, i.e., $p_1 \leq p_2 \leq \ldots \leq p_n$.

3 $P_{m-1,1} \mid r - a, pmtn \mid C_{\max}/\sum C_j$

In this section, we show that the problem $P_{m-1,1} \mid r - a, pmtn \mid C_{\max}/\sum C_j$ is in P by developing a polynomial time optimal algorithm. Let T^* be the optimal makespan among all the schedules with minimum total completion time. The basic idea of our optimal algorithm is to schedule the first $n - m$ jobs one by one using PSPT rule, which guarantees the optimal total completion time; compute the optimal makespan T^* based on the partial schedule and the last m jobs; and at last schedule all the jobs to minimize the total completion time subject to $C_{max} \leq T^*$ by calling the optimal algorithm for problem $P_{m-1,1} \mid r - a, pmtn \mid \sum C_j/C_{\max} \leq T^*$ which is developed by Huo and Zhao [8].

Let S be the partial schedule produced by scheduling the first $n-m$ jobs using PSPT rule. It is easy to see that jobs J_1, J_2, $\cdots$, J_{n-m} have the completion time $C_1 \leq C_2 \leq \cdots \leq C_{n-m}$. Since our schedule is preemptive, we can always exchange jobs or exchange jobs with unavailable intervals so that job J_{n-2m+1}, J_{n-2m+2}, $\cdots$, J_{n-m} are completed on machine M_1, M_2, $\cdots$, M_m, respectively and no jobs are scheduled after C_{n-2m+1}, C_{n-2m+2}, $\cdots$, C_{n-m} on machine M_1, M_2, $\cdots$, M_m, respectively. Let $f_i (1 \leq i \leq m)$ be the completion time of the last job on machine M_i. That is, $f_i = C_{n-2m+i}$ for all $1 \leq i \leq m$. Reschedule all the unavailable intervals such that all the unavailable intervals between f_i and $f_{i+1} (1 \leq i \leq m - 1)$ are on machine M_i and all the unavailable intervals after f_m are on machine M_m. And let $S_i (1 \leq i \leq m - 1)$ be the set of the unavailable intervals after f_i on M_i and $U_i (1 \leq i \leq m - 1)$ be the total length of all the intervals in S_i. For machine M_m, let S_m be the set of the unavailable intervals during $[f_m, f_1 + U_1 + p_{n-m+1}]$ and U_m be the total length of all the intervals in S_m.

Now let us consider the lowerbound of makespan by which jobs J_{n-m+1}, $\cdots$, J_n can be finished. Job J_n has the largest time, so in order to complete job J_n as early as possible, we should schedule J_n on machine M_1, that is, the lowerbound by which job J_n can be finished is $F_1 = f_1 + U_1 + p_n$ (Note that J_n can not be scheduled on more than one machines simultaneously at any time) and apparently this is also the lowerbound by which jobs J_{n-m+1}, $\cdots$, J_n can be finished and we use B_1 to represent this lowerbound. Similarly, J_{n-1} should be scheduled on M_2 and its completion time is $F_2 = f_2 + U_2 + p_{n-1}$. If $F_2 \leq B_1$, then the lowerbound by which job J_n and J_{n-1} can be finished is still B_1; otherwise, the lowerbound by which jobs J_n and J_{n-1} can be finished will be $B_1 + (F_2 - B_1)/2$, which is also the lowerbound by which jobs $J_{n-m+1}, \cdots, J_n$ can be finished and we use B_2 to represent this lowerbound. let $\Delta_1 = F_2 - B_1$ be the difference between the completion time of job J_{n-1} on machine M_2 and the lowerbound of makespan obtained so far, then $B_2 = \max\{B_1, B_1 + \Delta_1/2\}$. If $B_2 > B_1$, we update the $\Delta_1 = 0$ since the completion time of last job on M_2 and the so far obtained lowerbound of makespan are both B_2. We repeat this procedure and let B_{k-1} be the lowerbound of makespan by which jobs $J_{n-k+2}, \cdots, J_n$ can be finished, then we can get the lowerbound $B_k = \max\{B_{k-1}, B_{k-1} + (\sum_{j=1}^{k-1} \Delta_j)/k\}$ by which $J_{n-k+1}, \cdots, J_n$ can be finished, which is also the lowerbound by which J_{n-m+1}, $\cdots, J_n$ can be finished. And if $B_k > B_{k-1}$, we update $\Delta_j = 0$ for all $1 \leq j \leq k-1$. Now we consider job J_{n-m+1} and let B_{m-1} be the lowerbound of makespan by which jobs $J_{n-m+2}, \cdots, J_n$ can be finished. Job J_{n-m+1} will be scheduled on machine M_m and its completion time will be $F_m = f_m + U_m + p_{n-m+1}$ and the lowerbound for jobs $J_{n-m+1}, \cdots, J_n$ will be $B_m = \max\{B_{m-1}, B_{m-1} + (\sum_{j=1}^{m-1} \Delta_j)/m\}$. Note that U_m is the total length of unavailable intervals during $[f_m, f_1 + U_1 + p_{n-m+1}]$, and $f_1 + U_1 + p_{n-m+1}$ is the job J_{n-m+1}'s completion time if we schedule it by PSPT rule. Since all jobs $J_{n-m+2}, \cdots, J_n$ complete at or after $\min\{F_1, F_2, \cdots, F_{m-1}\} = \min\{f_1 + U_1 + p_n, f_2 + U_2 + p_{n-1}, \cdots, f_{m-1} + U_{m-1} + p_{n-m+2}\}$ which is larger than $f_1 + U_1 + p_{n-m+1}$, if there is an unavailable interval between $[f_1 + U_1 + p_{n-m+1}, \min\{F_m, B_m\}]$, then job J_{n-m+1} will be finished after a part of unavailable interval, which will increase the job J_{n-m+1}'s completion time, and thus the total completion time is not optimal any more. So in this case, we will fix job J_{n-m+1} so that it completes at the beginning of this unavailable interval and then compute the lowerbound of makespan for jobs $J_{n-m+2}, \cdots, J_n$ following the above procedure.

Formally, we compute $B_k (1 \leq k \leq m)$ which is the lowerbound of makespan by which jobs $J_{n-k+1}, \cdots, J_n$ can be finished as follows:

(1) $B_1 = F_1$ where $F_1 := f_1 + U_1 + p_n$.
(2) For $2 \leq k \leq m$, let $F_k = f_k + U_k + p_{n-k+1}$ and $\Delta_{k-1} = F_k - B_{k-1}$, compute $B_k = \max\{B_{k-1}, B_{k-1} + (\sum_{j=1}^{k-1} \Delta_j)/k\}$ and if $B_k > B_{k-1}$, update $\Delta_j = 0$ for all $1 \leq j \leq k - 1$.

Let s be the starting time of the earliest unavailable interval after $f_1 + U_1 + p_{n-m+1}$. If $F_m \leq s$ or $B_m \leq s$, set $T^* = B_m$. Otherwise, starting at time s, we feasibly backward schedule job J_{n-m+1} on machine M_m and available intervals on machines $M_{m-1}, M_{m-2}, \cdots$, in this order until job J_{n-m+1} is fully scheduled and then update f_i, S_i and U_i for $m - 1 \geq i \geq 1$ and we use f_i' and U_i' to represent the updated f_i and U_i. We compute $B_k' (1 \leq k \leq m - 1)$ same as B_k and set $T^* = B_{m-1}'$. That is,

(1) $B_1' = F_1$ where $F_1 := f_1' + U_1' + p_n$.
(2) For $2 \leq k \leq m - 1$, let $F_k = f_k' + U_k' + p_{n-k+1}$ and $\Delta_{k-1} = F_k - B_{k-1}$, compute $B_k' = \max\{B_{k-1}', B_{k-1}' + (\sum_{j=1}^{k-1} \Delta_j)/k\}$ and if $B_k' > B_{k-1}'$, update $\Delta_j = 0$ for all $1 \leq j \leq k - 1$.

we present the complete algorithm formally as follows:

Algorithm ALG-OPT

(1) Let S be the PSPT schedule of $J_1, \cdots, J_{n-m+1}$

(2) Exchange jobs or exchange jobs with unavailable intervals so that job $J_{n-2m+1}, J_{n-2m+2}, \cdots, J_{n-m}$ are completed on machine $M_1, M_2, \cdots,$ M_m, respectively and no jobs are scheduled after $C_{n-2m+1}, C_{n-2m+2}, \cdots,$ C_{n-m} on machine $M_1, M_2, \cdots, M_m$, respectively.

(3) Let f_i be the completion time of the last job scheduled on machine M_i, and assume that the machines are numbered so that $f_1 \leq f_2 \ldots \leq f_m$.

(4) Reschedule all the unavailable intervals such that all the unavailable intervals between f_i and $f_{i+1}(1 \leq i \leq m-1)$ are on machine M_i and all the unavailable intervals after f_m are on machine M_m. And let $S_i(1 \leq i \leq m-1)$ be the set of the unavailable intervals after f_i on M_i and $U_i(1 \leq i \leq m-1)$ be the total length of all the intervals in S_i. Let S_m be the set of the unavailable intervals during $[f_m, f_1 + U_1 + p_{n-m+1}]$ on machine M_m and U_m be the total length of all the intervals in S_m.

(5) Let $F_1 := f_1 + U_1 + p_n$, $B_1 := F_1$ and $k := 2$

(6) while $k \leq m$
  1. let $F_k = f_k + U_k + p_{n-k+1}$ and $\Delta_{k-1} = F_k - B_{k-1}$
  2. If $\sum_{j=1}^{k-1} \Delta_j \leq 0$
  3.   Set $B_k := B_{k-1}$
  4. Else
  5.   Set $B_k := B_{k-1} + (\sum_{j=1}^{k-1} \Delta_j)/k$ and update all $\Delta_j = 0$ for all $1 \leq j \leq k-1$

(7) Let s be the starting time of the earliest unavailable interval after $f_1 + U_1 + p_{n-m+1}$

(8) If $F_m \leq s$ Set $T^* = B_m$ and go to step (17)

(9) If $B_m \leq s$ Set $T^* = B_m$ and go to step (17)

(10) Let $p := p_{n-m+1}$

(11) $p := p - (s - f_m - U_m)$

(12) Set $k := m - 1$

(13) while $k \geq 1$
  1. If $p > 0$
  2.   If $p \leq U_{k+1}$ Set $U'_k := U_k + p$ and $p := 0$
  3.   Else if $p \leq U_{k+1} + f_{k+1} - f_k - U_k$ Set $U'_k := U_k + p$ and $p := 0$
  4.   Else set $f'_k := f_{k+1}, p := p - (U_{k+1} + f_{k+1} - f_k - U_k)$ and $U'_k := U_{k+1}$
  5. Else set $f'_k := f_k$ and $U'_k := U_k$
  6. $k = k - 1$

(14) Let $F'_1 := f'_1 + U'_1 + p_n$, $B'_1 := F'_1$ and $k := 2$

(15) while $k \leq m - 1$
  1. let $F'_k := f'_k + U'_k + p_{n-k+1}$ and $\Delta'_{k-1} = F'_k - B'_{k-1}$
  2. If $\sum_{j=1}^{k-1} \Delta'_j \leq 0$
  3.   Set $B'_k := B'_{k-1}$
  4. Else

5.   Set $B'_k := B'_{k-1} + (\sum_{j=1}^{k-1} \Delta'_j)/k$ and update all $\Delta'_j = 0$ for all $1 \le j \le k-1$

(16) Set $T^* = B'_{m-1}$

(17) Make machine M_m unavailable after s and other machines all available after s, and call the optimal algorithm for problem $P_{m-1,1} \mid r-a, pmtn \mid \sum C_j/C_{\max} \le T^*$ (See Appendix) to schedule all the jobs to minimize the total completion time subject to makespan less than or equal to T^* and let S' be the produced schedule

(18) Return S'

Lemma 3. *For any $2 \le x \le m$, we have either (1) $B_x = B_{x-1}$ or (2) $B_x = (F_x + F_{x-1} + \cdots + F_1)/x$ and $F_x - B_x = \frac{x-1}{x}(F_x - F_{x-1}) + \frac{x-2}{x}(F_{x-1} - F_{x-2}) + \cdots + \frac{1}{x}(F_2 - F_1)$*

Now, let us look at the schedule produced by PSPT rule, which, by [14], has the minimum total completion time. Jobs will be scheduled in increasing order of their processing times, that is, $J_1, \cdots, J_n$. And it is easy to see that job $J_i(1 \le i \le n)$ is scheduled and completes as early as possible by PSPT rule. We will show that for the problem of $P_{m-1,1} \mid r-a, pmtn \mid C_{\max}/\sum C_j$, there must exist an optimal schedule with the first $n-m$ jobs complete as early as possible.

Lemma 4. *For the problem of $P_{m-1,1} \mid r-a, pmtn \mid C_{\max}/\sum C_j$, there must exist an optimal schedule such that the first $n-m$ jobs complete as early as possible, which is same as the schedule produced by PSPT rule.*

Proof. Assume not. Let S^* be an optimal schedule such that the first $n-m$ jobs do not complete as early as possible and let job $J_i(i \le n-m)$ be the first job in S^* that is not completed as early as possible. Then there must exist a time $t < C_i$ and index $j > i$ such that J_j is scheduled at time t but job J_i is not scheduled at t. Let C^*_{max} be the makespan of S^*. We claim that the completion time of job J_j must be less than C^*_{max}. Assume not. That is, $C_j = C^*_{max}$. Since $i \le n-m$, by lemma 1, there must exist at least m jobs with the completion time larger than or equal to C_i, and among these m jobs, there must exist a job $J_k(k > i)$ not scheduled at time $[C_i - 1, C_i]$. If we do the following triple exchange: move job J_j at $[t, t+1]$ to $[C^*_{max}, C^*_{max} + 1]$, move job J_i at $[C_i - 1, C_i]$ to $[t, t+1]$, and move job J_k at $[C_k - 1, C_k]$ to $[C_i - 1, C_i]$. In this case, the completion time of job J_j is increased by 1, the completion time of job J_i and J_k are both decreased at least by 1, and the completion time of all other jobs are not changed, and thus the total completion time is decreased by 1, which is contradict to the fact that S^* has the minimum total completion time. So we must have $C_j < C^*_{max}$. In this case, there must exist a job J_k that is scheduled at time interval $[C_j, C_j + 1]$ but not at time interval $[C_i - 1, C_i]$ and we can convert S^* such that J_i is scheduled at time t without changing the total completion time and makespan by triple exchange: move J_i at $[C_i - 1, C_i]$ to time $[t, t+1]$, move job J_j at $[t, t+1]$ to $[C_j, C_j + 1]$, and move job J_k at $[C_j, C_j + 1]$ to $[C_i - 1, C_i]$.

We can repeat the above procedure and convert S^* such that the first $n-m$ jobs complete as early as possible, which is the same as the schedule produced by PSPT rule, that is, the schedule obtained after step 1 in Algorithm ALG-OPT.

Lemma 5. *Let S be the schedule obtained after step 4 in Algorithm ALG-OPT, for all $1 \leq k \leq m - 1$, there exists a feasible schedule to schedule the last k jobs on the first k machines by B_k based on S.*

Lemma 6. *Let S be the schedule obtained after step 4 in Algorithm ALG-OPT, there exists a feasible schedule to schedule jobs $J_{n-m+1}, \cdots, J_n$ by T^* on S.*

Lemma 7. *If $f_m + U_m + p_{n-m+1} > s$ and $s < B_m$ in Algorithm ALG-OPT, there must exists at least one job among $J_{n-m+1}, \cdots, J_n$ with completion time less than or equal to s in any schedule such that total completion time is minimized and the first $n - m$ jobs are scheduled in PSPT order.*

Lemma 8. *The schedule produced by Algorithm ALG-OPT, S, has the minimum total completion time.*

Proof. In Algorithm ALG-OPT, if $f_m + U_m + p_{n-m+1} \leq s$ or $s \geq B_m$, by Lemma 5 and Lemma 6, there exists a feasible schedule such that jobs $J_n, \cdots,$ J_{n-m+1} finish before B_m and no job finish after s on machine M_m, so the total completion time of jobs $J_n, \cdots, J_{n-m+1}$ in S is $\sum_{x=1}^{m}(f_x + U_x + p_{n-x+1})$ which is same as in the schedule produced by PSPT algorithm. On the other hand, if $f_m + U_m + p_{n-m+1} > s$ and $s < B_m$, from Algorithm ALG-OPT, we fix job J_{n-m+1} to finish at s and by Lemma 5, we can obtain a feasible schedule such that $J_n, \cdots, J_{n-m+2}$ finish before B'_m, so the total completion time is $\sum_{x=1}^{m}(f_x + U_x + p_{n-x+1})$, which is same as in the schedule produced by PSPT algorithm.

Lemma 9. *The schedule produced by Algorithm ALG-OPT, S, has the minimum makespan subject to the condition that the total completion time is minimized.*

Proof. In Algorithm ALG-OPT, we have two cases: (1) $f_m + U_m + p_{n-m+1} \leq s$ or $s \geq B_m$; (2) $f_m + U_m + p_{n-m+1} > s$ and $s < B_m$. Let S be the schedule obtained after step 4 in Algorithm ALG-OPT. For case (1), we claim that B_k is the earliest time that jobs $J_n, \cdots, J_{n-k+1}$ can be completed in S. We prove this by induction. If $k = 1$, apparently, $B_1 = f_1 + U_1 + p_n$ is the earliest time that job J_n can be completed in S. Assume when $k = x$, B_x is the earliest time that jobs $J_n, \cdots, J_{n-x+1}$ can be completed in S. Let us look at $k = x + 1$. If $\sum_{j=1}^{x} \Delta_j \leq 0$, we have $B_{x+1} := B_x$, and apparently it is true that B_{x+1} is the earliest time that $J_n, \cdots, J_{n-x}$ can be completed in S. Otherwise, we have $B_{x+1} := B_x + (\sum_{j=1}^{x} \Delta_j)/(x + 1)$ and apparently all the jobs $J_n, \cdots,$ J_{n-x} have the same completion time on machines $M_1, \cdots, M_{x+1}$. Since for any machine $M_{i_1}(1 \leq i_1 \leq x + 1)$ and any machine $M_{i_2}(x + 2 \leq i_2 \leq m)$ in S, $f_{i_1} + U_{i_1} \leq f_{i_2} \leq f_{i_2} + U_{i_2}$ and there is no idle time on any machines $M_{i_1}(1 \leq i_1 \leq x + 1)$ before B_{x+1}, B_{x+1} is the earliest time that jobs $J_n, \cdots,$ J_{n-x} can be completed in S.

For case (2), by Lemma 7, there must exist at least one job among $J_{n-m+1}, \cdots,$ J_n with completion time less than or equal to s in any schedule such that total completion time is minimized and the first $n - m$ jobs are scheduled in PSPT order. First, we show that there must exist an optimal schedule such that the first $n - m$ jobs completes as early as possible and J_{n-m+1} completes no later than s. Let S^* be an optimal schedule such that the first $n - m$ jobs completes as early as possible and let $J_y(n-m+1 \leq y \leq n)$ be the job that completes no later than s in S^*. Backward starting from the completion time of job J_{n-m+1} in S, for any time that job J_{n-m+1} is scheduled but J_y is not scheduled, we can always exchange job J_{n-m+1} at this time with J_y at the time that job J_y is scheduled but J_{n-m+1} is not scheduled. Since $p_y \geq p_{n-m+1}$, we must be able to exchange job J_{n-m+1} and job J_y without overlap and without changing the total completion time and the makespan of S^*. That is, we have convert S^* into an optimal schedule such that the first $n - m$ jobs completes as early as possible and J_{n-m+1} completes no later than s. Now by the same argument of case (1), we can show that B'_k is the earliest time that jobs $J_n, \cdots, J_{n-k+1}$ can be completed in S subject to the condition that job J_{n-m+1} completes no later than s.

So we have that T^* is the minimum makespan subject to the condition that the total completion time is minimized. So when the optimal algorithm for the problem $P_{m-1,1} \mid r - a, pmtn \mid \sum C_j / C_{\max} \leq T$ is called to produce the schedule such that all the jobs complete before T^* and no job is scheduled after t on machine M_m, the total completion time of all the jobs in the produced scheduled must be optimal, that is, the schedule produced by Algorithm ALG-OPT, S, has the minimum makespan subject to the minimized total completion time.

By Lemma 8 and Lemma 9, we have the following theorem.

Theorem 1. *Algorithm ALG-OPT is optimal for* $P_{m-1,1} \mid r - a, pmtn \mid C_{\max} / \sum C_j$.

4   Conclusion

In this paper, we study the bi-criteria scheduling problems subject to the machine unavailability constraint with total completion time and makespan as primary and secondary criteria respectively. Our focus is on the parallel machine environment where each machine can have multiple unavailable intervals, but at any time, there is at most one machine unavailable. We develop an optimal polynomial time algorithm. Both algorithm and the proof are quite involved and subtle. It is expected that the problem is still solvable when each machine has a release time after which the machine is always available. And it is interesting to know the problem's complexity when more than one machine are unavailable at some time.

References

1. Chen, C.L., Bulfin, R.L.: Complexity of single machine, multicriteria scheduling problems. European Journal of Operational Research 70, 115–125 (1993)
2. Diedrich, F., Jansen, K., Schwarz, U.M., Trystram, D.: A Survey on Approximation Algorithms for Scheduling with Machine Unavailability. In: Lerner, J., Wagner, D., Zweig, K.A. (eds.) Algorithmics. LNCS, vol. 5515, pp. 50–64. Springer, Heidelberg (2009)

3. Dileepan, P., Sen, T.: Bicriteria static scheduling research for a single machine. OMEGA 16, 53–59 (1988)
4. Graham, R.L., Lawler, E.L., Lenstra, J.K., Rinnooy Kan, A.H.G.: Optimization and approximation in deterministic sequencing and scheduling, a survey. Annals of Discrete Mathematics 5, 287–326 (1979)
5. Gupta, J.N.D., Ho, J.C., Webster, S.: Bicriteria optimisation of the makespan and mean flowtime on two identical parallel machines. Journal of Operational Research Society 51(11), 1330–1339 (2000)
6. Hoogeveen, J.A.: Multicriteria scheduling. European Journal of Operational Research 167(3), 592–623 (2005)
7. Huo, Y., Zhao, H.: Bicriteria Scheduling Concerned with Makespan and Total Completion Time Subject to Machine Availability Constraints. Theoretical Computer Science 412, 1081–1091 (2011)
8. Huo, Y., Zhao, H.: Bi-criteria Scheduling on Multiple Machines Subject to Machine Availability Constraints. In: Fellows, M., Tan, X., Zhu, B. (eds.) FAW-AAIM 2013. LNCS, vol. 7924, pp. 325–338. Springer, Heidelberg (2013) (The journal version of this paper is under review)
9. Lee, C.-Y., Liman, S.D.: Capacitated two-parallel machine scheduling to minimize sum of job completion times. Discrete Appl. Math. 41, 211–222 (1993)
10. Lee, C.-Y., Lei, L., Pinedo, M.L.: Current trends in deterministic scheduling. Annals of Operations Research 70, 1–41 (1997)
11. Lenstra, J.K., Rinnooy Kan, A.H.G., Brucker, P.: Complexity of machine scheduling problems. Annals of Discrete Math 1, 343–362 (1977)
12. Leung, J.Y.-T., Young, G.H.: Minimizing schedule length subject to minimum flow time. SIAM Journal on Computing 18(2), 314–326 (1989)
13. Leung, J.Y.-T., Pinedo, M.L.: Minimizing total completion time on parallel machines with deadline constraints. SIAM Journal on Computing 32, 1370–1388 (2003)
14. Leung, J.Y.-T., Pinedo, M.L.: A Note on the scheduling of parallel machines subject to breakdown and repair. Naval Research Logistics 51, 60–72 (2004)
15. Liu, Z., Sanlaville, E.: Preemptive scheduling with variable profile, precedence constraints and due dates. Discrete Applied Mathematics 58, 253–280 (1995)
16. Ma, Y., Chu, C., Zuo, C.: A survey of scheduling with deterministic machine availability constraints. Computers & Industrial Engineering 58(2), 199–211 (2010)
17. Panwalkar, S.S., Dudek, R.K., Smith, M.L.: Sequencing research and the industrial scheduling problem. In: Elmaghraby, S.E. (ed.) Proceedings of the Symposium on the Theory of Scheduling and Its Application, pp. 29–38. Springer, New York (1973)
18. Saidy, H., Taghvi-Fard, M.: Study of Scheduling Problems with Machine Availability Constraint. Journal of Industrial and Systems Engineering 1(4), 360–383 (2008)
19. Schmidt, G.: Scheduling with limited machine availability. European Journal of Operational Research 121(1), 1–15 (2000)
20. Multiple Criteria Optimization: Theory, Computation and Application. John Wiley, New York (1986)
21. T'kindt, V., Billaut, J.C.: Multicriteria Scheduling: Theory, Models and Algorithms. Springer, Heidelberg (2002)

Edge-Clique Covers of the Tensor Product

Wing-Kai Hon[1], Ton Kloks[1], Hsiang-Hsuan Liu[1,3], and Yue-Li Wang[2]

[1] National Tsing Hua University, Taiwan
{wkhon,kloks,hhliu}@cs.nthu.edu.tw
[2] National Taiwan University of Science and Technology, Taiwan
ylwang@cs.ntust.edu.tw
[3] University of Liverpool, UK
hhliu@liv.ac.uk

Abstract. In this paper we study the edge-clique cover number of the tensor product $K_n \times K_n$. We derive an easy lowerbound for the edge-clique number of graphs in general. We prove that, when n is prime $\theta_e(K_n \times K_n)$ matches the lowerbound. Moreover, we prove that $\theta_e(K_n \times K_n)$ matches the lowerbound if and only if a projective plane of order n exists. We also show an easy upperbound for $\theta_e(K_n \times K_n)$ in general, and give its limiting value when the Riemann hypothesis is true.

1 Introduction

The edge-clique cover problem is the problem of determining if the set of edges of a graph can be expressed as a union of k cliques (i.e., if k cliques in the graph can cover all the edges in the graph). We denote by $\theta_e(G)$ the minimum number of cliques that are necessary to cover all its edges. Finding a minimum edge-clique cover is NP-complete even in very restricted graph classes [1,8,13]. It is known that the edge-clique cover problem is equivalent to finding a set representation of a graph G with at most k elements in the universe [5,15]. This number is also known as the intersection number [5,9]. In this paper we concentrate on the edge-clique cover problem of the tensor product $K_n \times K_n$.

Unlike the clique cover problem, the edge-clique cover problem does not attract computer scientists' attention very much. However, the edge-clique cover problem is related to various applications in discrete mathematics, and more and more people started to conduct research on it [15]. For example, suppose that G is the *intersection graph* of a family of subsets of a set X. The minimal cardinality of X such that G is the intersection graph of a family of subset of X is equal to $\theta_e(G)$ [15].

Another problem related to edge-clique cover problem is about *competition graphs* (or *niche overlap graphs*). In ecology, we can use a competition graph to represent the competition between predators who prey on the same target. People started to ask what do the competition graphs of acyclic graphs look like. Roberts [14] found that by adding e isolated vertices to any graph G, where $e = |E(G)|$, the resulting graph becomes a competition graph of some acyclic graph. Further, Opsut [12] proved that the minimum number of isolated vertices

Q. Gu, P. Hell, and B. Yang (Eds.): AAIM 2014, LNCS 8546, pp. 66–74, 2014.

we need to add to G, denoted by $k(G)$, to make it a competition graph of an acyclic graph is bounded by

$$\theta_e(G) - n + 2 \leqslant k(G) \leqslant \theta_e(G),$$

where $n = |V(G)|$.

The research into edge-clique cover problem also made contributions to some other optimization problems. Kou et al. [10] showed that the optimization problem of keyword conflict is equivalent to the problem that finds the minimum number of cliques to cover the edges in a graph.

Finding the minimum edge-clique cover is hard. Gramm et al. [6] approached a method to reduce the number of vertices of G so that the edge-clique cover problem on G can be solved faster. Recently, Cygan et al. [3] proved that such an approach is optimal in running time by reducing 3-CNF-SAT formula with n variables and m clauses to an equivalent edge-clique cover instance (G, k) with $k = O(\log n)$ and $|V(G)| = O(n + m)$, provided that the Exponential Time Hypothesis holds. In this paper, we focus on the edge-clique cover problem of the tensor product of $K_n \times K_n$, which is one of the graph products that play an important role in graph decomposition into isomorphic subgraphs [11].

The rest of this paper is organized as follows. Section 2 introduces the tensor product $K_n \times K_n$. Section 3 derives a lowerbound for $\theta_e(K_n \times K_n)$. Sections 4 and 5, respectively, prove that $\theta_e(K_n \times K_n)$ matches the lowerbound when n is a prime and when a projective plane of order n exists. Section 6 derives an easy upperbound for $\theta_e(K_n \times K_n)$ in general, and gives its limiting value when the Riemann hypothesis is true.

2 The Tensor Product $K_n \times K_n$

We write $[n] = \{1, \ldots, n\}$. We refer to 'gridpoints' as the elements of the set

$$V = \{ (i, j) \mid i \in [n] \quad \text{and} \quad j \in [n] \}.$$

We also write the set V as $[n] \times [n]$.

Apart from the model suggested by the name we may also consider the gridpoints as the edges of the complete bipartite graph $K_{n,n}$. In the following we write $H = K_{n,n}$. The graph H has as color classes the sets R (for rows) and C (for columns). Thus we can write

$$V = E(H) = \{ (i, j) \mid i \in [n] \quad \text{and} \quad j \in [n] \}.$$

Definition 1. *The tensor product $K_n \times K_n$ has V as its set of vertices. Two vertices (i, j) and (k, ℓ) are adjacent if and only if*

$$i \neq k \quad \text{and} \quad j \neq \ell.$$

Notice that the tensor product is the complement of the Cartesian product $K_n \square K_n$ also known as rook's graph. In this product two pairs (i, j) and (k, ℓ) are adjacent if they lie in the same 'row' or 'column,' that is, the pairs are adjacent if either

$$(i = k \quad \text{and} \quad j \neq \ell) \quad \text{or} \quad (i \neq k \quad \text{and} \quad j = \ell).$$

Definition 2. *A matching in H is a subgraph M of H in which every vertex has exactly one neighbor.*

In other words, a matching in H is a collection of pairwise parallel edges. We call two edges 'parallel' if they have no endpoint in common. Notice that a matching in H is a clique of $K_n \times K_n$.

3   Edge-Clique Covers

We are interested in the edge-clique cover problem (especially in graph limits). In this paper we concentrate on the edge-clique cover problem of the tensor product $K_n \times K_n$.

Definition 3. *Let G be a graph. An edge-clique cover $\mathcal{C}$ is a set of cliques such that each edge of G has both its endpoints in at least one element of $\mathcal{C}$.*

We denote by $\theta_e(G)$ the minimum number of cliques that are necessary to cover all its edges. This number is also known as the intersection number.

Definition 4. *Let G be a graph. An edge-clique partition is a collection of cliques $\mathcal{C}$ such that each edge is in precisely one element of $\mathcal{C}$.*

Lemma 1. *If we denote by $\mathrm{ecp}(G)$ the minimal number of cliques in an edge-clique partition of G then we have*

$$\mathrm{ecp}(G) \geqslant \theta_e(G).$$

Proof. Obviously, the number $\mathrm{ecp}(G)$ exists, since there is a partition of the edges, namely where each clique is a single edge. Notice that any edge-clique partition is an edge-clique cover. Therefore,

$$\theta_e(G) \leqslant \mathrm{ecp}(G) \quad \text{for any graph.}$$

This proves the lemma.   □

Theorem 1. *For any graph G*

$$\theta_e(G) \geqslant \frac{2 \cdot e(G)}{w(G) \cdot (w(G) - 1)},$$

where we write $w(G)$ for the clique number of G and $e(G)$ for the number of edges of G.

Proof. Let $\mathcal{C}$ be an edge-clique cover. Every edge of G has both its endpoints in an element of $\mathcal{C}$. As each element of $\mathcal{C}$ has at most $\omega(G) \cdot (\omega(G) - 1)/2$ edges, thus we find that

$$e(G) \leqslant |\mathcal{C}| \cdot \omega(G) \cdot (\omega(G) - 1)/2. \tag{1}$$

This proves the theorem.   □

Notice that, to achieve this lowerbound we need to have an edge-clique partition of the edges into maximum cliques.

Theorem 2. *For the tensor product we find*

$$\theta_e(K_n \times K_n) \geqslant n(n - 1).$$

Proof. The number of edges in $K_n \times K_n$ is $n^2(n - 1)^2/2$, since there are n^2 vertices, and the degree of each vertex is exactly $(n - 1)^2$. The clique number $\omega(K_n \times K_n)$ is n, since the set of vertices $\{(i, i) \mid i \in [n]\}$ forms a clique, and any $n + 1$ vertices cannot form a clique (otherwise, by pigeonhole principle there exist two vertices (i, j) and (k, ℓ) with $i = k$, thus there are no edges between them). The theorem follows immediately from Theorem 1.   □

Remark 1. Notice that this lowerbound is much stronger than Gýarfás lowerbound [7], which gives, (since $K_n \times K_n$ has no twins)

$$\theta_e(K_n \times K_n) \geqslant \lceil \log_2(n^2 + 1) \rceil.$$

In the following section we show that the lowerbound of Theorem 2 holds for infinitely many $K_n \times K_n$ (when n is prime).

4   The Prime Case

In this section, we show that when n is prime then $\theta_e(K_n \times K_n)$ satisfies the lowerbound mentioned in Theorem 2. Indeed, we show a slightly more general result.

Theorem 3. *For any prime p and any m with $m \leqslant p$,*

$$\theta_e(K_p \times K_m) = p(p - 1).$$

Proof. We construct an edge-clique partition $\mathcal{C}$ with $p(p - 1)$ cliques, each clique containing exactly m vertices. For a clique $C_{x,y}$ in $\mathcal{C}$ with $x \in [p]$ and $y \in [p - 1]$, it contains the vertices

$$\{(x + (t - 1) \times y, t) \mid 1 \leqslant t \leqslant m\},$$

where the first coordinate is taken modulo p. As the total number of edges covered by all the cliques in $\mathcal{C}$ is at most $p(p - 1)m(m - 1)/2 = e(K_p \times K_m)$, to show that $\mathcal{C}$ is an edge-clique cover, it is sufficient to show that each edge in $K_p \times K_m$ is covered by some clique in $\mathcal{C}$. In other words, we want to show, for any two vertices (i, j) and (k, ℓ) with $i \neq j$ and $k \neq \ell$, there exists some clique containing both vertices.

Let u be the inverse of $j - \ell$ in modulo-p arithmetic. That is, $u \in [p - 1]$ such that $u(j - \ell) \equiv 1 \mod p$. Let $v \in [p - 1]$ be the value $u \times (i - k) \mod p$, and $z \in [p]$ be the value $i - (j - 1) \times v \mod p$. Then we can easily check that both (i, j) and (k, ℓ) belong to the clique $C_{z,v}$ in $\mathcal{C}$. This proves the theorem.   □

5   The Projective Plane Case

In this section, we further show that $\theta_e(K_n \times K_n)$ matches the lowerbound if and only if a projective plane of order n exists. Immediately, this implies the lowerbound is matched when n is a prime power.[1]

Definition 5 ([17]). *A finite projective plane of order n is defined as a set of $n^2 + n + 1$ points with the properties that:*

1. *Any two points determine a line;*
2. *Any two lines determine a point;*
3. *Every point has $n + 1$ lines on it, and*
4. *Every line contains $n + 1$ points.*

An example of a projective plane of order 2 is shown in Figure 1.

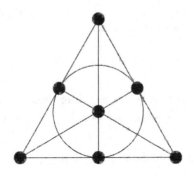

Fig. 1. A projective plane of order 2. Note that one of the lines is a circle.

We next show our main result of this section, which is inspired from the De Bruijn-Erdős theorem about $\theta_e(K_n)$ [4].

Theorem 4 (De Bruijn-Erdős for $K_n \times K_n$). *Assume that there exists a projective plane $\mathcal{P}$ of order n. Then there exists an edge-clique cover of $K_n \times K_n$ with $n(n-1)$ cliques.*

---

[1] There exists a projective plane of order n for each prime power n, and all known projective planes are of prime power order.

Proof. Let P denote the set of points and L denote the set of lines in the projective plane $\mathcal{P}$. Based on P and L, we choose a subset P' of n^2 points from P, and define a one-to-one correspondence to the $n \times n$ gridpoints of $K_n \times K_n$ as follows.

Take one line $L_0 \in L$. Take two points ∞ and ∞' on L_0. Through each of ∞ and ∞' go n lines other than L_0. In the case these n lines go through ∞, they are called 'columns.' Similarly, the n lines through ∞', except L_0, are called 'rows.'

Notice that each row intersects each column in exactly one point. Let P' denote the set of these intersection points, and we see that $|P'| = n^2$. If r_i is a row and c_j is a column then the unique intersection $r_i \cap c_j$ is the point of P corresponding to the gridpoint (r_i, c_j).

We next show that $n(n-1)$ lines of L correspond to an edge-clique cover of $K_n \times K_n$ with $n(n-1)$ cliques. Consider another point $p \in L_0 \setminus \{\infty, \infty'\}$. There are $n-1$ of those points and through each go n lines other than L_0. Each of these $n(n-1)$ lines hits every row and every column in exactly one point, so that the points on the line (except p) correspond to a maximum clique in $K_n \times K_n$.

Moreover, every pair of points in P' that are not on the same row or column, is on a unique line $q \in L$. Thus, each edge of $K_n \times K_n$ is covered by one of the $n(n-1)$ cliques defined above. The theorem thus follows.   □

As projective plane of each prime power order is known to exist, we have the following corollary.

Corollary 1. *For any* $n = p^k$ *where p is a prime and k is a positive integer,*

$$\theta_e(K_n \times K_n) = n(n-1).$$

Definition 6. *We define* diagonal *as a set of points in the grid that form a clique in the tensor product.*

Notice, if the projective plane-possibility occurs in Theorem 4, then all cliques A_i in the edge-clique cover of K_n have $n+1$ vertices. Each pair of vertices lies in exactly one clique (so we have an edge-clique partition in which each clique has cardinality $n+1$) and every pair of cliques intersect in exactly one point.

We next show the converse; if there exists an edge-clique covering $\mathcal{C}$ such that $|\mathcal{C}| = n(n-1)$, then there exists a projective plane of order n.

Definition 7. *An* affine plane $\mathcal{A}$ *is an ordered pair* $\mathcal{A} = (\mathcal{P}, \mathcal{L})$ *of sets. Say* $\mathcal{P}$ *is the set of 'points' and* $\mathcal{L}$ *is the set of 'lines.' There is an incidence relation between points and lines; we refer to this by saying that 'a point lies on a line' and 'a line goes through a point.' For the pair* $(\mathcal{P}, \mathcal{L})$ *to be an affine plane, the sets need to satisfy the following requirements.*

(1). *Every two points lie together on exactly one line.*
(2). *For every point-line pair* (P, L) *with* $P \notin L$ *there is exactly one line M that goes through P and such that M is parallel to L.*
(3). *There are three points that are not on one line.*

Definition 8. *An* affine plane of order n *is an affine plane that*

(1). *All lines contain n points.*
(2). *Each point is contained in $n + 1$ lines.*
(3). *There are n^2 points in $\mathcal{P}$.*
(4). *There are $n^2 + n$ lines in $\mathcal{L}$.*

Part of the converse is proved by the following result.

Theorem 5 ([18]). *For every affine plane $\mathcal{A}$ of order n, there exists a unique, up to isomorphism, projective plane $\mathcal{P}$ of order n.*

Theorem 6 (The converse theorem). *Assume that*

$$\theta_e(K_n \times K_n) = n(n - 1).$$

Then there exists a projective plane of order n.

Proof. Since $\mathcal{C}$ is an edge-clique cover that reaches the lowerbound, the cliques of $\mathcal{C}$ *partition* the edges of $K_n \times K_n$ into *maximum* cliques. Each clique has cardinality n (it is a maximal diagonal of $[n] \times [n]$). We refer to these cliques as 'lines.'

We add two sets of lines, namely the set of rows and the set of columns. We refer to $\mathcal{A} = (V, \mathcal{C}')$ as the incidence structure of points and lines, where

$$V = \text{The gridpoints of the grid with size } n \times n$$

$$\mathcal{C}' = \mathcal{C} \cup \{\text{ rows, columns }\}.$$

We prove that $\mathcal{A} = (V, \mathcal{C}')$ is an affine plane.

Notice that the rows and columns are sets of *parallel* lines, that is, their pairwise intersection is empty. Of course, we also have that every line of $\mathcal{C}$ hits every row and column exactly once.

Notice that the first requirement for an affine plane is satisfied. We prove the second.

Let (P, L) be a point-line pair of $\mathcal{A}$ such that P lies *not* on L. We prove that there is exactly one line M, parallel to L, that goes through P. (Case 1) If L is a row (or a column), we set M to be the row (or the column) that contains P; also, all the other lines that go through P must intersect L. (Case 2) Else, if L is a line of $\mathcal{C}$, then for each point $\ell \in L$, there is exactly one line through P that contains ℓ; also, for any distinct points ℓ and ℓ' in L, the line through P and ℓ and the line through P and ℓ' must be distinct (else, the edge $\{\ell, \ell'\}$ are covered twice in $\mathcal{C}$). Since $|L| = n$, this determines n lines through P. On the other hand, the point P lies on $n - 1$ diagonals of $\mathcal{C}$, and on one row and one column, so that P is in $n + 1$ lines, thus implying that *exactly one* line through P does not contain any point of L. Then this line M through P must be parallel to L.

The above shows that the second requirement of an affine plane is satisfied. For the third requirement, we can easily check that $(1, 1), (1, 2), (2, 1)$ are not on the same line, for any line of $\mathcal{C}'$. Thus, $\mathcal{A}$ is an affine plane. Furthermore, we can easily check that $\mathcal{A}$ is an affine plane of order n. By Theorem 5, a projective plane of order n exists. This proves the theorem. $\square$

We obtain the following theorem as an immediate corollary.

Theorem 7. *The computational complexity of the edge-clique partition problem into maximal cliques is equivalent to the time-complexity of checking if there exists a projective plane of order n.*

Remark 2. We are not aware of a time-complexity class defined by the equivalence of a problem to the existence of a projective plane of order n. There are many problems in this class. The De Bruijn -Erdös theorem plays also an important role in the proof of Cygan et al.

6   An Upperbound and a Limiting Value for $\theta_e(K_n \times K_n)$

In this section we derive an easy upperbound for $\theta_e(K_n \times K_n)$ in general, and give its limiting value when the Riemann hypothesis is true.

A good strategy to find a small edge-clique cover of $K_n \times K_n$ for general n, seems to be to go to the next prime number $n' > n$, and to 'truncate' the optimal solution of $K_{n'} \times K_{n'}$ (which has $n'(n'-1)$ cliques). By Bertrand's postulate [16], it is known that $n' \leqslant 2n - 1$ for all $n \geqslant 2$, so immediately we have the following theorem and corollary.

Theorem 8. *For any integer $n \geqslant 2$,*

$$\theta_e(K_n \times K_n) \leqslant (2n - 1)(2n - 2).$$

Corollary 2. *For any integers m and n with $n \geqslant m \geqslant 2$,*

$$\theta_e(K_n \times K_m) \leqslant (2n - 1)(2n - 2).$$

Using the same idea, we show the following result.

Theorem 9. *For any integers m and n with $n \geqslant m \geqslant 2$,*

$$\lim_{n \to \infty} \frac{\theta_e(K_n \times K_m)}{n(n - 1)} = 1$$

if the Riemann hypothesis is true.

Proof. Let p_i denote the ith smallest prime. Next, assume that the Riemann hypothesis is true; then, Cramér [2] showed that $p_{i+1} - p_i = O(\sqrt{p_i} \ln p_i)$. For our case, let us consider the largest prime p, say p_j, that is smaller than n. Then, we have $p_j = \Theta(n)$ (as $n/2 \leqslant p_j \leqslant n$ by Bertrand's postulate), so that

$$n' - n = p_{j+1} - n < p_{j+1} - p_j = O(\sqrt{n} \ln n).$$

Consequently, by Theorem 2 and Corollary 2 we get

$$1 \leqslant \lim_{n \to \infty} \frac{\theta_e(K_n \times K_m)}{n(n - 1)}$$

$$\leqslant \lim_{n \to \infty} \frac{\theta_e(K_{n'} \times K_{n'})}{n(n - 1)}$$

$$\leqslant \lim_{n \to \infty} \frac{(n + O(\sqrt{n} \ln n))(n - 1 + O(\sqrt{n} \ln n))}{n(n - 1)} = 1.$$

This completes the proof of the theorem.   □

References

1. Chang, M.-S., Müller, H.: On the tree-degree of graphs. In: Brandstädt, A., Le, V.B. (eds.) WG 2001. LNCS, vol. 2204, pp. 44–54. Springer, Heidelberg (2001)
2. Cramér, H.: On the order of magnitude of the difference between consecutive prime numbers. Acta Arithmetica 2, 23–46 (1936)
3. Cygan, M., Pilipczuk, M., Pilipczuk, M.: Known algorithms for EDGE CLIQUE COVER are probably optimal. In: Proceedings of SODA 2013, pp. 1044–1053 (2013)
4. Bruijn, N.G., Erdős, D., On, P.: a combinatorial problem. Proceedings Koninklijke Nederlandse Akademie van Wetenschappen 51, 1277–1279 (1948)
5. Erdős, P., Goodman, A., Pósa, L.: The representation of a graph by set intersections. Canadian Journal of Mathematics 18, 106–112 (1966)
6. Gramm, J., Guo, J., Hüffner, F., Niedermeier, R.: Data reduction and exact algorithms for clique cover. ACM Journal of Experimental Algorithmics 13, article 2 (2009)
7. Gyárfás, A.: A simple lowerbound on edge covering by cliques. Discrete Mathematics 85, 103–104 (1990)
8. Hoover, D.N.: Complexity of graph covering problems for graphs of low degree. Journal of Combinatorial Mathematics and Combinatorial Computing 11, 187–208 (1992)
9. Kong, J., Wu, Y.: On economical set representations of graphs. Discrete Mathematics and Theoretical Computer Science 11, 71–96 (2009)
10. Kou, L.T., Stockmeyer, L.J., Wong, C.K.: Covering edges by cliques with regard to keyword conflicts and intersection graphs. Communications of the ACM 21, 135–139 (1978)
11. Moradi, S.: A note on tensor product of graphs. Iranian Journal of Mathematical Sciences and Informatics 7, 73–81 (2012)
12. Opsut, R.J.: On the computation of the competition number of a graph. SIAM Journal on Algebraic Discrete Methods 3, 420–428 (1982)
13. Orlin, J.B.: Contentment in graph theory: Covering graphs with cliques. Indagationes Mathematicae 80, 406–424 (1977)
14. Roberts, F.S.: Food webs, competition graphs, and the boxicity of ecological phase space. In: Theory and Applications of Graphs, pp. 477–490 (1978)
15. Roberts, F.S.: Applications of edge coverings by cliques. Discrete Applied Mathematics 10, 93–109 (1985)
16. Weisstein, E.: Bertrand's postulate. Wolfram Mathworld
17. Weisstein, E.: Projective plane. Wolfram Mathworld
18. Cameron, P.J.: Projective and Polar Spaces. QMW Maths Notes 13 (1991)

Protein Name Recognition
Based on Dictionary Mining and Heuristics

Shian-Hua Lin, Shao-Hong Ding, and Wei-Sheng Zeng

Department of Computer Science and Information Engineering,
National Chi Nan University, Puli, Nantou Hsien, 545, Taiwan
{Shlin,s95321526}@ncnu.edu.tw, wilson7126@gmail.com

Abstract. We propose a novel method that integrates dictionary, heuristics and data mining approaches to efficiently and effectively recognize exact protein names from the literature. According to the protein name dictionary and heuristic rules published in related studies, core tokens of protein names can be efficiently detected. However, exact boundaries of protein names are hard to be identified. By regarding tokens of a protein name as items within a transaction, we apply mining associations to discover significant sequential patterns (SSPs) from the protein name dictionary. Based on SSPs, protein name parts are extended from core tokens to left and right boundaries for correctly recognizing the protein name. Based on Yapex101 corpus, Protein Name Recognition System (PNRS) achieves the F-score (74.49%) better than existing systems and papers.

Keywords: protein name recognition, association mining, dictionary mining, heuristics.

1   Introduction

The huge amount of literature data drives an urgent demand for integrating *information/knowledge extraction* (IKE) techniques with bioinformatics systems. One prerequisite for knowledge extraction from scientific literature is accurately recognizing and mapping biological entity names in free text to corresponding entries in biological databases [16]. Several text and data mining conferences are continuously focusing on the research topic: *applying IKE methods to bioinformatics systems*. The Genomics Track competition of TREC (http://trec.nist.gov/) and BIOKDD (http://www.acm.org/) competition of ACM repeatedly hold competitions of extracting biological information and knowledge. However, these studies and bioinformatics systems suffer from a common problem, *name recognition.*

Recognizing protein names is the first step while biomedical researchers attempt to investigate some proteins through computer-aided extractions of protein-protein interactions (PPI), protein-disease relations and protein-function annotations [18]. However, new and unknown protein names are frequently created in the literature, IKE systems cannot effectively extract protein-related information without identifying protein names first. Consequently, *protein name recognition* (**PNR**) is an important issue for protein-related IKE systems.

Q. Gu, P. Hell, and B. Yang (Eds.): AAIM 2014, LNCS 8546, pp. 75–87, 2014.
© Springer International Publishing Switzerland 2014

Conventional name recognition systems aim to identify names of people, locations or organizations from general documents like news articles. In these applications, the best systems published in MUC (Message Understanding Conference) have reached the perfect performance almost comparable to human annotators [21]. Thus, traditional name recognition problems are regarded as solved problems. However, in the topic of molecular biology, recognizing protein names from biomedical literature is a real challenge since there is no standard nomenclature for naming protein entities. Many examples of irregularities and ambiguities of the protein naming are summarized in [7]. Following cases also make PNR problems more complex.

- Protein names may be extremely short or extremely long (compound words), in terms of number of characters or number of words. For example, the Swiss-Prot dictionary has protein names varying from 2 characters (AR) to 192 characters (including space characters). Consequently, finding exact left and right boundaries of a protein name is very difficult.
- Names of chemical compounds, genes, cells, or PPI may also be identified as protein names. E.g., "CTC box binding factor 75 kDa subunit" is a protein name but looks like the PPI information.
- Divergent recognitions of humans: One recognizes "TFPI-C-factor Xa" and the other prefers "TFPI" and "factor Xa".

Above examples list only some PNR problems, there are still many exceptions for naming proteins. Consequently, there are no rules for biomedical researchers while giving names to new proteins, recognizing name entities from the literatures is therefore more difficult than identifying general names like locations, people, news events, etc.

2   Related Works

Although many systems combine several approaches to identify protein names, PNR methods can be roughly classified as follows: dictionary-, learning-, rule- and alignment-based methods. Dictionary-based methods [6][9][21] match each n-gram text of a sentence with patterns appearing in dictionaries or curated patterns. The effectiveness is therefore limited to proteins included in dictionaries so that the recall rate is low. Machine learning-based methods [4][5][11][13][18][23][24][25] reduce the PNR problem to classify tokens into positive or negative parts of protein names based on the curated training corpus. Then, positive tokens are merged to form a complete name. The performance is influenced by the annotated corpus and may be overestimated due to the bias problem of mixing training and testing data. Rule-based methods [7][8][21][22] obviously suffer from the manual cost of experts' heuristics, though rule systems are efficient and easy to be implemented. Alignment-based methods [12][15] translate each character of both protein names (within dictionaries) and sentences of articles into a symbol so that protein names and sentences are regarded as symbolic sequences. By running the local alignment method to align the sentence (query) and dictionary (target) sequences, acceptable recall rate may be achieved.

However, the precision rate is low due to many meaningless alignments like "page 20" aligned with "protein p 20".

Due to different test sets and evaluation criteria, these methods are hard to be compared under the same benchmark. According to the widely used Yapex test corpus, the best F-score values under the STRICT evaluation are 60.95% and 67.1%, which are published in the Yapex website[1] and the literature [7], respectively. Therefore, the Yapex test corpus with proposed evaluation criteria is employed to evaluate our system. We also follow the definition of protein name used in the Yapex naming policy: *"A protein name semantically denotes a single biological entity composed of one or more amino acid chains."* This definition excludes the names of genes, protein families, domains, fragments and organisms, as well as, unspecific references to single protein molecules [18].

3   PNRS

The system, *Protein Name Recognition System* (**PNRS**), has developed and provides services on the Web. We propose a novel dictionary-mining approach to find sequential patterns for extending protein name boundaries from core tokens detected by heuristic rules and dictionaries. Several studies [7][8][10][21] were cited to manually build heuristic rules of protein naming features. These rules are applied to detect protein name fragments as *core tokens* since many protein name fragments share common morphological features like upper/lower case English alphabets, digits, Greek letters or Roman numbers, etc. Based on heuristic rules, core tokens can be efficiently identified, but exact left and right boundaries of protein names are hard to be correctly recognized.

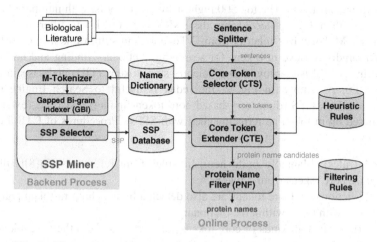

Fig. 1. The system architecture of PNRS: backend and online processes

---

[1] http://www.sics.se/humle/projects/prothalt/yapex_prestanda.html

Based on these ideas, the system architecture of PNRS is shown in Fig. 1. PNRS reads biomedical articles (XML format) and extract metadata through our PubMed Crawler module. The title and abstract are partitioned into sentences by the Sentence Splitter module. Implementations of both modules are trivial and omitted in this paper. In the **CTS** module, heuristic rules and the Swiss-Prot protein name dictionary[2] are applied to extract core tokens within sentences. Once core tokens are detected, **CTE** module extracts candidates of protein names by performing left-and-right extension (**L-R-Ext**) from core tokens based on significant sequential patterns (**SSPs**) mined from the dictionary in the backend process (**SSP Miner**). Finally, protein name candidates pass filtering rules (verified by the **PNF** module) are recognized as exact protein names.

3.1   Core Token Selector (CTS)

Given a sentence string, Tokenizer modules first split the string into tokens. Both types of Tokenizers are employed to extract different kinds of tokens for CTS and following SSP Miner. *C-Tokenizer* splits a sentence into tokens (*c-tokens*) according to user-defined delimiter characters: space, ' , ', ' . ', ' / ', ' ; ' and ' ? '. Morphological information like uppercase/lowercase phenomena of tokens remains unchanged to keep the primitive tokens so that heuristic rules can be applied to detect core tokens. Using c-tokens extracted by C-Tokenizer as the input, CTS applies heuristic rules to select core tokens. However, some rules tend to select common words as core tokens. The rule "starts with one uppercase character and number of tokens is no more than 4" [8] will select irrelevant tokens like "Dual", "Two", "Like", etc. To remedy this drawback, the PubMed Crawler automatically collects more than 18 millions abstracts from NCBI PubMed and counts each token frequency appearing in these titles and abstracts. The top 500 highest-frequent words, with number of characters no more than 4, are selected into the PNRS stopword list to filter off irrelevant core tokens. M-Tokenizer inherits from C-Tokenizer to split a token into useful patterns. To rapidly retrieve m-tokens for searching, dictionary finding and mining; the inverted index [20] is employed to efficiently store tokens (TID) and associated sentences (SID). As shown in Fig. 2, the index is useful to Search Engine (search PubMed articles), CTS (dictionary-based core token finding) and CTE (extend core tokens based on m-token patterns matched with SSPs). Procedures of CTS are listed as follows:

- Step 1 (rules): Manually build heuristic rules from references [7][8][10][21] to recognize core tokens from c-tokens.
- Step 2 (dictionary): Core tokens are also detected by matching m-token patterns of a sentence with those within the dictionary.
- Step 3 (union): Both kinds of core tokens are merged to form the core token set of the sentence.

---

[2] ftp://ftp.uniprot.org/pub/databases/uniprot/
current_release/knowledgebase/complete/uniprot_sprot.xml.gz

- Step 4 (filtering): To filter off irrelevant tokens that are not likely to be protein name fragments, tokens appearing in the PNRS stopword list (including the NCBI stopword list[3]) or containing special symbols ('#', '@', '~', '!', '$', '%', '<', '>', '&', '*' and '\') are omitted. Some general words (gain, contain, increase, decrease, phase, etc.) that match the rule (ends with "-in" and "-ase") are also removed from the core token set.

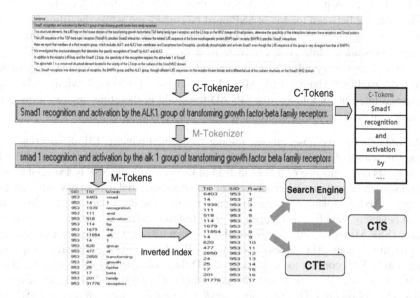

Fig. 2. Tokenize sentences into tokens stored in the inverted index for reusing

3.2   Significant Sequential Pattern (SSP) Miner

Heuristic rules are useful to detect core tokens, but hard to determine the exact protein name. Rule-based approaches [8][21] also use rules to perform left-and-right extensions from core tokens so that both boundaries of the protein name are probably recognized. However, rules are not effective in finding name boundaries due to no standard nomenclature for naming proteins. To extend core tokens and find protein name boundaries, the semantic relationship between tokens (co-occurrence of words), referred to as *"significant pattern* (**SP**)" of tokens, is considered. Numerous protein names were curated in the dictionary, mining SP tokens from dictionary seems feasible. Mining association [1][2] attempts to discover co-occurrence of items in the same transactions within the database. Based on this idea, we successfully mine term associations from documents to improve the performance of document classification [14]. However, following observations make mining SP different from mining association.

---

[3] http://www.ncbi.nlm.nih.gov/entrez/query/
static/help/pmhelp.html

- The sequences of tokens within SPs are meaningful. For example, SP <alpha crystallin> differs from <crystallin alpha> so mining significant pattern is revised as mining *"Significant **Sequential** Pattern (**SSP**)"*.

- Tokens within a sequential pattern can be *gapped* since <A chain>, <B chain>, <chain 1>, <chain 2>, and other similar patterns are frequently used in the dictionary. Therefore, SSP tokens should tolerate one or more *"don't care tokens* (X-tokens)". E.g. "Alpha crystallin A chain" may derive <crystallin X chain>.

- The *distance* between tokens of SSP must be considered while regarding a sequential pattern as SSP. *AD* (Average Distance) measure is proposed to estimate the average distance while grouping patterns with same begin-end tokens. E.g. <crystallin X chain>, <crystallin X X chain> and <crystallin X X X chain> are grouped into <crystallin * chain> with AD = 2 (6 X-tokens in 3 SPs). Trivially, AD is the average tolerable X-tokens of a SP; AD-threshold is therefore applied to assess a sequential pattern, within tolerable X-token counts (*gaps*), as a SSP.

According to above observations, mining "sequential pattern" [3], discovering a sequence of itemsets frequently appearing in transactions, seems useful to mine SSP. The SSP consists of a sequence of tokens instead of itemsets; mining SSP is a special case of mining sequential pattern since each itemset within a sequential pattern has only one item (token). To extract tokens of sequential patterns from the dictionary, n-gram index approach is simply modified as Gapped Bi-gram Indexer (**GBI**), as shown in the backend process (**SSP Miner**) of Fig. 1. All possible bi-gram tokens with gaps are extracted and grouped to calculate AD values. Then, the SSP selector qualified bi-gram sequential patterns as SSPs according to the predefined AD-threshold (default value = 1, i.e., allowing one X-tokens). Without losing the generality, only bi-gram SSP is considered in PNRS since following CTE *gradually extending core tokens through bi-gram SSPs* will cover tri-gram extensions, and vice versa. Consequently, the problem of mining SSP is translated into mining sequential pattern with the extra AD measure.

- The Swiss-Prot dictionary is regarded as the transaction database.
- A protein name entry in the dictionary is mapped to a transaction of the database.
- Protein name m-tokens extracted by M-Tokenizer are sequences of itemsets.

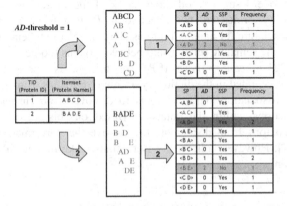

Fig. 3. Mining SSP with the AD measure

As shown in Fig. 3, tokens (itemsets) of the first protein name (transaction) indexed by GBI generate 6 patterns, in which <A D> will be filtered off by AD-threshold = 1. After processing the second transaction, <A D> becomes SSP since the AD value is reduced to 1. All sequential patterns with AD values are stored in the database so that SSP selector is able to rapidly extract SSPs with different AD-threshold values for various experiments or purposes. Currently, support and confidence values considered by mining association to find qualified association rules or sequential patterns are not used in SSP Miner due to low-frequencies of SSPs. For each protein name entry in the dictionary, GBI invokes the SPAD algorithm to extract all sequential patterns and stores information of occurrence frequency (F) and accumulated gaps (G, i.e., the accumulated X-token counts) into the database table. After parsing all the dictionary's entries, SSP Miner removes patterns that contain tokens included in the PNRS stopword list. Finally, the AD value of a sequential pattern is set to G/F (columns are defined as REAL data type in the database table).

Algorithm. SPAD /* store SP and AD into the database */
 Input: p // a protein name entry
 Output: Write each pattern with AD into SPTable (in database)
   Build a m-token array S for p by M-Tokenizer
   Set L to the array length of S
   for $i = 0$ to $L - 2$ **do**
     for $j = i + 1$ to $L - 1$ **do**
       $psp = S[i] + `\ ` + S[j]$ /* a possible sequential pattern */
       $AD = j - i - 1$ /* X-token count: gap = 0, 1, 2, ... */
       Update the database table by the SQL statement[*]
     end for
   end for
 Update each pattern's AD value by one SQL statement[+]

[*]: update SPTable set G = G + AD, F = F + 1 where pattern = SP
[+]: update SPTable set AD = G/F

3.3   Core Token Extender (CTE)

By defining the AD-threshold value (default = 1), SSPs can be efficiently retrieved from the database. Given a sentence with core token set, the Core Token Extender (**CTE**) module applies SSPs to perform left-and-right extension (L-R-Ext) processes from core tokens. Assuming the sentence "A B C D E" has a core token "C" and the AD-threshold is set to 1, L-Ext and R-Ext start from "C" and verify if <A C>, <A B>, <B C>, <C E>, <D E> and <C D> are SSPs. If <A C> is SSP, the protein name is left extended to "A B C". Obviously, CTE tries to extend more tokens through the farthest SSPs based on the AD-threshold. For the example "Tail-anchored protein of 66 amino acid residues that is homologous to the yeast YSY6 protein", CTS generates the core token set: {protein, yeast, YSY6 protein}. Former two core tokens are based on rules and the last one is matched by the Swiss-Prot dictionary. After mining SSPs from the dictionary, CTE extends the first core token "protein" to find a candidate "tail-anchored protein" through L-Ext with SSP <tail protein>. Another candidate "yeast YSY6 protein" can be obtained by R-Ext with the core token "YSY6 protein". Finally, CTE discovers two protein name candidates, "Tail-anchored protein" and "yeast YSY6 protein". The implementation of L-Ext and R-Ext algorithm is trivial and simple.

3.4   Protein Name Filter

Protein Name Filter (**PNF**) applies following rules [21] to remove noisy protein name.

- Candidates only contain following words: factor(s), protein(s), kinase(s), receptor(s), chain(s), etc.
- Candidates only contain Greek letters: alpha, beta, gamma, etc.
- Candidates only contain Roman characters like I, II, etc.

We follow the strict definition of protein name [7] used in Yapex tagging policy to develop our system. By consulting with domain experts, following filtering rules are additionally included:

- Protein names end with "-ed", "-ing", "-tion", "-rial", "-ent", etc.
- In the sentence, if the succeeding token of the protein name is: cell(s), domain(s) or folding(s).
- Groups, but not specific proteins: growth factor, nuclear factor, transcription factor, vitamin, etc.
- Contains 20 amino acid code abbreviations: ARG, ALA, LEU, etc.

4   Experiment and Evaluation

To compare the performance of PNRS with other systems, Yapex test corpus (Yapex101), is employed as the benchmark. The Yapex webpage also lists current best results of precision, recall and F-score measures based on different evaluation methods summarized as followings [7]:

- STRICT: Only count exact matches to the Yapex answers. Of course, the answer count is 1,966.
- PNP (Protein Name Parts): Count each matched token as one match, the answer count is the total number of c-tokens.
- SLOOPY: Consult with domain experts to revise PNP hits. This evaluation method is not considered since it may have biases due to different domain knowledge of experts.
- L BOUNDARY: Exactly match the left boundary of an answer. The answer count is the same with that of STRICT.
- R BOUNDARY: Exactly match the right boundary of an answer. The answer count is the same with that of STRICT.
- ANY BOUNDARY: Exactly match left or right boundaries of an answer. The answer count doubles that of STRICT.

Based on the "dictionary + rule" method and experiment results, CTE applies SSPs (AD-threshold = 1, i.e. one X-token) to enhance the rule-based CTE module and extend the protein name parts from core tokens. L-Ext and R-Ext are separately used to observe the effectiveness. As shown in Fig. 4, the performance of R-Ext (F-score = 65.82%) is slightly better than that of L-Ext (F-score = 65.61%) since most derived protein names share the same prefix tokens (left boundaries). However, PNF works better for R-Ext process due to filtering off more irrelevant candidates of protein names. Since filtering rules tend to focus on right extension variances, the F-score improvements of PNF for R-Ext and L-Ext are 7.44% and 5.97%, respectively. Obviously, the best performance (F-score = 74.49%) is simultaneously applying L- and R-Ext processes. Consequently, SSP Miner achieves 11.12% increments on the F-score in par with our "dictionary + rule" method.

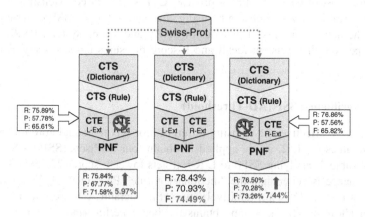

Fig. 4. Results of applying SSP to L-R-Ext (L-Ext vs. R-Ext)

Table 1. Systems usually combine several approaches denoted by: Rule (**R**), Dictionary (**D**), Machine Learning (**ML**), Data Mining (**DM**), Natural Language Processing (**NLP**)

	ML/D [13]	ML/R/D [18]	R [21]	R/D [21]	ML [13]	NLP/R/D [7]	DM/R/D PNRS
Precision (%)	45.1	61.0	57.6	58.3	**76.0**	67.8	70.9
Recall (%)	69.7	59.0	65.3	66.8	59.5	66.4	78.4
F-score (%)	54.8	60.0*	61.2	**62.3**	66.7	67.1	74.5

*: The F-score is measured without the case of induction bias. Using training data as the testing set results in the bias of machine learning and gets the higher F-score (75%).

Table 2. Comparison of the Yapex results with PNRS

Evaluation Methods	Systems	Answers	Analyses	Hits	Recall	Precision	F-score
STRICT	**PNRS**	**1,966**	**2,174**	**1,542**	**78.43**	**70.93**	**74.49**
	Yapex	1,966	1,899	1,178	59.91	62.03	60.95
PNP	**PNRS**	**2,704***	**2,918**	**2,219**	**82.06**	**76.05**	**78.94**
	Yapex	2,705	2,720	2,062	76.22	75.80	76.01
L BOUNDARY	**PNRS**	**1,966**	**2,174**	**1,618**	**82.30**	**74.42**	**78.16**
	Yapex	1,966	1,899	1,340	68.15	70.56	69.34
R BOUNDARY	**PNRS**	**1,966**	**2,174**	**1,621**	**82.45**	**74.56**	**78.31**
	Yapex	1,966	1,899	1,422	72.32	74.88	73.58
ANY BOUNDARY	**PNRS**	**3,932**	**4,348**	**3,239**	**82.38**	**74.49**	**78.24**
	Yapex	3,932	3,798	2,762	70.24	72.72	71.46

*: There is one token match difference between PNRS and Yapex under the PNP evaluation. We have carefully verified the data and parsing programs to make sure without bugs.

Comparing with the current best result [7] and other methods, PNRS outperform other systems, as shown in Table 1. In par with the primitive "rule + dictionary" method [21] and the complex NLP-based approach [7], PNRS improves the F-score with 12.2% and 7.7%, respectively. More comparisons of systems and evaluations on the Yapex benchmark are summarized in [13]. Obviously, PNRS is as simple as the "rule + dictionary" method since we apply backend dictionary mining to discover SSPs so that many exceptions of rule-based extension method can be cured. Without considering the time cost of backend mining process, CTE's SSP-based extension process is as efficient as rule-based extension by using clustered index (on SSP patterns). Table 2 shows the detail measures of matches of PNRS and Yapex system. PNRS outperforms Yapex on all precision, recall and F-score measures based on any evaluation methods.

4.1   Experiments: SSP (AD-Threshold)

To evaluate the effectiveness of using different SSP sets to extend core tokens, AD-threshold values (0, 1, 2, 3) are applied to obtain four SSP sets (SSP0 – SSP4). As shown in Table 3, applying SSP0 to L-R-Ext gets lower F-score (72.35%); however, the performance is better than the "dictionary + rule" PNR method (Fig. 4) and the current best F-score (67.1%). Intuitively, we set the default AD-threshold to 1. Using SSPs that allow one X-token gap obtains the better performance, but not the best. However, the deviations among SSP1 – SSP3 can be neglected and using more SSPs does not guarantee better results. To find the best F-score of the optimal SSP set, we use fine-grain (step = 0.1) AD-thresholds to verify the F-score vs. the size of SSP set. As shown in Fig. 5, the number of SSPs is slightly increased during every 0.1 increment on the AD-threshold. Dramatic promotions of SSP counts happen in integer thresholds. This phenomenon certifies our observations:

Table 3. Evaluate precision, recall and F-score with AD-threshold (0, 1, 2, 3) in SSP Miner

AD-Threshold	R (%)	R (%)	F-score (%)
0 (SSP0)	75.33	69.60	72.35
1 (SSP1)	78.43	70.93	74.49
2 (SSP2)	78.69	70.83	74.55

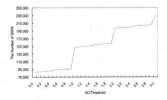

Fig. 5.   SSP counts vs. AD-thresholds

- Gapped sequential patterns are widely used in naming proteins like "Alpha crystallin A chain" named from the gapped pattern <crystallin * chain>.
- The frequency of SSP, such as <crystallin * chain>, if very low so that the integer AD-threshold has striking increment.

Applying each SSP set to evaluate matches (Analyses, Hits) with corresponding to precision, recall and F-score, results are summarized in Table 4. In general, larger AD-threshold values (looser SSP filtering constraints) identify more protein names.

Hits of protein names are basically increased following larger thresholds, though some analyzed names are not matched with answers in equal ratios with increments of AD-Threshold values. The best result (F-score = 74.85%) happen in two thresholds: 0.4 and 0.5. However, the difference between largest and smallest (74.37%) F-score is very low (0.48%); finding the optimal AD-Threshold seems not important in the range [0.1, 3]. As our observation, considering gaps within begin-end tokens of the SSP is useful to improve the performance. Consequently, mining SSP from dictionary is effective to discover extension features to enhance the extension process and find exact protein names.

Table 4. Detail measures on matches, precision, recall and F-score with AD-threshold from 0 to 3 stepped by 0.1. Rows with the same column values are merged into one row.

AD-Th	Analysis	Hits	R (%)	P (%)	F-Score
0	2,128	1,481	75.33	69.60	**72.35**
Best (0.4, 0.5)	**2,162**	**1,545**	**78.59**	**71.46**	**74.85**
Worst (3)	**2,189**	1,545	78.59	70.58	**74.37**

Detail measures are available on http://www.bios.csie.ncnu.edu.tw/PNRS/Table4.htm.

By re-implementing dictionary- and rule-based PNR methods, we observed that one of the hardest PNR problems is extending core tokens to find exact protein name boundaries. We propose a novel dictionary mining approach to discover SSPs for extending protein name fragments so that exact protein names can be correctly identified. Based on our "dictionary + rule" method, SSP-based CTE improves the F-score with 11.12% increments (Fig. 4 and 5). In par with the "rule + dictionary" method (Seki and Mostafa, 2003), F-score = 62.3%, PNRS reaches 12.19% improvements on F-score. Accordingly, the novel SSP mining approach is effective to find more exact protein names.

5   Conclusion

We developed the Protein Name Recognition System that applies mining significant sequential patterns from the dictionary to deal with the problem of protein name extension. We propose a flexible architecture to efficiently do several experiments. CTS module applies C-Tokenizer and M-Tokenizer to parse a sentence into different types of tokens, c-tokens and m-tokens, for rule-matching, dictionary-finding and pattern-mining. We carefully verified both dictionary- and rule-based methods (with CTS, CTE and PNF modules) and obtained similar results (F-score = 63.37%) in par with past studies. Based on the result, CTE applies SSPs to perform L-R-Ext process and find exact protein names. Based on the STRICT evaluation criterion on Yapex101, SSP-based extension approach achieves F-score = 74.49%, with 7.4% increments on the current best result (67.1%). The general architecture of PNRS is flexible to be customized for recognizing miscellaneous name entities like diseases, genes, functions, protein-protein interactions, etc. Based on mining SSPs from various domains dictionaries, the challenge of name extension may be efficiently and effectively solved in the future.

References

1. Agrawal, R., Imielinski, T., Swami, A.: Mining association rules between sets of items in large databases. In: Proceedings of the ACM SIGMOD Conference, pp. 207–216 (1993)
2. Agrawal, R., Srikant, R.: Fast algorithms for mining association rules. In: Proceedings of International Conference on Very Large Databases, Santiago, Chile, pp. 487–499 (September 1994)
3. Agrawal, R., Srikant, R.: Mining sequential patterns. In: Proceedings of International Conference on Data Engineering, Taipei, Taiwan, pp. 3–14 (March 1995)
4. Chang, J.T., Schutze, H., Altman, R.: GAPSCORE: finding gene and protein names one word at a time. Bioinformatics 20, 216–225 (2004)
5. Collier, N., Nobata, C., Tsujii, J.: Extracting the names of genes and gene products with a hidden markov model. In: Proceedings of the 18th International Conference on Computational Linguistics, pp. 201–207 (2000)
6. Egorov, S., Yuryev, A., Daraselia, N.: A simple and practical dictionary-based approach for identification of proteins in MEDLINE abstracts. Journal of the American Medical Informatics Association 11(3), 174–178 (2004)
7. Franzen, K., Eriksson, G., Olsson, F., Asker, L., Liden, P., Cöster, J.: Protein names and how to find them. International Journal of Medical Informatics 67(3), 49–61 (2002)
8. Fukuda, K., Tsunoda, T., Tamura, A., Takagi, T.: Toward information extraction: identifying protein names from biological papers. In: Proceedings of the 3rd Pacific Symposium on Biocomputing, pp. 707–718 (1998)
9. Hanisch, D., Fluck, J., Mevissen, H., Zimmer, R.: Playing biology's name game: Identifying protein names in scientific text. In: Proceedings of the 8th Pacific Symposium on Biocomputing, pp. 403–414 (2003)
10. Huang, M.L., Zhu, X.Y., Hao, Y., Payan, D.G., Qu, K.B., Li, M.: Discovering patterns to extract protein-protein interactions from full texts. Bioinformatics 20, 3604–3612 (2004)
11. Kazama, J., Makino, T., Ohta, Y., Tsujii, J.: Tuning support vector machines for biomedical named entity recognition. In: Proceedings of the ACL 2002 Workshop on Natural Language Processing in the Biomedical Domain, pp. 1–8 (2002)
12. Krauthammer, M., Rzhetsky, A., Morozov, P., Friedman, C.: Using BLAST for identifying gene and protein names in journal articles. Gene 259(1-2), 245–252 (2000)
13. Kou, Z., Cohen, W.W., Murphy, R.F.: High-recall protein entity recognition using a dictionary. Bioinformatics 21, i266–i273 (2005)
14. Lin, S.-H., Shih, C.-S., Chen, M.C., Ho, J.-M., Ko, M.-T., Huang, Y.-M.: Extracting Classification Knowledge of Internet Documents: A Semantics Approach. In: Proceedings of the 21st ACM SIGIR Conference, pp. 241–249 (1998)
15. Lipman, D.J., Pearson, W.R.: Rapid and sensitive protein similarity searches. Science 227, 1435–1441 (1985)
16. Liu, H., Hu, Z.-Z., Zhang, J., Wu, C.: BioThesaurus: a web-based thesaurus of protein and gene names. Bioinformatics 22(1), 103–105 (2006)
17. Malik, R., Franke, L., Siebes, A.: Combination of text-mining algorithms increases the performance. Bioinformatics 22, 2151–2157 (2006)
18. Mika, S., Rost, B.: Protein names precisely peeled off free text. Bioinformatics 20, 241–247 (2004)
19. Nobata, C., Collier, N., Tsujii, J.: Automatic term identification and classification in biology texts. In: Proceedings of the 5th Natural Language Pacific Rim Symposium, pp. 369–375 (1999)

20. Salton, G., McGill, M.J. (1983) Introduction to Modern Information Retrieval. McGraw-Hill (1983)
21. Seki, K., Mostafa, J.: An approach to protein name extraction using heuristics and a dictionary. In: Proceedings of the American Society for Information Science and Technology Annual Conference, ASIST (2003)
22. Tanabe, L., Wilbur, W.J.: Tagging gene and protein names in biomedical texts. Bioinformatics 18, 1124–1132 (2003)
23. Tsai, T.-H., Chou, W.-C., Wu, S.-H., Sung, T.-Y., Hsiang, J., Hsu, W.-L.: Integrating linguistic knowledge into a conditional random field framework to identify biomedical named entities. Expert Systems with Applications 30, 117–128 (2006)
24. Yeganova, L., Smith, L., Wilbur, W.J.: Identification of related gene/protein names based on an HMM of name variations. Computational Biology and Chemistry 28(2), 97–107 (2004)
25. Zhou, G.D., Zhang, J., Su, J., Shen, D., Tan, C.L.: Recognizing names in biomedical texts: A machine learning approach. Bioinformatics 20(7), 1178–1190 (2004)

A Facility Coloring Problem in 1-D[*]

Sandip Das[1], Anil Maheshwari[2], Ayan Nandy[1], and Michiel Smid[2]

[1] Indian Statistical Institute, Kolkata 700 108, India
[2] School of Computer Science, Carleton University, Ottawa, Canada

Abstract. Consider a line segment R consisting of n facilities. Each facility is a point on R and it needs to be assigned exactly one of the colors from a given palette of c colors. At an instant of time only the facilities of one particular color are 'active' and all other facilities are 'dormant'. For the set of facilities of a particular color, we compute the one dimensional Voronoi diagram, and find the cell, i.e, a segment of maximum length. The users are assumed to be uniformly distributed over R and they travel to the nearest among the facilities of that particular color that is active. Our objective is to assign colors to the facilities in such a way that the length of the longest cell is minimized. We solve this optimization problem for various values of n and c. We propose an optimal coloring scheme for the number of facilities n being a multiple of c as well as for the general case where n is not a multiple of c. When n is a multiple of c, we compute an optimal scheme in $\Theta(n)$ time. For the general case, we propose a coloring scheme that returns the optimal in $O(n^2 \log n)$ time.

1 Introduction

In this paper we study a facility location problem. There are n facilities to be distributed between c classes of service providers. Each class of service provider should be assigned at least one facility and no facility should be assigned to more than one class. Moreover, when one class of service provider is active, all other classes are dormant. Our objective is to partition the set of facilities in c classes such that the users are served as equitably as possible, i.e., the maximum length amongst the regions covered by any facility of any class is minimized.

In the area of wireless sensor networks, an effective approach for energy conservation is scheduling sleep intervals for sensors [1]. One can assign a color to each sensor, each color representing a set of sensors which would be active at a given time when the rest are in the sleep mode. Here the objective would be to color the nodes such that the maximum area covered by any active node is minimized. Lin et al [2] have explored the problem of maximizing the lifetime of wireless sensor networks. Their study is based on finding the maximum number of disjoint connected covers that satisfy both sensing coverage and network connectivity.

---

[*] This research is partially supported by NSERC and the Commonwealth Scholarship Program of DFAIT, Canada.

Q. Gu, P. Hell, and B. Yang (Eds.): AAIM 2014, LNCS 8546, pp. 88–99, 2014.

Here we study the MinVor problem. Let n_η be the number of facilities of color class η, where $\eta = 0, \ldots, c-1$ and $n = \Sigma_{\eta=0}^{c-1} n_\eta$. For each η, we draw the Voronoi diagram considering the corresponding n_η facilities which represents the active sensors while the remaining $n - n_\eta$ facilities represent the sensors in sleep mode. A Voronoi diagram of a set of k sites partitions the Euclidean space into k regions such that the region of each site consists of all points that are closer to it than to any other site. For our problem, the region R is a horizontal line segment. Let $\gamma_{\eta,j}$ denote the length of the Voronoi zones of the j-th facility from the left end of some color η.

Formally, the MinVor problem is to devise a coloring scheme that minimizes $\max_{\eta=0,1,\ldots,c-1} \max_{j=1,2,\ldots,j_\eta} \gamma_{\eta,j}$ where j_η is the total number of facilities assigned the color η.

Problems similar to the MinVor problem in the plane have been considered in [1, 3–5]. Funke et al [1] presented a greedy algorithm that provides complete coverage with an approximation factor no better than $\Omega(\log n)$, where n is the number of sensor nodes. An algorithm is said to provide complete coverage if the set of the selected sensors always covers the region R, provided that there exists a feasible solution. The communication graph is an undirected graph in which sensors are represented as nodes and there is an edge between two nodes if they can talk to each other. Attempts have been made to cover the Communication Graph using a connected dominating set (CDS) S', which is a subset of the set of sensors, S, such that each node in $S \setminus S'$ is adjacent to some node in S' and the communication subgraph induced by S' is connected. Clark et al [3] have shown that the problem of finding a minimum CDS for unit-disk graphs is NP-hard. An 8-approximation algorithm with $O(n)$ time complexity was suggested by Wan et al [4] which was later improved to a 6.91 approximation factor [5].

In this paper we consider the case where R is a horizontal line segment and provide optimal solutions. As input to the problem we have the location of n facilities, specified by a distance vector $\bar{d} = \langle d_1, d_2, \ldots, d_{n+1} \rangle$, where the i-th facility is placed at a distance of $\sum_{j=1}^{i} d_j$ from the left end of R, for $i = 1, \ldots, n$ and d_{n+1} is the distance of the n-th facility from the right end of R. We consider various cases depending on the values of n and c. Note that each facility is assigned exactly one of the colors from $\{0, 1, \ldots, c-1\}$ and each color is assigned to at least one facility. We assume that the density of users is uniform over R and hence the Voronoi length reflects the proportional user volume.

For $n < 2c$ observe that there has to be a color with only one facility whose Voronoi zone is the whole space. In Section 3 we prove that C_1 (see Definition 1) is an optimal coloring for any distance vector if the number of facilities is twice the number of colors. In Section 4 we show that if the number of facilities is any multiple of c, the same coloring scheme provides an optimal solution. Section 5 suggests a coloring scheme for the general case where n is not a multiple of c which produces an optimal coloring in $O(n^2 \log n)$ time.

2   Notations and Definitions

The facilities on the horizontal line segment R are to be assigned colors from the set $\{0, 1, \ldots, c-1\}$. A facility at *position* f means that among the n facilities on the line segment R, it is the f-th one from the left and hence at a distance $\Sigma_{k=1}^{f} d_k$ from the left boundary of the line segment R. $P_{i,j}$ is the *position* of the j-th facility from the left belonging to the color class i, i.e., there are exactly $P_{i,j} - 1$ facilities to the left of this facility among which $j - 1$ are of color i. We define $M(a, b) = \frac{1}{2}\Sigma_{\ell=a+1}^{b} d_\ell$, where $M(a, b)$ is the Voronoi length of a facility at some position f whose immediate left neighbor of the same color is at position a and whose immediate right neighbor of the same color is at position b, $a < f < b$ (see Figure 1 for an illustration). We define $L(a, b) = \Sigma_{\ell=1}^{a} d_\ell + \frac{1}{2}\Sigma_{\ell=a+1}^{b} d_\ell$, where $L(a, b)$ is the Voronoi length of a facility at position a which is the leftmost of its color and whose immediate right neighbor (of the same color) is at position b. Analogously, we define $R(a, b) = \frac{1}{2}\Sigma_{\ell=a+1}^{b} d_\ell + \Sigma_{\ell=b+1}^{n+1} d_\ell$, where $R(a, b)$ is the Voronoi length of a facility at position b which is the rightmost of its color and whose immediate left neighbor (of the same color) is at position a.

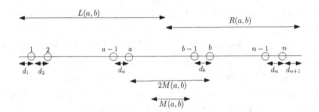

Fig. 1. $M(a, b)$, $L(a, b)$ and $R(a, b)$

We define the objective function of a coloring C as $\Delta(C)$, which is the largest Voronoi length among all the facilities corresponding to all the colors.

In this paper we use certain fixed coloring schemes, as they turn out to be optimal for specific configurations. One such scheme is as follows.

Definition 1. (*Coloring C_1*): *Consider a coloring of the facilities from the left in the order $0, 1, \ldots, c-1, c-1, \ldots, 0, 0, 1, \ldots, c-1, c-1, \ldots, 0, \ldots$. We define this assignment of colors as the coloring C_1. The position of the jth facility of color i for C_1 is denoted by $F_{i,j}$. Note that $F_{i,j} = 2c\lfloor\frac{j-1}{2}\rfloor + (2c-i)(1-\chi_j) + (i+1)\chi_j$, where $\chi_j = 1$ if j is odd and 0 otherwise.*

For example, for $n = 9$ and $c = 3$, C_1 will color the facilities by the colors $0, 1, 2, 2, 1, 0, 0, 1, 2$, from left to right on R (see Figure 2). Let $\alpha_{i,j}$ be the *length* of the Voronoi cell of the j-th facility from the left belonging to the color class i, where facilities are colored by C_1. Note that

$$\alpha_{i,j} = \begin{cases} M(F_{i,j-1}, F_{i,j+1}) & \text{if } j \neq 1, j_i \\ L(F_{i,j}, F_{i,j+1}) & \text{if } j = 1 \\ R(F_{i,j-1}, F_{i,j}) & \text{if } j = j_i \end{cases}$$

Fig. 2. An example for $n = 9, c = 3$ using C_1. The facilities are represented by circles and the numbers 0, 1 or 2 within the facility represent the color of the facility. The distance vector is $\bar{d} = \langle 3, 2, 5, 7, 3, 5, 1, 2, 5, 7 \rangle$. The length of Voronoi cells of the facilities are as follows: $\alpha_{0,1} = L(1,6) = 3 + \frac{(2+5+7+3+5)}{2} = 14$, $\alpha_{0,2} = M(1,7) = \frac{(2+5+7+3+5+1)}{2} = 11.5$, $\alpha_{0,3} = R(6,7) = \frac{1}{2} + (2+5+7) = 14.5$, $\alpha_{1,1} = L(2,5) = (3+2) + \frac{(5+7+3)}{2} = 12.5$, $\alpha_{1,2} = M(2,8) = \frac{(5+7+3+5+1+2)}{2} = 11.5$, $\alpha_{1,3} = R(5,8) = \frac{(5+1+2)}{2} + (5+7) = 16$, $\alpha_{2,1} = L(3,4) = (3+2+5) + \frac{7}{2} = 13.5$, $\alpha_{2,2} = M(3,9) = \frac{(7+3+5+1+2+5)}{2} = 11.5$, $\alpha_{2,3} = R(4,9) = \frac{(3+5+1+2+5)}{2} + 7 = 15$. Therefore $\Delta(C_1) = \alpha_{1,3} = 16$.

$\Delta(C_1) = \max\{\alpha_{i,j}\}$, for all $i = 0, \ldots, c - 1$ and $j = 1, 2, \ldots, j_i$, where j_i is the total number of facilities assigned to color i.

The value of the objective function for the optimum coloring is denoted by Opt.

3 $n = 2c$

In Theorem 1 we show that C_1 is an optimal coloring when we have $n = 2c$ facilities. It is obvious that for each color there will be exactly two facilities, otherwise the Voronoi cell for at least one of the colors will be whole of R, which is clearly non-optimal. In Case 1 of Theorem 1, we show that if the objective function for C_1 returns the Voronoi length of the 1st facility of some color i (i.e. $\Delta(C_1) = \alpha_{i,1}$), then any attempt to get a new coloring to reduce $\alpha_{i,1}$ will ensure that in the new coloring there will be some color i', such that the length of the Voronoi cell corresponding to its 1st facility will be at least $\alpha_{i,1}$. The analogous result for the case where the objective function for C_1 returns the Voronoi length of 2nd facility of some color i is shown in Case 2 of Theorem 1. Using these two cases, we show in Theorem 1 that for $n = 2c$, C_1 is an optimal coloring.

We consider C' to be a coloring scheme different from C_1 introduced as a candidate for possible improvement over C_1. Let $\alpha'_{i,j}$ denote the length of the Voronoi cell corresponding to the j-th facility from the left belonging to the color class i in C'. Let $F'_{i,j}$ be the position of the j-th facility from the left of color i using C'.

Theorem 1. *For $n = 2c$, C_1 is an optimal coloring for the* MinVor *problem.*

Proof. Suppose $\Delta(C_1) = \alpha_{i,\ell}$ for some $i \in \{0, \ldots, c - 1\}$ and some $\ell \in \{1, 2\}$ for C_1.

Case 1: $\ell = 1$: We investigate if it is possible to achieve an objective function whose value is smaller than $\Delta(C_1)$ by any alternate coloring scheme C'. $F'_{j,k}$ is the position of the k-th facility of color j for the coloring C' and $\alpha'_{j,k}$ be the Voronoi length of the facility at $F'_{j,k}$ in C'. Note that each color in C' is assigned to exactly two facilities. Let $S_1 = \{i' : F'_{i',1} \geq i+1\}$, i.e., S_1 is the set of colors whose first facility for coloring C' is at position $i+1$ or higher. Clearly $|S_1| \geq c - i$. Note $F'_{i',1} \geq F_{i,2}$ implies $\alpha'_{i',1} = L(F'_{i',1}, F'_{i',2}) \geq L(F_{i,1}, F_{i,2}) \geq \alpha_{i,1} = \Delta(C_1)$. Hence $F'_{i',1} < F_{i,2} = 2c-i \ \forall \ i' \in \{0, \ldots, c-1\}$ (from Definition 1). Let $S_2 = \{i' : F'_{i',2} \geq 2c-i\}$ i.e., S_2 is the set of colors whose second facility for coloring C' is at position $2c - i$ or higher. Since $F'_{i',1} < 2c - i \ \forall \ i'$, $|S_2| = i + 1$. Therefore $|S_1| + |S_2| > c$, i.e., S_1 and S_2 are not disjoint and there exists some i' such that $F'_{i',1} \geq F_{i,1}$ and $F'_{i',2} \geq F_{i,2} \Rightarrow L(F'_{i',1}, F'_{i',2}) \geq L(i+1, 2c-i) = L(F_{i,1}, F_{i,2}) \Rightarrow \alpha'_{i',1} \geq \alpha_{i,1}$.

Case 2: $\ell = 2$: Let $d'_i = d_{n+2-i} \ \forall \ i = 1, \ldots, n+1$. Then, for the distance vector $\bar{d}' = \langle d'_1, \ldots, d'_{n+1} \rangle$, the value of the objective function is $\alpha_{i,1}$. This is because if the Voronoi length of the 2nd facility for color i is the value of the objective function for coloring C_1, if we look at the mirror image such that the ℓ-th facility from the left is now the ℓ-th facility from the right, then the Voronoi length of the 1st facility for color i is the value of the objective function for coloring C_1. So, as demonstrated in Case 1, for the distance vector $\bar{d}'$, for any new coloring C', $\exists$ some i' such that $F'_{i',1} \geq F_{i,1}$ and $F'_{i',2} \geq F_{i,2}$. Hence for the distance vector $\bar{d} = \langle d_1, \ldots, d_{n+1} \rangle$, for any coloring C', there exists some i' such that $F'_{i',1} \leq F_{i,1}$ and $F'_{i',2} \leq F_{i,2}$. Therefore, there exists some i' such that $F'_{i',1} \leq F_{i,1}$ and $F'_{i',2} \leq F_{i,2} \Rightarrow R(F'_{i',1}, F'_{i',2}) \geq R(F_{i,1}, F_{i,2}) \Rightarrow \alpha'_{i',2} \geq \alpha_{i,2}$.

Hence, no coloring C' can result in an objective function with value less than $\Delta(C_1)$. □

4 $n = kc$, $k > 2$

In this section, we extend the C_1-coloring result of the previous section to the cases where $n = kc$ and $k > 2$.

Theorem 2. For $n = kc$ and any integer $k > 2$, C_1 is an optimal coloring for the MinVor problem.

Proof. As the facilities are colored by the coloring C_1, let Voronoi length of the j-th facility of the color η achieve maximum amongst all the facilities, i.e. $\Delta(C_1) = \alpha_{\eta,j}$. If $j = 1$ or k, then we are looking at facilities at the left or right end among all the facilities of color η. Else, we are dealing with some intermediate facility of color η.

Consider first the case $j \notin \{1, k\}$. Let $S = \{F_{\eta,j-1}+1, F_{\eta,j-1}+2, \ldots, F_{\eta,j+1} - 1\}$. We investigate if it is possible to achieve an objective function with value smaller than $\Delta(C_1)$ by the alternate coloring scheme C'. There exists an i such that for the coloring C' there is at most one facility of color i in S because there are c colors and $|S| = 2c - 1$. Let there be one facility, say the j_1-th of color i with a position in S and j_{max} is the number of facilities of color i. Then for $1 < j_1 < j_{max}$, $F'_{i,j_1-1} \leq F_{\eta,j-1} < F'_{i,j_1} < F_{\eta,j+1} \leq F'_{i,j_1+1}$ which implies $\alpha'_{i,j_1} \geq \alpha_{\eta,j} = \Delta(C_1)$. If $j_1 = 1$, $F_{\eta,j-1} < F'_{i,1} < F_{\eta,j+1} \leq F'_{i,2}$ which implies $\alpha'_{i,j_1} = L(F'_{i,1}, F'_{i,2}) \geq \alpha_{\eta,j} = \Delta(C_1)$. If $j_1 = j_{max}$, $F'_{i,j_{max}-1} \leq F_{\eta,j-1} < F'_{i,j_{max}} < F_{\eta,j+1}$ which implies $\alpha'_{i,j_1} = R(F'_{i,j_{max}-1}, F'_{i,j_{max}}) \geq \alpha_{\eta,j} = \Delta(C_1)$.

For $j = 1$, the situation is similar to the Case 1 of Theorem 1. We have $\Delta(C_1) = \alpha_{\eta,1} = L(\eta + 1, 2c - \eta)$. We investigate if it is possible to achieve an objective function with value smaller than $\Delta(C_1)$ by C'. Let $S_1 = \{i' : F'_{i',1} \geq F_{\eta,1}\}$. Since $F_{\eta,1} = \eta + 1$, $|S_1| \geq c - \eta$. Let $S_2 = \{i' : F'_{i',2} \geq F_{\eta,2}\}$. Since $F_{\eta,2} = 2c - \eta$ and $F'_{i',1} < F_{\eta,2} \ \forall \ i'$, $|S_2| \geq \eta + 1$. Therefore $|S_1| + |S_2| > c$, i.e., S_1 and S_2 are not disjoint and $\exists$ some i' such that $F'_{i',1} \geq F_{\eta,1}$ and $F'_{i',2} \geq F_{\eta,2} \Rightarrow L(F'_{i',1}, F'_{i',2}) \geq L(F_{\eta,1}, F_{\eta,2}) \Rightarrow \alpha'_{i',1} \geq \alpha_{\eta,1}$.

For $j = k$, we have $\Delta(C_1) = \alpha_{\eta,k} = L(n - 2c + u, n - u + 1)$ for some $u \in \{1, \ldots, c\}$. Let $j_{i'}$ be the number of facilities assigned to the color i' by the new coloring scheme C' for which we investigate if it is possible to achieve an objective function with value smaller than $\Delta(C_1)$. Note that if $F'_{i',j_{i'}} \leq F_{\eta,k-1}$ for some i', then $R(F'_{i',j_{i'}-1}, F'_{i',j_{i'}}) \geq R(F_{\eta,k-1}, F_{\eta,k})$ since $F'_{i',j_{i'}-1} < F'_{i',j_{i'}} \leq F_{\eta,k-1} < F_{\eta,k}$. Hence we assume $F'_{i',j_{i'}} > F_{\eta,k-1} \ \forall \ i'$. Let $S'_1 = \{i' : F'_{i',j_{i'}} \leq F_{\eta,k}\}$. Since $F_{\eta,k} = n - u + 1$, $|S'_1| \geq c - u + 1$. Let $S'_2 = \{i' : F'_{i',j_{i'}-1} \leq F_{\eta,k-1}\}$. Since $F_{\eta,k-1} = n - 2c + u$ and $F'_{i',j_{i'}} > F_{\eta,k-1} \ \forall \ i'$, $|S'_2| \geq u$. Therefore $|S'_1| + |S'_2| > c$, i.e., S'_1 and S'_2 are not disjoint and $\exists$ some i' such that $F'_{i',j_{i'}} \leq F_{\eta,k}$ and $F'_{i',j_{i'}-1} \leq F_{\eta,k-1} \Rightarrow R(F'_{i',j_{i'}-1}, F'_{i',j_{i'}}) \geq R(F_{\eta,k-1}, F_{\eta,k}) \Rightarrow \alpha'_{i',j_{i'}} \geq \alpha_{\eta,k}$. □

5 $n = kc + m$

In this section we consider the general case where n is not a multiple of c. Note that $0 < m < c$. In Section 5.1, we introduce a coloring C_2 where the facility of any color and its next to next neighbouring facilities of the same color have exactly $2c - 1$ facilities in between them. If the maximum Voronoi length among all the facilities in this coloring corresponds to an interior facility of some color, i.e, a facility which is neither the leftmost nor the rightmost of its color, then C_2 is the optimal coloring. Otherwise, for a given Δ, we define a coloring C_Δ in Section 5.2. We denote the Voronoi length of the j-th facility of the i-th color for coloring C_2 as $\beta_{i,j}$ and for coloring C_Δ as $\gamma_{i,j}$. In Theorem 3 we show that there exists an optimal C_Δ which can be obtained in $O(n^2 \log n)$ time.

5.1   Coloring C_2

Definition 2. *(Coloring C_2): Let $S_1 = 0, 1, \ldots, m - 1$, $\bar{S}_1 = m - 1, m - 2, \ldots, 1, 0$, $S_2 = m, m + 1, \ldots, c - 1$, $\bar{S}_2 = c - 1, c - 2, \ldots, m + 1, m$. Consider a coloring of the facilities from the left in the order $S_1, S_2, \bar{S}_1, \bar{S}_2, S_1, S_2, \bar{S}_1, \bar{S}_2, \ldots$. We define this assignment of colors as the coloring C_2.*

Lemma 1. *If the value of the objective function for C_2 is not equal to $\beta_{i,1}$ or $\beta_{i,k}$ for some $i \in \{m, m + 1, \ldots, c - 1\}$, then C_2 is an optimal coloring.*

Proof. The value of the objective function for C_2 can not be equal to $\beta_{i',1}$ or $\beta_{i',k+1}$ for some $i' \in \{0, \ldots, m - 1\}$. By Definition 2, $\forall \ i' \in \{0, 1, \ldots, m - 1\}$ and $\forall \ i \in \{m, m + 1, \ldots, c - 1\}$, $P_{i',1} = i' + 1 < i + 1 = P_{i,1} < P_{i,2} = c + i' + 1 < c + i + 1 = P_{i,2}$, which implies $\beta_{i',1} = L(P_{i',1}, P_{i',2}) < L(P_{i,1}, P_{i,2}) = \beta_{i,1}$.

Similarly $P_{i',k} < P_{i,k-1} < P_{i',k+1} < P_{i,k}$ and hence $\beta_{i',k} = R(P_{i',k}, P_{i',k+1}) < R(P_{i,k-1}, P_{i,k}) = \beta_{i,k}$.

Consider the case where the value of the objective function for C_2 is $\beta_{i,j}$ for some intermediate facility j of any color i. We call j-th facility of a color i as an intermediate facility when $2 \leq j \leq k-1$ for $i \in \{m, m+1, \ldots, c-1\}$ or when $2 \leq j \leq k$ for $i \in \{0, 1, \ldots, m-1\}$). Then, by Definition 2, $P_{i,j+1} - P_{i,j-1} = 2c$. Any alternate coloring scheme C' will have at least one color i' with at most one facility in $S = \{P_{i,j-1}+1, \ldots, P_{i,j+1}-1\}$. If there is one facility of i' in S, then its Voronoi length is at least $\beta_{i,j}$ and if there is no facility of i' in S, then the Voronoi length of a facility of color i' nearest to S is greater that $\beta_{i,j}$. Hence the value of the objective function can not be reduced from $\beta_{i,j}$ by any different coloring scheme C'. □

5.2 Coloring C_Δ

If the objective function for C_2 returns the Voronoi length of the 1st or last facility of some color in $\{m, m+1, \ldots, c-1\}$, C_2 need not be optimal. For example if $n = 18$, $c = 5$ and the distance vector $d = (2, 10, 1, 1, 1, 8, 1, 1, 1, 1, 1, 1, 1, 1, 1, 1, 1, 1)$, then C_2 returns an objective function 20.5 where as $Opt = 19$ for the coloring scheme $0, 1, 2, 3, 4, 4, 3, 2, 1, 0, 3, 0, 1, 2, 4, 2, 1, 0$. To handle such cases, let us consider a number Δ such that $\max(\max_{u=1,\ldots,n-2c} M(u, u+2c), \max_{u=1,\ldots,c} L(u, 2c - u + 1), \max_{u=1,\ldots,c} R(n - 2c + u, n - u + 1)) \leq \Delta \leq \max(M(1, n), \max_{u=1,\ldots,c} L(u, n - u + 1), \max_{u=1,\ldots,c} R(u, n - u + 1))$. The lower bound of Δ will be clear from Observation 1 and the upper bound is arrived considering the extreme cases where the 1st and 3rd facilities of a colour are at positions 1 and n respectively, or the 1st and 2nd facilities of a colour are at positions u and $n - u + 1$ respectively for some $u \in \{1, 2, \ldots, c\}$, or the last but one and last but one facilities of a colour are at positions u and $n - u + 1$ respectively for some $u \in \{1, 2, \ldots, c\}$. We intend to define a Coloring C_Δ where the value of the objective function will be lesser than or equal to Δ. If Δ is less than the optimum, clearly the coloring doesn't exist. There are $O(n^2)$ possible choices for Δ since the optimum can be of the form $L(a, b)$, $M(a, b)$ or $R(a, b)$, where $1 \leq a < b \leq n$.

Observation 1. *If* $\max_{u=1,\ldots,n-2c} M(u, u+2c) > \Delta$ *or* $\max_{u=1,\ldots,c} L(u, 2c - u + 1) > \Delta$ *or* $\max_{u=1,\ldots,c} R(n - 2c + u, n - u + 1) > \Delta$, *then* C_Δ *does not exist.*

Proof. Proof omitted due to paucity of space. □

After choosing a Δ at least as large as the maximum among $\max_{u=1,\ldots,n-2c} M(u, u+2c)$, $\max_{u=1,\ldots,c} L(u, 2c - u + 1)$ and $\max_{u=1,\ldots,c} R(n - 2c + u, n - u + 1)$, our scheme to obtain C_Δ can be divided into two parts. The first part assigns k facilities to each of the colors $m, \ldots, c - 1$, ensuring that the Voronoi length of each of these $k(c - m)$ facilities is lesser than or equal to Δ. The second part assigns the remaining colors $0, \ldots, m - 1$ to the remaining $m(k + 1)$ facilities, each of these colors being assigned to $k + 1$ facilities.

In the first part of the coloring scheme, we identify the facilities to be assigned the colors $m, \ldots, c-1$. Initially we define the positions of the first facilities of these colors as $P_{i,1} = i+1 \; \forall \; i \in \{m, \ldots, c-1\}$. $P_{c-1,2}$ is such that $L(P_{c-1,1}, P_{c-1,2}) \leq \Delta < L(P_{c-1,1}, P_{c-1,2}+1)$, i.e., the farthest positon to ensure that $\gamma_{c-1,1} \leq \Delta$.

Observation 2. $P_{c-1,2} \geq c+1$

Proof. Δ is chosen such that $\Delta > \max_{u=1,\ldots,c} L(u, 2c-u+1) \geq L(c, c+1)$. As $P_{c-1,1} = c$ and $P_{c-1,2}$ is such that $L(P_{c-1,1}, P_{c-1,2}) \leq \Delta < L(P_{c-1,1}, P_{c-1,2}+1)$, the observation follows. □

Now $\forall \; i = c-2, \ldots, m$, we define $P_{i,2}$ such that $L(i+1, P_{i,2}) \leq \Delta < L(i+1, P_{i,2}+1)$. If k is odd and $P_{i,2} > 2c - i + 2m$, we set $P_{i,2} = 2c - i + 2m$. If k is even and $P_{i,2} > 2c - i + m$, we set $P_{i,2} = 2c - i + m$. If $P_{i,2} \leq P_{i+1,2}$, we backtrack by reducing $P_{i+1,2}$ and if necessary even earlier defined assignments of colors. Essentially we reduce $P_{i+1,2}$ by one, but if that conflicts with $P_{i+2,2}$, we reduce that by 1 and so on. Please note that the operation of backtracking will stop at $P_{c-1,2}$ or before because of the lower bound discussed in Observation 2. We observe, similar to the Observation 2, the following:

Observation 3. $P_{i,2} \geq 2c - i$

To assign the j-th facility of each of these colors, $\forall \; j = 3, \ldots, k-1$:

1. If j is odd, we define $P_{i,j}$, where the order of i is $m, \ldots, c-1$, i.e, first $P_{m,j}$ is identified, then $P_{m+1,j}$ and so on. If k is odd, $P_{i,j} = (j-1)c + i + 1$. If k is even, $P_{i,j}$ is such that $M(P_{i,j-2}, P_{i,j}) \leq \Delta < M(P_{i,j-2}, P_{i,j}+1)$. If $i > m$ and $P_{i,j} \leq P_{i-1,j}$, we backtrack starting with $P_{i-1,j} = P_{i,j} - 1$. If $i = m$ and $P_{i,j} \leq P_{i,j-1}$, we backtrack starting with $P_{i,j-1} = P_{i,j} - 1$. For even k and $P_{i,j} > jc + i + m + 1$, we set $P_{i,j} = jc + i + m + 1$.
2. If j is even, the order of i is $c-1, \ldots, m$ while defining $P_{i,j}$, i.e, first $P_{c-1,j}$ is identified, then $P_{c-2,j}$ and so on. We ensure that $M(P_{i,j-2}, P_{i,j}) \leq \Delta < M(P_{i,j-2}, P_{i,j}+1)$. If $i < c-1$ and $P_{i,j} \leq P_{i+1,j}$, we backtrack starting with $P_{i+1,j} = P_{i,j} - 1$. If $i = c-1$ and $P_{i,j} \leq P_{i,j-1}$, we backtrack starting with $P_{i,j-1} = P_{i,j} - 1$. If k is odd and $P_{i,j} > jc - i + 2m$, we set $P_{i,j} = jc - i + 2m$. If k is even and $P_{i,j} > jc - i + m$, we set $P_{i,j} = jc - i + m$.

Observation 4. $P_{i,j+1} - P_{i,j-1} \geq 2c \; \forall \; j = 2, 3, \ldots, k-1$.

Proof. Δ is chosen such that $\Delta \geq \max_{u=1,\ldots,n-2c} M(u, u+2c)$. If $P_{i,j+1} - P_{i,j-1} < 2c$ for any $j \in \{2, 3, \ldots, k-1\}$, then $\gamma_{i,j} = M(P_{i,j-1}, P_{i,j+1}) < M(u, u+2c)$ for some u. □

Observation 5. $P_{i,j} \geq (j-1)c + i + 1$ for odd j and $P_{i,j} \geq jc - i$ for even j.

Proof. $P_{i,1} = i+1$ and by Observation 4 for odd j, $P_{i,j} - P_{i,1} \geq \frac{j-1}{2}2c \Rightarrow P_{i,j} \geq (j-1)c + i + 1$. By Observations 2 and 3, $P_{i,2} \geq 2c - i$ for all $i \in \{m, m+1, \ldots, c-1\}$. Therefore for even j, $P_{i,j} - P_{i,2} \geq \frac{j-2}{2}2c \Rightarrow P_{i,j} \geq jc - i$. □

Now we define $P_{i,k}$, i.e., the last position for the colors $i = m, \ldots, c - 1$. If k is odd, we define $P_{i,k} = n - c - m + i + 1 \; \forall \; i = m, \ldots, c - 1$. If k is even, we define $P_{i,k} = n - i \; \forall \; i = m, \ldots, c - 1$. Irrespective of whether k is odd or even, if $M(P_{i,k-2}, P_{i,k}) > \Delta$ for any $i \in \{m, \ldots, c - 1\}$ or $R(P_{i,k-1}, P_{i,k}) > \Delta$ for any $i \in \{m, \ldots, c - 1\}$, then C_Δ does not exist for the given Δ. Otherwise we proceed to the second part of the coloring scheme.

In the second part of the scheme, we assign facilties to the colors $0, \ldots, m - 1$. We have remaining $m(k + 1)$ facilities to be colored $0, \ldots, m - 1$. We color them using Coloring C_1 for m colors. It is easy to note that,

Observation 6. For $i = 0, 1, \ldots, m - 1$, we have $P_{i,1} = i + 1$, $P_{i,k+1} = n - i$ for odd k and $P_{i,k+1} = n - m + i + 1$ for even k.

5.3   Optimality of C_Δ

In this section we prove that an optimal coloring C_Δ and the corresponding $\Delta = Opt$ can be identified in $O(n^2 \log n)$ time. Proof of Lemma 4 is omitted because of paucity of space.

Lemma 2. For all $i' \in \{0, \ldots, m - 1\}$

1. $\gamma_{i',1} \leq L(u, 2c - u + 1)$ for some $u \in \{1, \ldots, c\}$
2. $\gamma_{i',k+1} \leq R(n - 2c + u, n - u + 1)$ for some $u \in \{1, \ldots, c\}$
3. $P_{i',j'+1} - P_{i',j'-1} \leq 2c \; \forall \; j' \in \{2, \ldots, k\}$.

Proof. If $2c - i' < P_{i',2} \leq 2c$ for some $i' \in S_1 = \{0, \ldots, m - 1\}$, then to its left there are at least $2c - i'$ facilities of which the first c are the first facilities of each color (by the description of C_Δ and Observation 6). Moreover, there are $m - i' - 1$ more facilities with colors in S_1 with position less than $P_{i',2}$. So in $\{c + 1, \ldots, P_{i',2} - 1\}$ there are atleast $2c - i' - (m - i' - 1) = c - m + 1$ facilities with colors in $S_2 = \{m, \ldots, c - 1\}$, which implies $P_{i'',3} < P_{i',2} < 2c$ for some $i'' \in S_2$. Then $P_{i'',3} - P_{i'',1} < 2c$ violating Observation 4. If $P_{i',2} > 2c$, then there are at most $2m - 1$ of the facilities with position less than $P_{i',2}$ are from S_1. Therefore there are at least $2c - (2m - 1) = 2(c - m) + 1$ facilities assigned colors from $m, m + 1, \ldots, c - 1$ with positions in $\{1, 2, \ldots, c\}$, i.e., there is a color $i \in \{m, m + 1, \ldots, c - 1\}$ such that $P_{i,3} \leq 2c$. But $P_{i,1} > m$ for all $i \in \{m, m + 1, \ldots, c - 1\}$ and hence $P_{i,3} - P_{i,1} < 2c$ violating Observation 4. So $P_{i',2} \leq 2c - i'$ for all $i' \in S_1$ and hence $\gamma_{i',1} \leq L(u, 2c - u + 1)$ for some $u \in \{1, \ldots, c\}$.

Similarly $\gamma_{i',k+1} \leq R(n - 2c + u, n - u + 1)$ for some $u \in \{1, \ldots, c\}$.

Suppose $\exists \; i' \in S_1$ such that $P_{i',j'+1} - P_{i',j'-1} > 2c$. Let $T = \{P_{i',j'-1} + 1, \ldots, P_{i',j'+1} - 1\}$ and $|T| = 2c - 1 + u$ where $u > 0$. Since the colors in S_1 form an C_1 of m colors for the $m(k + 1)$ facilities they are assigned to, $\forall \; i_1 \in S_1 - \{i'\}$ there are exactly two facilities in T. Hence there are $2c - 1 + u - (2m - 1) = 2(c - m) + u$ facilities in T with colors in S_2. Let colors $i_1, \ldots, i_u \in S_2$ have 3 facilities each in T. If the leftmost facility in T of any color $i_a \in \{i_1, \ldots, i_u\}$ is positioned at $P_{i_a,j_a} > P_{i_1,j_1-1} + u$, then $P_{i_a,j_a+2} - P_{i_a,j_a} <$

$2c$, violating Observation 4. Similarly, the rightmost facility in T of any color $i_a \in \{i_1,\ldots,i_u\}$ can not positioned at $P_{i_a,j_a+2} < P_{i_1,j_1+1} - u$. Therefore, positions $P_{i_1,j_1-1}+1,\ldots,P_{i_1,j_1-1}+u$ are assigned to colors from $\{i_1,\ldots,i_u\}$ and positions $P_{i_1,j_1+1}-u,\ldots,P_{i_1,j_1+1}-1$ are assigned to colors from $\{i_1,\ldots,i_u\}$. Let $P_{i_a,j_a} = P_{i_1,j_1-1}+1$. If $P_{i_a,j_a+2} = P_{i_1,j_1+1} - u$, then Theorem$P_{i_a,j_a+2} - P_{i_a,j_a} = 2c - 1 < 2c$, violating Observation 4. If $P_{i_a,j_a+2} > P_{i_1,j_1+1} - u$, then there exists an i_b such that $P_{i_b,j_b+2} = P_{i_1,j_1+1} - u$ and $P_{i_b,j_b} > P_{i_1,j_1-1} + 1$. Hence $P_{i_b,j_b+2} - P_{i_b,j_b} < (P_{i_1,j_1+1} - u) - (P_{i_1,j_1-1}+1) = 2c - 1 < 2c$, again violating Observation 4. □

Lemma 3. *If C_Δ exists and $\gamma_{i,j}$ is the Voronoi length of the j-th facility of color i, then $\forall i$ and $\forall j$, $\gamma_{i,j} \leq \Delta$.*

Proof. By the definition of C_Δ, if it exists, $\gamma_{i,j} \leq \Delta \ \forall i \in \{m,\ldots,c-1\}$, $\forall j \in \{1,\ldots,k\}$.

From Lemma 2 we have for all $i' \in \{0,\ldots,m-1\}$

1. $\gamma_{i',1} \leq L(u, 2c-u+1)$ for some $u \in \{1,\ldots,c\}$
2. $\gamma_{i',k+1} \leq R(n-2c+u, n-u+1)$ for some $u \in \{1,\ldots,c\}$
3. $P_{i',j'+1} - P_{i',j'-1} \leq 2c \ \forall j' \in \{2,\ldots,k\}$, i.e., $\forall j \in \{2,\ldots,k\}$, $\gamma_{i',j} \leq M(u, u+2c)$ for some $u \in \{1,\ldots,n-2c\}$

Our coloring scheme for C_Δ chooses a Δ at least as large as the maximum among $\max_{u=1,\ldots,n-2c} M(u, u+2c)$, $\max_{u=1,\ldots,c} L(u, 2c-u+1)$ and $\max_{u=1,\ldots,c} R(n-2c+u, n-u+1)$. Therefore $\gamma_{i',j} \leq \Delta \ \forall i' \in \{0,\ldots,m-1\}$, $\forall j \in \{1,\ldots,k+1\}$. □

Lemma 4. *If C_a is an optimal coloring where the colors of the facilities $1,2,\ldots,c$ are not distinct or the colors of the facilities $n-c+1, n-c+2,\ldots,n$ are not distinct, there exists another optimal coloring C_b where the colors of the facilities $1,2,\ldots,c$ are distinct and the colors of the facilities $n-c+1, n-c+2,\ldots,n$ are distinct.*

Lemma 5. *If $Opt \leq \Delta$, there exists a C_Δ where the value of the objective function is Δ.*

Proof. Suppose, if possible, there exists some $\Delta \geq Opt$ and the coloring scheme announced that C_Δ does not exist for such Δ. This announcement is made if, for some $i \in \{m,\ldots,c-1\}$, at least one of the following occurs:

1. $M(P_{i,k-2}, P_{i,k}) > \Delta$
2. $R(P_{i,k-1}, P_{i,k}) > \Delta$

If $M(P_{i,k-2}, P_{i,k}) > \Delta$ and there would exist a coloring C' where the j-th facility for color η is at position $P'_{\eta,j}$, has Voronoi length $\gamma'_{\eta,j}$, $\max_{i=0,\ldots,c-1} \max_{j=1,\ldots,j_{max_\eta}} \gamma'_{\eta,j} \leq \Delta$ and at least one of the following is true:

1. $P'_{i,k-2} > P_{i,k-2}$
2. $P'_{i,k} < P_{i,k}$

As is obvious from the definition of C_Δ, $P_{i,k-2}$ could not be increased without increasing $\gamma_{i,k-1}$ or $\gamma_{\eta,u}$ for some η and some $u \leq k-1$. So if $P'_{i,1} = P_{i,1} = i+1$, then $P'_{i,k-2} > P_{i,k-2}$ would imply that the value of the objective function for C' is greater than Δ. Lemma 4 suggests for every optimal coloring we can have a recoloring such that $P'_{i,1} = P_{i,1} = i+1 \ \forall \ i \ \in \ \{0,1,\ldots,c-1\}$. $P'_{i,k} < P_{i,k}$ can not be achieved because by Lemma 4 there exists an optimal coloring such that the colors of the facilities $n-c+1, n-c+2, \ldots, n$ are distinct and if $P'_{i,1} = P_{i,1} = i+1 \ \forall \ i \ \in \ \{0,1,\ldots,c-1\}$, the maximum Voronoi length will only increase if $P'_{i,k} < P_{i,k}$ since it would violate the C_1 coloring for the colors $m, m+1, \ldots, c-1$.

If $R(P_{i,k-1}, P_{i,k}) > \Delta$ and there would exist a coloring C' where the j-th facility for color η is at position $P'_{\eta,j}$, has Voronoi length $\gamma'_{\eta,j}$, $\max_{i=0,\ldots,c-1} \max_{j=1,\ldots,j_{max\eta}} \gamma'_{\eta,j} \leq \Delta$ and at least one of the following is true:

1. $P'_{i,k-1} > P_{i,k-1}$
2. $P'_{i,k} > P_{i,k}$

But, as explained above while considering the case $P'_{i,k-2} > P_{i,k-2}$, we can not have $P'_{i,k-1} > P_{i,k-1}$ if $P'_{i,1} = P_{i,1} = i+1 \ \forall \ i \ \in \ \{0,1,\ldots,c-1\}$. $P'_{i,k} > P_{i,k}$ would violate the C_1 coloring for the colors $m, m+1, \ldots, c-1$ as it would for $P'_{i,k} < P_{i,k}$. □

Theorem 3. *For $n = kc + m, 0 < m < c$, an optimal coloring scheme can be obtained in $O(n^2 \log n)$ time.*

Proof. From Lemma 5 it is clear that there exists a Δ such that C_Δ is optimal. A candidate for Δ is any of the following:

1. $L(a,b)$ where $1 \leq a < b \leq n$
2. $M(a,b)$ where $1 \leq a < b \leq n$
3. $R(a,b)$ where $1 \leq a < b \leq n$

There are $O(n^2)$ candidates for Δ, which can be sorted in ascending order in $O(n^2 \log n)$ time. For a given Δ, we can identify C_Δ in $O(cn)$ time as follows. We have a list of n cumulative distances $D_j = \Sigma_{i=1}^{j} d_i \ \forall \ j = 1, 2, \ldots, n$. One can calculate $L(a,b) = \frac{D_a + D_b}{2}$, $M(a,b) = \frac{D_b - D_a}{2}$ and $R(a,b) = d_{n+1} + D_n - \frac{D_a + D_b}{2}$ for any pair of a and b in constant time. For odd k, $P_{i,j}$ is fixed for odd j and, for even j, $jc - i \leq P_{i,j} \leq jc - i + 2m$. For even k, $jc + i \leq P_{i,j} \leq jc + i + m + 1$ for odd j and $jc - i \leq P_{i,j} \leq jc - i + m$ for even j. So, for identifying $P_{i,j}$ for $i \ \in \ \{m, m+1, \ldots, c-1\}$ using the rule $M(P_{i,j-2}, P_{i,j}) \leq \Delta \leq M(P_{i,j-2}, P_{i,j}+1)$ for $j > 2$ or $L(P_{i,1}, P_{i,2}) \leq \Delta \leq L(P_{i,1}, P_{i,2}+1)$, we need $O(m)$ time for each i, j and hence $O(cn)$ time for all the facilities as $m < c$. For backtracking we need:

1. At most $c - i$ time for odd k and even j for each $i \ \in \ \{m, m+1, \ldots, c-1\}$
2. No backtracking for odd k and odd j
3. At most $c - i$ time for even k and even j for each $i \ \in \ \{m, m+1, \ldots, c-1\}$
4. At most m time for even k and odd j for each $i \ \in \ \{m, m+1, \ldots, c-1\}$

So for backtracking we need $O(cn)$ time and altogether we need $O(cn)$ time for each Δ. Using binary search over the sorted list of $O(n^2)$ candidates for Δ, we obtain the optimum coloring by trying atmost $O(\log n)$ candidates. Hence the total time complexity is $O(cn \log n) + O(n^2 \log n)$, i.e., $O(n^2 \log n)$. $\square$

6   Conclusions

With the objective of coloring the available facilities to a population of users distributed uniformly in a line segment R such that the load of the different facilities are distributed as equitably as possible (MinVor problem) we obtained some interesting results. We observed that when $n = kc$, C_1 offers us an optimal coloring for any distance vector, while for $n = kc + m$ facilities with c colors for the MinVor problem, we have C_2 is the optimal coloring in some special cases and otherwise C_Δ is the optimal coloring with Δ being the optimal value which can be obtained in $O(n^2 \log n)$ time.

If $\Delta < \max_{u=1,\ldots,n-2c} M(u, u + 2c) = M(u_1, u_1 + 2c)$, there must be a color i with only one facility in $\{u_1 + 1, \ldots, u_1 + 2c - 1\}$. The Voronoi length of that facility of color i is clearly atleast $M(u_1, u_1 + 2c) > \Delta$.

Let $\Delta < \max_{u=1,\ldots,c} L(u, 2c - u + 1) = L(u_2, 2c - u_2 + 1)$. If the first facility of color i is at position u_2 or higher, we say that it follows Property 1. If the 2nd facility of color i is at position $(2c - u_2 + 1)$ or higher, we say that it follows Property 2. Clearly at least $c - u_2 + 1$ colors follow Property 1 and u_2 colors follow Property 2. So, there is at least one color, say i_1, whose 1st facility is at position u_2 or higher and 2nd facility at $2c - u_2 + 1$ or higher. Clearly $L(P_{i_1,1}, P_{i_1,2}) \geq L(u_2, 2c - u_2 + 1) > \Delta$.

Similarly if $\Delta < \max_{u=1,\ldots,c} R(n - 2c + u, n - u + 1) = R(n - 2c + u_3, n - u_3 + 1)$, there is at least one color, say i_2, whose last facility is at position $n - u_3 + 1$ or lower and the 2nd last facility at $n - 2c + u_3$ or lower, which implies $R(P_{i_2,j-1}, P_{i_2,j}) \geq R(n - 2c + u_3, n - u_3 + 1) > \Delta$, where j is the total number of facilities of color i_2.

References

1. Funke, S., Kesselman, A., Kuhn, F., Lotker, Z., Segal, M.: Improved approximation algorithms for connected sensor cover. Wireless Networks 13(2), 153–164 (2007)
2. Lin, Y., Zhang, J., Chung, H.S.-H., Ip, W.-H., Li, Y., Shi, Y.-H.: An Ant Colony Optimization Approach for Maximizing the Lifetime of Heterogeneous Wireless Sensor Networks. IEEE Transactions on Systems, Man, and Cybernetics, Part C 42(3), 408–420 (2012)
3. Clark, B.N., Colbourn, C.J., Johnson, D.S.: Unit disk graphs. Discrete Mathematics 86(1-3), 165–177 (1990)
4. Wan, P.-J., Alzoubi, K.M.: Ophir Frieder: Distributed Construction of Connected Dominating Set in Wireless Ad Hoc Networks. MONET 9(2), 141–149 (2004)
5. Funke, S., Kesselman, A., Meyer, U., Segal, M.: A simple improved distributed algorithm for minimum CDS in unit disk graphs. TOSN 2(3), 444–453 (2006)

Approximation Algorithms for Packing Element-Disjoint Steiner Trees on Bounded Terminal Nodes

Daiki Hoshika and Eiji Miyano

Department of Systems Design and Informatics, Kyushu Institute of Technology,
Iizuka, Fukuoka 820-8502, Japan
hoshika@theory.ces.kyutech.ac.jp, miyano@ces.kyutech.ac.jp

Abstract. In this paper we discuss approximation algorithms for the
ELEMENT-DISJOINT STEINER TREE PACKING problem (Element-STP for
short). For a graph $G = (V, E)$ and a subset of nodes $T \subseteq V$, called
terminal nodes, a Steiner tree is a connected, acyclic subgraph that contains all the terminal nodes in T. The goal of Element-STP is to find as
many element-disjoint Steiner trees as possible. Element-STP is known
to be $\mathcal{APX}$-hard even for $|T| = 3$ [1]. It is also known that Element-STP
is $\mathcal{NP}$-hard to approximate within a factor of $\Omega(\log |V|)$ [3] and there
is an $O(\log |V|)$-approximation algorithm for Element-STP [2,4]. In this
paper, we provide a $\lceil \frac{|T|}{2} \rceil$-approximation algorithm for Element-STP on
graphs with $|T|$ terminal nodes. Furthermore, we show that the approx-
imation ratio of 3 for Element-STP on graphs with five terminal nodes
can be improved to 2.

1 Introduction

In this paper we study a popular variant of the STEINER TREE PACKING problem on undirected graphs (STP for short): The instance of STP is a pair of an
undirected graph $G = (V, E)$ and a subset of nodes $T \subseteq V$, where each node in
T is called a *terminal* node, and each non-terminal node, i.e., node in $V \setminus T$ is
called a *Steiner* node. Throughout the paper, we use n to denote the number
of nodes $|V|$ in the input graph. A Steiner node or an edge in G is called an
element. For an instance (G, T), a *Steiner tree* is a connected, acyclic subgraph
in G that contains all the terminal nodes in T, but Steiner nodes are optional.
A Steiner tree that contains all terminal nodes in T is called a T-Steiner tree.
Given a graph $G = (V, E)$ and a set T of terminal nodes, we say that the set T
is k-edge connected if there exist k edge-disjoint paths between any two nodes
$u, v \in T$. Also, we say that the set T is k-element connected if there exist k
element-disjoint paths between any two nodes $u, v \in T$. If all the nodes in G are
terminal nodes, i.e., $T = V$, then we simply say that the graph G is k-edge or
k-element connected.

The STP problem was originally formulated by Grötschel, Martin, and Weismantel [7,8]. Then, many researchers have been interested in the computational

Q. Gu, P. Hell, and B. Yang (Eds.): AAIM 2014, LNCS 8546, pp. 100–111, 2014.

complexity and (in)approximability of STP since it has applications in VLSI circuit design [9], wireless networks, and data broadcasting [11]. One of the most popular variants of STP in the literature is the ELEMENT-DISJOINT STEINER TREE PACKING problem (Element-STP for short). The objective of Element-STP is to find as many element-disjoint T-Steiner trees as possible. Element-STP was first considered by Hind and Oellermann [10]. (Also, the element-connectivity is independently reintroduced as a connectivity measure intermediate to edge and vertex connectivities by Jain, Măndoiu, Vazirani, and Williamson [12].) For Element-STP, Cheriyan and Salavatipour [4] provide a randomized algorithm achieving an approximation ratio of $O(\log n)$. Subsequently, Calinescu, Chekuri, and Vondrak [2] give a derandomized $O(\log n)$-approximation algorithm. As for inapproximability of Element-STP, in [3], Cheriyan and Salavatipour prove that Element-STP is hard to approximate within a factor of $\Omega(\log n)$.

If the input is a non-restricted, general unweighted graph, then Element-STP has a high computational complexity, even from the viewpoint of the approximability; the $O(\log n)$-approximation algorithm is the best possible one. Thus, several researchers have investigated the (in)tractability and the (in)approximability of Element-STP when its input graphs are restricted to some special classes of graphs and/or the number of terminal nodes is bounded. It is shown [3] that the inapproximability of $\Omega(\log n)$ holds even if the input graphs are restricted to bipartite graphs. On the other hand, Aazami, Cheriyan, and Jampani [1] design an approximation algorithm for Element-STP on planar graphs that achieves an approximation ratio of almost 2. In the same paper, they show that the problem of finding two element-disjoint T-Steiner trees in an input planar graph is $\mathcal{NP}$-hard. This implies that one cannot improve the approximability of 2 for Element-STP if we impose no further restrictions on the inputs.

Our Contributions. The goal of this paper is to design an approximation algorithm whose approximation ratio does not depend on the number n of nodes, but depends only on the number $|T|$ of terminal nodes; we provide a $\lceil |T|/2 \rceil$-approximation algorithm for Element-STP on graphs with $|T|$ terminal nodes by proving the ratio between the element-connectivity of T and the maximum number of element-disjoint T-Steiner trees is bounded by $\lceil |T|/2 \rceil$. Furthermore, we show that the approximation ratio of 3 for Element-STP on graphs with five terminal nodes can be improved to 2. On the other hand, it is known [1] that Element-STP is $\mathcal{APX}$-hard even for graphs with three terminal nodes, and the standard LP relaxation of Element-STP even on planar graphs with $|T|$ terminal nodes has an integrality gap $\geq 2 - 2/|T|$. It is important to note that in [4], the authors mention that their randomized algorithm achieving an approximation ratio of $O(\log n)$ can be modified to $O(\log |T|)$ by sacrificing the success probability of the Monte Carlo randomized algorithm. Unfortunately, however, the constant coefficient hidden by the big-O notation is clearly large (it would be at least 6 by a careful reading). This implies that when $|T| = 5$, the previous approximation ratio is at least 12. Therefore, even our deterministic $\lceil |T|/2 \rceil$-approximation algorithm can outperform the previous randomized algorithm in [4] (or its derandomized version in [2]) from the viewpoint of the approximation ratio at least if the number of terminal nodes is bounded by a small constant. (Due to the page limitation, we omit some proofs from this extended abstract.)

Related Work. Another variant of STP which has been energetically studied is the EDGE-DISJOINT STEINER TREE PACKING problem (Edge-STP for short). Kaski [13] shows that Edge-STP is $\mathcal{NP}$-hard even on graphs with seven terminal nodes. Chriyan and Salavatipour [3] prove the $\mathcal{APX}$-hardness of Edge-STP on graphs with four terminal nodes, and subsequently, Aazami, Cheriyan, and Jampani [1] prove the $\mathcal{APX}$-hardness of Edge-STP on graphs with three terminal nodes. As for the approximability of Edge-STP, Jain, Mahdian, and Salavatipour [11] provide an $O(|T|)$-approximation algorithm for Edge-STP on graphs with $|T|$ terminal nodes. The constant-factor approximation algorithm is first provided by Lau [16,17]; he show that if the terminal set T is $24k$-edge connected, then there exist k edge-disjoint T-Steiner trees in the input graph, and provide a 24-approximation algorithm by using the approximation algorithm of Jain et al. [11] for a linear programming relaxation of Edge-STP. Later, West and Wu [20] improve the approximation ratio to 6.5 by showing that if the terminal set T is $6.5k$-edge connected, then there exist k edge-disjoint T-Steiner trees. Very recently, DeVos, McDonald, and Pivotto [5] make a further improvement to $5k+4$ from $6.5k$. Furthermore, Kriesell [15] designs a 1.5-approximation algorithm for $|T| = 4$ terminal nodes.

2   Preliminaries

In this paper, we only consider simple, undirected, unweighted and connected graphs. Let $G = (V, E)$ be a graph; we sometimes denote by $V(G)$ and $E(G)$ the node set and edge set of G, respectively. For a subset V' of $V(G)$, we denote by $G[V']$ the subgraph of G induced by V'. We denote simply by $G \setminus V'$ the induced subgraph $G[V \setminus V']$. For a subgraph G' of G, let $G \setminus G' = G \setminus V(G')$. Let $N(v)$ ($N(S)$, respectively) be the set of neighbor nodes of the node v (the set of neighbor nodes of the set S, respectively).

In this paper we focus on Element-STP, which is formulated as follows:

ELEMENT-DISJOINT STEINER TREE PACKING (Element-STP)

Instance: Graph $G = (V, E)$ and terminal nodes $T \subseteq V$.
Goal: Find a maximum cardinality set of element-disjoint T-Steiner trees.

We often call the terminal nodes *black* nodes, and the non-terminal nodes *white* nodes. For example, b_1 and w_1 represent a terminal node and a Steiner node, respectively, in Figure 1-(a). Also, an edge between two white nodes is called a *white* edge. A *cut* $C(T_1, T_2)$ is a partition of a set $T = T_1 \cup T_2$, where $T_1 \cap T_2 = \emptyset$. An *element cut-set* of a cut $C(T_1, T_2)$ is defined as a set of Steiner nodes and edges whose removal disconnects $T = T_1 \cup T_2$ into two subsets T_1 and T_2. The *element connectivity* of the set T of terminal nodes is the minimum size of all the element cut-sets of T. The set T of terminal nodes is said to be k-*element connected* if the element connectivity of T is at least k.

For a pair of a graph G and a terminal set T as an instance of Element-STP, let $ALG(G, T)$ and $OPT(G, T)$ be the numbers of obtained element-disjoint T-Steiner trees by a polynomial-time algorithm ALG and an optimal algorithm, respectively. Then, the algorithm ALG is called an α-approximation algorithm and the approximation ration of ALG is α where $\alpha \geq 1$ if the inequality $OPT(G, T)/ALG(G, T) \leq \alpha$ holds.

Recall that the problem of finding a set of edge-disjoint T-Steiner trees of maximum cardinality is denoted by Edge-STP. For a while, we consider Edge-STP. If all of the nodes are terminal nodes, i.e., $T = V$, then the problem is the same as the problem of finding a maximum cardinality set of edge-disjoint spanning trees. For this special case of Edge-STP, Tutte [19] and Nash-Williams [18] independently show a necessary and sufficient condition for an input graph to have k edge-disjoint spanning trees. Moreover, as a corollary of their result, every $2k$-edge connected graph contains k edge-disjoint spanning trees. The matroid intersection algorithm can find a largest set of edge-disjoint spanning trees in polynomial time. Kriesell [14] conjectures that the corollary of Tutte and Nash-Williams's result can generalize to Edge-STP:

Conjecture 1 (Kriesell's conjecture [14]). If $T \subseteq V(G)$ is $2k$-edge connected, then there is a polynomial-time algorithm which finds at least k edge-disjoint T-Steiner trees in G.

If this conjecture is settled by a constructive proof, then we may obtain a 2-approximation algorithm for Edge-STP. As mentioned above, DeVos, McDonald, and Pivotto [5] recently have made a major advance on Kriesell's conjecture by presenting the following result:

Proposition 1 ([5]). *If $T \subseteq V(G)$ is $5k + 4$ edge-connected, then there is a polynomial-time algorithm which finds at least k edge-disjoint T-Steiner trees in G.*

Now we go back to Element-STP. Note that if $T = V$, then Element-STP is equivalent to Edge-STP since there are no Steiner nodes. Thus, both problems can be solved in polynomial time if $T = V$. However, Element-STP is quite different from Edge-STP if $T \neq V$; for example, Edge-STP has a constant-factor approximation algorithm, but Element-STP is hard to approximate within a factor of $\Omega(\log n)$ as mentioned before. On the other hand, similarly to Edge-STP, an $O(\log n)$-approximation algorithm for Element-STP is provided by showing the relationship between the element-connectivity and the number of element-disjoint T-Steiner trees in a graph.

Proposition 2 ([4,2]). *If $T \subseteq V(G)$ is $O(k \cdot \log n)$-element connected, then there is a polynomial-time algorithm which finds at least k element-disjoint T-Steiner trees in G.*

Now we define the ELEMENT DISJOINT BIPARTITE STEINER TREE PACKING problem as a subproblem of Element-STP: The input is restricted to be a bipartite graph such that the set T of terminal nodes is one part of the bipartition, and the set $V \setminus T$ of Steiner nodes is the other part. Let the input bipartite graph be denoted by $G = (T \cup (V \setminus T), E)$. It is known [10,4] that we can efficiently transform any graph $G = (V, E)$ with the set T of terminal nodes into a bipartite graph G' with the same terminal set T such that T is one part of the bipartition in G' and the element-connectivity of T is unchanged.

Proposition 3 ([10,4]). *Consider a graph $G = (V, E)$ that has a set T of terminal nodes such that T is k-element connected. There is a polynomial-time algorithm that repeatedly deletes or contracts white edges to obtain a bipartite graph G' from G such that T stays k-element connected, and moreover, T forms one part of the bipartition in G'.*

It is important to note that the element-connectivity of the terminal set T is an upper bound on the maximum number of element-disjoint T-Steiner trees. Thus, for a bipartite graph $G = (T \cup (V \setminus T), E)$ such that T is k-element connected, if we can design a polynomial-time algorithm which can find at least $\lceil \frac{k}{\alpha} \rceil$ element-disjoint T-Steiner trees, then the approximation ratio of the algorithm is α. Indeed, for a special case where the degree of every Steiner node in an input bipartite graph is at most Δ, Frank, Király, and Kriesell [6] present a Δ-approximation algorithm by using the matroid intersection theorem and algorithm:

Proposition 4 ([6]). *Consider a bipartite graph $G = (T \cup (V \setminus T), E)$, where T is a set of terminal nodes and $V \setminus T$ is a set of Steiner nodes. Then, if the degree of every Steiner node is at most Δ and T is $\Delta \cdot \ell$-element connected, then there is a polynomial-time algorithm which finds at least ℓ element-disjoint T-Steiner trees in G.*

The following observation is very simple but plays an important role in the next two sections: Suppose that the number of element-disjoint T-Steiner trees in a bipartite graph $G = (T \cup (V \setminus T), E)$ is at least ℓ. Also suppose that a subset S_i of Steiner nodes for each $i = 1, 2, \cdots, \ell$ forms the ith element-disjoint T-Steiner tree (in some order), where $S_i \cap S_j = \emptyset$ for $i \neq j$. Then, $G[T \cup S_1]$ through $G[T \cup S_\ell]$ can be thought as ℓ element-disjoint T-Steiner trees. Furthermore, one can easily see that, for example, for some $i \in \{1, 2, \cdots, \ell\}$, the number of element-disjoint T-Steiner trees in $G[V \setminus (S_1 \cup \cdots \cup S_i)]$ is at least $\ell - i$. In the following sections, our approximation algorithms construct such element-disjoint T-Steiner trees, one by one.

3 $\lceil |T|/2 \rceil$-Approximation Algorithm for Graphs with $|T|$ Terminal Nodes

In this section we show that there is a $\lceil |T|/2 \rceil$-approximation algorithm for Element-STP on graphs with a set T of terminal nodes. Assume that the set T of terminal nodes in the input graph G is k-element connected. Then, we achieve the $\lceil |T|/2 \rceil$-approximability of Element-STP by showing that we can find at least $\lfloor 2k/|T| \rfloor$ element-disjoint T-Steiner trees from G such that T is k-element connected. Our algorithm consists of the following four phases, (Phase 1) through (Phase 4):

(Phase 1). We transform the input graph G with the set T of terminal nodes into the bipartite graph, say, $G^0 = (T \cup (V \setminus T), E)$ such that G^0 has the same set T of terminal nodes such that T remains k-element connected by using the algorithm described in Proposition 3, deleting or contracting white edges.

Note that the maximum degree Δ of Steiner nodes in $V \setminus T$ is at most $|T|$. Therefore, a trivial consequence of Proposition 4 is that we can design a $|T|$-approximation algorithm on graphs with $|T|$ terminal nodes. For $d = 1, \cdots, |T|$, let the subset of Steiner nodes of exactly degree d be denoted by W_d, i.e., $W_d \cap W_{d'} = \emptyset$ for $d \neq d'$, and $V \setminus T = \bigcup_{d=1,2,\cdots,|T|} W_d$. In the following we can show that the approximability can be improved to $\lceil |T|/2 \rceil$ by removing repeatedly all the Steiner nodes in $\bigcup_{d=\lceil |T|/2 \rceil+1,\cdots,|T|} W_d$, while keeping the element-connectivity of T in the resulting graph sufficiently high:

Lemma 1. *Consider a bipartite graph $G = (T \cup (V \setminus T), E)$, where T is a set of terminal nodes, $V \setminus T$ is a set of Steiner nodes, and T is k-element connected $(k > 0)$. Suppose that there is a Steiner node w_d of the degree $d \geq \lceil |T|/2 \rceil + 1$ in G. Then, there exists a T-Steiner tree in G such that (i) a subset S of Steiner nodes containing w_d forms the T-Steiner tree, and (ii) in the induced graph $G[V \setminus S]$, the element connectivity of T is at least $(k - \lceil |T|/2 \rceil)$. (The proof is omitted.)* □

(Phase 2). Let w_Δ be a Steiner node of the maximum degree $\Delta \geq \lceil |T|/2 \rceil + 1$ in the bipartite graph G^0. From Lemma 1, we can find a T-Steiner tree in G^0 such that a subset S of Steiner nodes containing w_Δ forms the T-Steiner tree, and moreover T in $G^0 \setminus S$ is $(k - \lceil |T|/2 \rceil)$-element connected. Then, we remove S from G^0 and obtain $G^1 = G^0 \setminus S$. If $\bigcup_{d=\lceil |T|/2 \rceil+1,\cdots,|T|} W_d \neq \emptyset$ in G^1, then set $G^0 = G^1$ and repeat this phase; otherwise, i.e., if $\bigcup_{d=\lceil |T|/2 \rceil+1,\cdots,|T|} W_d = \emptyset$, then go to the next phase.

Suppose that eventually we can obtain L element-disjoint T-Steiner trees in total from the initial G^0 after the Lth iteration in (Phase 2), and the maximum degree of Steiner nodes in the obtained graph G^1 is at most $\lceil |T|/2 \rceil$. From Lemma 1, we can assume that G^1 is (at least) $(k - L\lceil |T|/2 \rceil)$-element connected.

(Phase 3). By setting the maximum degree of Steiner nodes $\Delta = \lceil |T|/2 \rceil$ and thus $\ell = (k - L\lceil |T|/2 \rceil)/(\lceil |T|/2 \rceil) = k/(\lceil |T|/2 \rceil) - L$ in Proposition 4, we can find $(k/(\lceil |T|/2 \rceil) - L)$ element-disjoint T-Steiner trees from G^1, i.e., in total $k/(\lceil |T|/2 \rceil)$ element-disjoint T-Steiner trees can be found from the initial bipartite graph G^0.

(Phase 4). By "uncontracting" the edges that were contracted in (Phase 1), we obtain a set of $k/(\lceil|T|/2\rceil)$ element-disjoint T-Steiner trees in the input graph G.

Theorem 1. *There is a $\lceil|T|/2\rceil$-approximation algorithm for* Element-STP *on graphs with a set T of terminal nodes where $|T| \geq 2$.*

4   2-Approximation Algorithm for Graphs with at Most Five Terminal Nodes

In this section, we consider a special case where the number of terminal nodes is at most five. From Theorem 1, we have a 3-approximation algorithm for Element-STP if the terminal nodes are bounded by five. In the following we show that the approximation ratio of 3 can be improved to 2 for graphs with at most five terminal nodes.

Theorem 2. *There is a 2-approximation algorithm for* Element-STP *on graphs with a set of at most five terminal nodes.*

Proof. Without loss of generality, we can assume that the input is a bipartite graph $G = (T \cup (T \setminus V), E)$ with a terminal set T from Proposition 3. Let $T = \{b_1, b_2, b_3, b_4, b_5\}$ be the set of five terminal nodes. Also, let the subset of Steiner nodes of exactly degree d be denoted by W_d, i.e., $W_d \cap W_{d'} = \emptyset$ for $d \neq d'$ and $V \setminus T = W_1 \cup W_2 \cup W_3 \cup W_4 \cup W_5$. Here, we can assume that $W_1 = \emptyset$ since Steiner nodes of degree one do not connect any two terminal nodes and thus they are not contained in any element-disjoint T-Steiner trees.

In the following we show that if T is k-element connected, then there is a polynomial-time algorithm which finds at least $\lceil\frac{k}{2}\rceil$ element-disjoint T-Steiner trees in G by using very similar ideas to the proof of the previous section. We show that, repeatedly, we can choose a T-Steiner tree in G such that T is k-element connected, and then remove all Steiner nodes in the T-Steiner tree from G, while preserving the $(k-2)$-element connectivity of T in the resulting graph, until the maximum degree of Steiner nodes decreases to two. Our algorithm consists of the following four phases: (Phase 1) $W_5 \neq \emptyset$, (Phase 2) $W_5 = \emptyset$ but $W_4 \neq \emptyset$, (Phase 3) $W_5 = W_4 = \emptyset$ but $W_3 \neq \emptyset$, and (Phase 4) $W_5 = W_4 = W_3 = \emptyset$. If T is 2-element connected, then it is sufficient for us to find just one T-Steiner tree in order to obtain the approximability of 2. Thus, we always assume that $k \geq 3$ in the following.

(Phase 1). $W_5 \neq \emptyset$. Every Steiner node w in W_5 and its adjacent five terminal nodes in T forms a T-Steiner tree, say, ST, where every terminal node is a leaf in the tree ST, and hence its degree is one. Therefore, if all the Steiner nodes in ST are removed from the graph, then the element-connectivity of the set T of terminal nodes decreases exactly by one. As a result, we can find $|W_5|$ element-disjoint T-Steiner trees, each of which contains one Steiner node in W_5, and thus, by removing all the Steiner nodes in W_5 from G, we can obtain a new graph G' such that T in G' is k'-element connected, where $k' = k - |W_5| \geq k - 2|W_5|$.

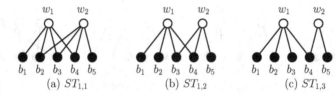

Fig. 1. Three types of T-Steiner trees in Phase 2

(Phase 2). $W_5 = \emptyset$ but $W_4 \neq \emptyset$. By renaming G' and k' in Phase 1 as G and k, respectively, we again consider a graph G such that T is k-element connected. See Figure 1. Consider a Steiner node, say, w_1 of degree four, assuming $N(w_1) = \{b_1, b_2, b_3, b_4\}$, without loss of generality. If T is k-element connected, then there is at least one Steiner node, say, w_2, which is adjacent to the rightmost terminal node b_5 and at least one black node of $\{b_1, b_2, b_3, b_4\}$. One can see that there are only three types of T-Steiner trees, $ST_{1,1}$, $ST_{1,2}$ and $ST_{1,3}$ as shown in Figures 1-(a), (b), and (c), respectively. Therefore, two Steiner nodes w_1 and w_2, and all the terminal nodes in T forms one T-Steiner tree, say, ST. If we remove w_1 and w_2 from the graph, the element-connectivity of the set T of terminal nodes decreases by at most two. By repeating this phase, we remove all Steiner nodes in W_4 from $V(G)$; W_4 eventually gets empty. When we find ℓ T-Steiner trees in this phase, T is still $(k - 2\ell)$-element connected.

(Phase 3). $W_5 = W_4 = \emptyset$ but $W_3 \neq \emptyset$. Our goal in this phase is to remove all Steiner nodes of degree three in the selected T-Steiner trees, and reduce into the graph which only contains Steiner nodes of degree at most two. Phase 3 is further divided into the following two phases (more detailed descriptions of them are given later):

(Phase 3-1). We first choose a T-Steiner tree, say, ST in G such that the maximum degree of *terminal* nodes in ST is at most *two*. Then, we remove the set S of all the Steiner nodes in ST and obtain the graph $G' = G \backslash S$. If G' contains another T-Steiner tree such that the maximum degree of terminal nodes in the T-Steiner tree is at most two, then we repeat this phase; otherwise, go to (Phase 3-2).

(Phase 3-2) We first choose a T-Steiner tree, say, ST' in G' such that the maximum degree of *terminal* nodes in ST' is at most *three*. Then, we remove the set S' of all the Steiner nodes in ST' and obtain the graph $G'' = G' \backslash S'$. If $W_3 \neq \emptyset$, then we repeat this phase; otherwise, go to (Phase 4).

Take a look at Figure 2. It can be easily verified that there are only four different types of T-Steiner trees such that the maximum degree of terminal nodes is at most two, (a) $ST_{2,1}$, (b) $ST_{2,2}$, (c) $ST_{2,3}$, and (d) $ST_{2,4}$. Actually, however, we do not need to consider $ST_{2,2}$-type T-Steiner trees. The reason is as

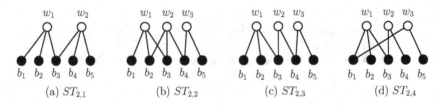

Fig. 2. Four types of T-Steiner trees in Phase 3-1, (a) $ST_{2,1}$, (b) $ST_{2,2}$, (c) $ST_{2,3}$, and (d) $ST_{2,4}$

follows: Now we assume that T is k-element connected and $k \geq 3$. Then, another white node, say, w_4 must connect the leftmost black node b_1 to at least one node of $\{b_2, b_3, b_4, b_5\}$. For example, if w_4 connects b_1 to b_2, then $G[T \cup \{w_2, w_3, w_4\}]$ is identical to $ST_{2,4}$. As another example, if w_4 connects b_1 to b_5, then $G[T \cup \{w_1, w_3, w_4\}]$ is identical to $ST_{2,3}$, and so on.

Lemma 2. *Consider a bipartite graph $G = (T \cup (V \setminus T), E)$, where T is a set of terminal nodes, $V \setminus T$ is a set of Steiner nodes, and T is k-element connected ($k \geq 3$). Suppose that G contains a T-Steiner tree $ST_{2,1}$ illustrated in Figure 2-(a) ($ST_{2,3}$ illustrated in Figure 2-(c) and $ST_{2,4}$ illustrated in Figure 2-(d), respectively). Then, the element connectivity of T in $G \setminus \{w_1, w_2\}$ ($G \setminus \{w_1, w_2, w_3\}$ and $G \setminus \{w_1, w_2, w_3\}$, respectively) is at least $(k-2)$. (The proof is omitted.)* □

The detailed description of (Phase 3-1) is as follows:

(Phase 3-1) (i) We find an $ST_{2,1}$-type T-Steiner tree in a graph G, remove the set, say, S_1 of all the Steiner nodes in the T-Steiner tree, and obtain the graph $G \setminus S_1$. If $G \setminus S_1$ contains another $ST_{2,1}$-type T-Steiner tree, then repeat (i); otherwise go to the next. (ii) Let the obtained graph after the iterations of (i) be denoted by G_1. We find an $ST_{2,3}$-type T-Steiner tree in G_1, remove the set, say, S_2 of all the Steiner nodes in the T-Steiner tree, and obtain the graph $G_2 = G_1 \setminus S_2$. If $G_1 \setminus S_2$ contains another $ST_{2,3}$-type T-Steiner tree, then repeat (ii); otherwise go to the next. (iii) The very similar operations to (i) and (ii) are iterated for $ST_{2,4}$-type T-Steiner trees if there is an $ST_{2,4}$-type T-Steiner tree in G_2.

Now we can assume that a (current) bipartite graph $G = (T \cup (V \setminus T), E)$ contains no T-Steiner trees illustrated in Figures 2-(a) through (d), where the set T of terminal nodes is k-element connected ($k \geq 3$). For the case where $W_3 \neq \emptyset$, the remaining types of T-Steiner trees are only three ones illustrated in Figure 3, (a) $ST_{3,1}$, (b) $ST_{3,2}$, and (c) $ST_{3,3}$. Recall that the maximum degree of black nodes is three. Roughly speaking, we first remove all $ST_{3,1}$-type T-Steiner trees, then remove all $ST_{3,2}$-type T-Steiner trees, and finally remove all $ST_{3,3}$-type T-Steiner trees. Here, we view $ST_{3,1}$ and $ST_{3,2}$:

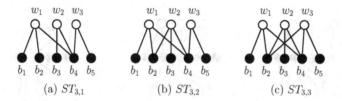

Fig. 3. Three types of T-Steiner trees in Phase 3-2, (a) $ST_{3,1}$, (b) $ST_{3,2}$, and (c) $ST_{3,3}$

Lemma 3. *Consider a bipartite graph $G = (T \cup (V \setminus T), E)$ which does not contain any T-Steiner tree illustrated in Figures 2-(a) through (d), where T is a set of terminal nodes, $V \setminus T$ is a set of Steiner nodes, and T is k-element connected ($k \geq 3$). Suppose that G contains a T-Steiner tree $ST_{3,1}$ illustrated in Figure 3-(a) (or $ST_{3,2}$ illustrated in Figure 3-(b)). Then, the element connectivity of T in $G \setminus \{w_1, w_2, w_3\}$ is at least $(k-2)$.*

Proof. **(a)** First consider an $ST_{3,1}$-type T-Steiner tree. We can show that from the assumptions that T in G is k-element connected and G does not contain any T-Steiner tree illustrated in Figure 2, if we remove any set W' of $(k-3)$ white nodes in $V \setminus \{w_1, w_2, w_3\}$ from G, then the resulted graph $G \setminus W'$, renamed by G', always contains a subgraph whose shape is identical to $ST_{3,1}^+$ illustrated in Figure 4-(a) although the reasons are omitted here. Consider the graph $G \setminus (W' \cup \{w_1, w_2, w_3\})$, which contains the connected graph $G[T \cup \{w_4, w_5, w_6, w_7\}]$. Namely, T in $G[T \cup \{w_4, w_5, w_6, w_7\}]$, or equivalently T in $G \setminus (W' \cup \{w_1, w_2, w_3\})$ is still (at least) 1-element connected. Thus, it can be shown that the element connectivity of T in $G \setminus \{w_1, w_2, w_3\}$ is at least $(k-2)$ as follows: Suppose for contradiction that the element connectivity of T in $G \setminus \{w_1, w_2, w_3\}$ is at most $(k-3)$. It follows that there must exist a set W' of $(k-3)$ white nodes such that T in $(G \setminus \{w_1, w_2, w_3\}) \setminus W'$ is disconnected, and so we have arrived at a contradiction. Therefore, the element connectivity of T in $G \setminus \{w_1, w_2, w_3\}$ must be at least $(k-2)$.

(b) Similarly to the above, we can show that G contains a subgraph whose structure is identical to $ST_{3,2}^+$ illustrated in Figure 4-(b) such that the element connectivity T in $ST_{3,2}^+$ is at least three, and the element connectivity of T in $G \setminus \{w_1, w_2, w_3\}$ is at least $(k-2)$.   [End of the proof of Lemma 3]   □

Observation. The remaining T-Steiner trees we have to consider are identical to $ST_{3,3}$ in Figure 3-(c). Without loss of generality, we consider a bipartite graph $G = (T \cup (V \setminus T), E)$ such that T is 3-element connected and G contains a T-Steiner tree $ST_{3,3}$ illustrated in Figure 3-(c). From the assumption that T is 3-element connected and similar arguments as in the proof of Lemma 3, G must have further three Steiner nodes, say, w_4, w_5 and w_6; namely, G must contain a subgraph which is identical to $ST_{3,3}^+$ in Figure 4-(c). Therefore, if T in G is 3-element connected, then we can find at least *two* element-disjoint T-Steiner trees, $G[T \cup \{w_1, w_5, w_6\}]$ and $G[T \cup \{w_2, w_3, w_4\}]$.

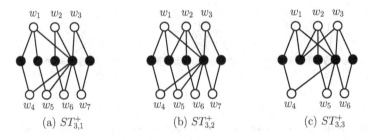

Fig. 4. T-Steiner trees in the proof of Lemma 3 and Observation

The detailed description of (Phase 3-2) is as follows:

(Phase 3-2) (i) We find an $ST_{3,1}$-type T-Steiner tree in a graph G, remove the set, say, S_1 of all the Steiner nodes in the T-Steiner tree, and obtain the graph $G \setminus S_1$. If $G \setminus S_1$ contains another $ST_{3,1}$-type T-Steiner tree, then repeat (i); otherwise go to the next. (ii) Let the obtained graph after the iterations of the previous (i) be denoted by G_1. We find an $ST_{3,2}$-type T-Steiner tree in G_1, remove the set, say, S_2 of all the Steiner nodes in the T-Steiner tree, and obtain the graph $G_2 = G_1 \setminus S_2$. If $G_1 \setminus S_2$ contains another $ST_{3,2}$-type T-Steiner tree, then repeat (ii); otherwise go to the next. (iii) If G_2 contains an $ST_{3,3}$-type subgraph, then we find two element-disjoint T-Steiner trees, $G[T \cup \{w_1, w_5, w_6\}]$ and $G[T \cup \{w_2, w_3, w_4\}]$ in G_2. Then, we remove the set $\{w_1, w_2, \cdots, w_6\}$ of six Steiner nodes and obtain the graph $G \setminus \{w_1, w_2, \cdots, w_6\}$. If the obtained graph contains another $ST_{3,3}$-type subgraph, then we repeat (iii); otherwise go to (Phase 4).

Consider the current bipartite graph after (Phase 3) and rename it as $G' = (T \cup (V' \setminus T), E')$. Suppose that we have already found L element-disjoint T-Steiner trees. Then, from Lemma 2, Lemma 3 and the above observations, the element connectivity of T in G' is at least $(k - 2L)$ if the initial element connectivity of T in the input graph G is at least k. Note that at this moment, $W_5 = W_4 = W_3 = \emptyset$.

(Phase 4) By using the algorithm in Proposition 4, we can find at least $(\lceil k/2 \rceil - L)$ element-disjoint T-Steiner trees from G'.

In total, at least $\lceil k/2 \rceil$ element-disjoint T-Steiner trees can be found from the input bipartite graph G. The algorithm obviously runs in polynomial time. This completes the proof of Theorem 2.   □

Acknowledgments. This work is partially supported by KAKENHI Grant Number 26330017.

References

1. Aazami, A., Cheriyan, J., Jampani, K.R.: Approximation algorithms and hardness results for packing element-disjoint Steiner trees in planar graphs. Algorithmica 63, 425–456 (2012)
2. Calinescu, G., Chekuri, C., Vondrak, J.: Disjoint bases in a polymatroid. Random Structures & Algorithms 35(4), 418–430 (2009)
3. Cheriyan, J., Salavatipour, M.R.: Hardness and approximation results for packing Steiner trees. Algorithmica 45(1), 21–43 (2006)
4. Cheriyan, J., Salavatipour, M.R.: Packing element-disjoint Steiner trees. ACM Trans. Algorithms 34(4), Article 47 (2007)
5. DeVos, M., McDonald, J., Pivotto, I.: Packing Steiner Trees. arXiv:1307.7621 (2013)
6. Frank, A., Király, T., Kriesell, M.: On decomposing a hypergraph into k connected sub-hypergraphs. Discrete Applied Math. 131(2), 373–383 (2003)
7. Grötschel, M., Martin, A., Weismantel, R.: Packing Steiner trees: polyhedral investigations. Mathematical Programming 72, 101–123 (1996)
8. Grötschel, M., Martin, A., Weismantel, R.: Packing Steiner trees: A cutting plane algorithm and computational results. Mathematical Programming 72, 125–145 (1996)
9. Grötschel, M., Martin, A., Weismantel, R.: The Steiner tree packing problem in VLSI design. Mathematical Programming 78, 265–281 (1997)
10. Hind, H.R., Oellermann, O.: Menger-type results for three or more vertices. Congr. Number 113, 179–204 (1996)
11. Jain, K., Mhdian, M., Salavatipour, M.R.: Packing Steiner trees. In: Proc. ACM-SIAM SODA, pp. 266–274 (2003)
12. Jain, K., Măndoiu, I.I., Vazirani, V.V., Williamson, D.P.: A Primal-Dual Schemes Based Approximation Algorithm for the Element Connectivity Problem. In: Proc. ACM-SIAM SODA, pp. 484–489 (1999)
13. Kaski, P.: Packing Steiner trees with identical terminal sets. Information Processing Letters 91(1), 1–5 (2004)
14. Kriesell, M.: Edge-disjoint trees containing some given vertices in a graph. J. Combinatorial Theory, Series B 88, 53–65 (2003)
15. Kriesell, M.: Packing Steiner trees on four terminals. J. Combinatorial Theory, Series B 100, 546–553 (2010)
16. Lau, L.C.: On approximate min-max theorems for graph connectivity problems. PhD thesis, University of Toronto (2006)
17. Lau, L.C.: An approximate max-Steiner-tree-packing min-Steiner-cut theorem. Combinatorica 27, 71–90 (2007)
18. Nash-Williams, S.J.A.: Edge disjoint spanning trees of finite graphs. J. London Math. Soc. 36, 445–450 (1961)
19. Tutte, W.T.: On the problem of decomposing a graph into n connected factors. J. London Math. Soc. 36, 221–230 (1961)
20. West, D.B., Wu, H.: Packing of Steiner trees and S-connectors in graphs. J. Combinatorial Theory, Series B 102, 186–205 (2012)

The Garden Hose Complexity
for the Equality Function

Well Y. Chiu[1], Mario Szegedy[2], Chengu Wang[3,*], and Yixin Xu[4]

[1] Department of Applied Mathematics,
National Chiao Tung University 1001 University Road, Hsinchu 300, Taiwan
well.am94g@nctu.edu.tw
[2] Department of Computer Science, Rutgers,
The State University of New Jersey 110 Frelinghuysen Road,
Piscataway, NJ 08854, USA
szegedy@cs.rutgers.edu
[3] Google Inc. 1600 Amphitheatre Parkway, CA 94043, USA
wangchengu@gmail.com
[4] Department of Computer Science, Rutgers,
The State University of New Jersey 110 Frelinghuysen Road,
Piscataway, NJ 08854, USA
yixinxu@cs.rutgers.edu

Abstract. The garden hose complexity is a new communication complexity introduced by H. Buhrman, S. Fehr, C. Schaffner and F. Speelman [BFSS13] to analyze position-based cryptography protocols in the quantum setting. We focus on the garden hose complexity of the equality function, and improve on the bounds of O. Margalit and A. Matsliah [MM12] with the help of a new approach and of our handmade simulated annealing based solver. We have also found beautiful symmetries of the solutions that have lead us to develop the notion of *garden hose permutation groups*. Then, exploiting this new concept, we get even further, although several interesting open problems remain.

1 Introduction

1.1 Quantum Position Based Cryptography and the Garden Hose Model

Position based cryptography was first introduced in [CGMO09], while its quantum setting was introduced in [BFS11] and also [BCF+11]. The basic idea is to use the geographical location as its only credential. For example, one message might be decrypted only if the receiver is in a specified location. In the setting of *position verification*, a special application of position based cryptography, Alice wants to convince Bob that she is in a particular position. In [CGMO09] it

---

* The work was done when this author was a Ph.D. student in Tsinghua University. He is now in Google.

Q. Gu, P. Hell, and B. Yang (Eds.): AAIM 2014, LNCS 8546, pp. 112–123, 2014.

has been shown that position verification using classical protocols is impossible against colluding adversaries (who control all positions except the prover's claimed position). In the quantum setting [BFS11], a general impossibility has been shown: using an enormous amount of quantum entanglement, colluding adversaries are always able to make it look to the verifiers as if they were at the claimed position.

In [BFSS13], the authors proposed a protocol PV_{qubit}^f for one dimensional quantum position verification, while the basic ideas generalize to higher dimensions. There are two verifiers V_0, V_1 and one prover P in between them and f is a fixed publicly known Boolean function $f : \{0,1\}^n \times \{0,1\}^n \to \{0,1\}$. Again, the above protocol PV_{qubit}^f is insecure under two adversarial attackers Alice and Bob if they share enough number of EPR pairs. It has been shown that there is a one-one correspondence between attacking the position-verification scheme PV_{qubit}^f and computing the function f in the *garden-hose model*[BFSS13]. More generally, we can translate any strategy of Alice and Bob in the garden-hose model to a perfect quantum attack of PV_{qubit}^f by using one EPR pair per pipe and performing Bell measurements where the players connect the pipes. We omit the details of the connection between them here while emphasizing on the new communication complexity model: the garden hose model.

Alice and Bob want to compute $f(x,y)$ of a Boolean function $f : \{0,1\}^n \times \{0,1\}^n \to \{0,1\}$ together where Alice gets x and Bob gets y. They have m water pipes numbered by $1, 2, \cdots, m$ between them and in addition Alice has a water tap 0. When Alice gets input x, she makes a *configuration* $A(x)$: a (nontrivial) partial matching in $\{0, 1, \cdots, m\}$, i.e. Alice uses hoses to connect those pairs of pipes according to the matching. When Bob gets input y, he makes a configuration $B(y)$: a (nontrivial) partial matching in $\{1, \cdots, m\}$. Then Alice opens the water tap, when water comes out from Alice's side, we say the output is 1, otherwise it is 0. Say that $f : \{0,1\}^n \times \{0,1\}^n \to \{0,1\}$ can be computed by m pipes in the garden hose model if for every possible input (x,y) there is a configuration pair $(A(x), B(y))$ such that water comes out in correct side. The *garden hose complexity* of f, denoted by $GH(f)$, is defined as the minimum number of pipes that computes f.

We will focus on computing $GH(EQ_n)$ for the equality function EQ_n in this paper, where $EQ_n(x,y)$ is 1 if $x = y$ and 0 if $x \neq y$.

Example 1. Let $n = 1$. Then three pipes suffice for Alice and Bob in computing EQ_1, i.e. $GH(EQ_1) \leq 3$. Here is one solution:

$$A(0) = \{01\}, \quad A(1) = \{02\}, \quad B(0) = \{13\}, \quad B(1) = \{23\}.$$

Figure 1 describes the configuration pictorially.

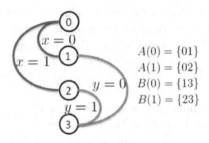

$$A(0) = \{01\}$$
$$A(1) = \{02\}$$
$$B(0) = \{13\}$$
$$B(1) = \{23\}$$

Fig. 1. Implementation of EQ_1 function

1.2   Prior Results and Our Results

It has been known that $GH(EQ_n)$ is lower bounded by $n+1$. Actually computing upper bound for $GH(EQ_n)$ featured as April 2012's "Ponder This" puzzle on the IBM website[1]. The best solution there gives

$$GH(EQ_n) \leq \frac{8}{\log 46} \cdot n + O(1) \approx 1.448n + O(1)$$

by applying IBM SAT-Solver[MM12]. With the help of a new approach and of our handmade simulated annealing based solver, we have improved their bounds. We have also found beautiful symmetries of the solutions that have lead us to develop the notion of *garden hose permutation groups*. Then, exploiting this new concept, we push the upper bound to the following.

Theorem 1. *The garden hose complexity of the equality function:*

$$GH(EQ_n) \leq \frac{28}{\log 3^{13}} \cdot n + O(1) \approx 1.359n + O(1).$$

2   Matrix Idea

Similar to the role of matrix in communication complexity, here we introduce the *configuration matrix* M_m for garden hose model where m is the number of water pipes. Rows (columns) in M_m are indexed by all possible configurations for Alice (Bob). Each entry is either 0 or 1 determined by which direction the water comes out according to its corresponding row and column. Below explicitly describes all elements in M_3 and M_4, two smallest nontrivial configuration matrices.

$$M_3 = \begin{array}{c} \\ 01 \\ 02 \\ 03 \end{array} \begin{array}{ccc} 12 & 13 & 23 \\ \left[\begin{array}{ccc} 1 & 1 & 0 \\ 1 & 0 & 1 \\ 0 & 1 & 1 \end{array}\right] \end{array}$$

---

[1] http://ibm.co/I7yvMz

$$M_4 = \begin{array}{c} \\ 01 \\ 02 \\ 03 \\ 04 \\ 01,23 \\ 01,24 \\ 01,34 \\ 02,13 \\ 02,14 \\ 02,34 \\ 03,12 \\ 03,14 \\ 03,24 \\ 04,12 \\ 04,13 \\ 04,23 \end{array} \begin{array}{cccccc} 12 & 13 & 14 & 23 & 24 & 34 \\ \left[\begin{array}{cccccc} 1 & 1 & 1 & 0 & 0 & 0 \\ 1 & 0 & 0 & 1 & 1 & 0 \\ 0 & 1 & 0 & 1 & 0 & 1 \\ 0 & 0 & 1 & 0 & 1 & 1 \\ 0 & 0 & 1 & 0 & 0 & 0 \\ 0 & 1 & 0 & 0 & 0 & 0 \\ 1 & 0 & 0 & 0 & 0 & 0 \\ 0 & 0 & 0 & 0 & 1 & 0 \\ 0 & 0 & 0 & 1 & 0 & 0 \\ 1 & 0 & 0 & 0 & 0 & 0 \\ 0 & 0 & 0 & 0 & 0 & 1 \\ 0 & 0 & 0 & 1 & 0 & 0 \\ 0 & 1 & 0 & 0 & 0 & 0 \\ 0 & 0 & 0 & 0 & 0 & 1 \\ 0 & 0 & 0 & 0 & 1 & 0 \\ 0 & 0 & 1 & 0 & 0 & 0 \end{array}\right] \end{array}$$

We observe that:

Lemma 1. $GH(EQ_n) \leq m$ *if and only if* M_m *contains a permutation submatrix of size* 2^n.

We will delay the proof to the full version of the paper.

We already gain some information from above M_3 and M_4: $GH(EQ_1) = 3$, $GH(EQ_2) = 4$, $GH(EQ_3) \geq 5$.

To make the configuration matrix smaller, lemma 2 says we can assume Alice has only one open pipe for even m. Before the lemma, we define some notations first.

For a configuration $A(x)$, we call the pipe which connects the water tap (number 0) the *water-in pipe,* and we call the pipe where the water comes out (from Alice's side) in the configuration $(A(x), B(x))$ the *water-out pipe.*

We divide M_m into blocks $M_m = [M_m^{i,j}]_{1 \leq i \leq m/2, 1 \leq j \leq (m-1)/2}$ where $M_m^{i,j}$ consists of those rows where Alice has i hoses and those columns where Bob has j hoses. For example, $M_3 = M_3^{1,1}, M_4 = \begin{bmatrix} M_4^{1,1} \\ M_4^{2,1} \end{bmatrix}$.

Lemma 2. *For even* $m \geq 4$, *if* M_m *contains a permutation submatrix of size* k, *then the last-row-block of* M_m, $[M_m^{m/2,1}, M_m^{m/2,2}, \cdots, M_m^{m/2,(m-2)/2}]$, *also contains a permutation submatrix of size* k.

Proof. We take the permutation submatrix of size k, denoted by $S \times T$. If S is not contained in the last-row-block, then we take a row $A(x) \in S$, which is outside of the last-row-block. In configuration $A(x)$, Alice has m pipes: a water-in pipe, a water-out pipe, some pairs of pipes connected with hoses, and others. Because m is even, there are even "other" pipes. We connect them by hoses arbitrarily. After that, every pipe is connected with a hose, except the water-out pipe. Therefore, this configuration, denoted by $A'(x)$, is in the last-row-block. We replace $A(x)$ by $A'(x)$. We can prove that it is still a permutation submatrix. Then, we repeat the replacement many times, and we can move $S \times T$ into the last-row-block finally.   □

By the lemma above, we know that if m is even, in order to search maximum size of permutation submatrix in M_m, we can restrict ourselves to the last-row-block of M_m, which means on Alice's side there is only one pipe without any hose connection. In the following construction, we assume that m is always even.

Actually we do not have to restrict ourselves to permutation matrices of size power two for the following lemma.

Lemma 3. *If M_m contains a permutation submatrix of size k, then M_{mt} contains a permutation submatrix of size k^t for every $t = 1, 2, \cdots$.*

Proof. In M_m, we denote the rows and columns of the permutation submatrix of size k by $\{A(x)|x \in [k]\}$ and $\{B(y)|y \in [k]\}$, respectively, where $A(x)$ and $B(y)$ intersect at entry 1 if and only if $x = y$. In M_{mt}, we will define A' and B' s.t. $\{A'(x')|x' \in [k]^t\} \times \{B'(y')|y' \in [k]^t\}$ is permutation submatrix of size k^t.

We group mt pipes into t blocks, where each block has m pipes. B' is just the product of B in each block, i.e.

$$B'(y') = \{\{im + a, im + b\}|\{a, b\} \in B(y_i'), i \in [t]\}.$$

The construction of A' is almost the same as B', but we connect the water-out pipe to the water-in pipe in the next block. Formally speaking,

$$\begin{aligned} A'(x') = \quad & \{\{im + a, im + b\}|\{a, b\} \in A(x_i'),\ a, b \neq 0,\ i \in [t]\} \\ & \cup \{\{0, \mathrm{in}(A(x_0'))\}\} \\ & \cup \{\{im + \mathrm{out}(A(x_i)), (i+1)m + \mathrm{in}(A(x_{i+1}))\}|i = 0, 1, \cdots, t-2\}. \end{aligned}$$

□

By the two lemmas above, we have the following corollary.

Corollary 1. *If there exist m and k such that M_m contains a permutation submatrix of size k, then the garden hose complexity of the equality function:*

$$\mathrm{GH}(\mathrm{EQ}_n) \leq \frac{m}{\log k} n + O(1).$$

For example when $m = 4$, $k = 6$ thus $\frac{m}{\log k} = \frac{4}{\log 6} \approx 1.547$. In [MM12] it is shown that when $m = 8$, $k = 46$ thus $\frac{m}{\log k} = \frac{8}{\log 46} \approx 1.448$ which is the best known result before this paper.

For larger m, we are devising computer programs to search for the maximum size of permutation submatrix in M_m. We need more properties to search faster.

Lemma 4. *Bob's one configuration cannot cover another one, i.e. $B(y) \subseteq B(y') \Rightarrow y = y'$.*

Proof. The water comes out from Alice's side in configuration $(A(y), B(y))$. If we add more hoses on Bob's side, the water path does not change. So, the water comes out from the same pipe on Alice's side in configuration $(A(y), B(y'))$. Therefore, $y = y'$.   □

To find a large permutation submatrix, we first want to find a large set of Bob's configurations such that one doesn't cover another. In the program, we assume each configuration in the set has the same number of pipes, so we don't need to worry about the covering problem. Moreover, to maximize the size of the set, we assume Bob's hoses covers almost half of the pipes. Appendix A gives more ideas of our simulated annealing program.

3   Some Symmetric Solutions

From previous section we know that when $m = 12$, $k = 395$, $\frac{m}{\log k} = \frac{12}{\log 395} \approx$ 1.391 which is already better than 1.448 in [MM12]. However, if we want better, unstructured searching program does not help any more. One possible hope is to study the structure of solutions, in particular we are interested in those symmetries behind some solutions. In this section we use one example to explain the basic idea then generalize it in next section.

It is known that M_4 contains a permutation submatrix of size 6. Here we write down one solution(the permutation submatrix).

$$
\begin{array}{c}
\begin{array}{cccccc} 12 & 13 & 14 & 23 & 24 & 34 \end{array} \\
\begin{array}{c} 01,23 \\ 01,24 \\ 01,34 \\ 02,13 \\ 02,14 \\ 03,12 \end{array}
\left[\begin{array}{cccccc}
0 & 0 & 1 & 0 & 0 & 0 \\
0 & 1 & 0 & 0 & 0 & 0 \\
1 & 0 & 0 & 0 & 0 & 0 \\
0 & 0 & 0 & 0 & 1 & 0 \\
0 & 0 & 0 & 1 & 0 & 0 \\
0 & 0 & 0 & 0 & 0 & 1
\end{array}\right]
\end{array}
$$

Here Alice and Bob are computing the equality function $f(x, y)$ where $x, y \in$ [6]. On input x, Alice makes a configuration $A(x)$, while on input y, Bob makes a configuration $B(y)$. Thus we can also write the above solution in the following form.

$$A(1) = \{01, 23\}, B(1) = \{14\}$$

$$A(2) = \{01, 24\}, B(2) = \{13\}$$

$$A(3) = \{01, 34\}, B(3) = \{12\}$$

$$A(4) = \{02, 13\}, B(4) = \{24\}$$

$$A(5) = \{02, 14\}, B(5) = \{23\}$$

$$A(6) = \{03, 12\}, B(6) = \{34\}$$

The symmetry of the above solution can be seen in two ways, although equivalent.

First let us assume that Alice and Bob only know how to connect pipes on input $(1, 1)$, i.e. they only have

$$A(1) = \{01, 23\}, B(1) = \{14\}$$

now the question is: how can they get other configurations from $(A(1), B(1))$? One possible way is to use a group to act on $(A(1), B(1))$. By doing this, hopefully they can get all other configurations. In other words, the final solution is invariant under some group action.

Two groups fit for our setting here, the alternating group A_4 and the symmetric group S_4. Take A_4 for example.

$$A_4 = \{(1),(123),(132),(124),(142),(134),(143),(234),(243),(12)(34),(13)(24),(14)(23)\}$$

For an element $\sigma \in A_4$, define its action on $A(1), B(1)$ as

$$A(\sigma) = \{0\sigma(1), \sigma(2)\sigma(3)\}, B(\sigma) = \{\sigma(1)\sigma(4)\}$$

then we have

$$
\begin{aligned}
A((1)) &= \{01, 23\}, B((1)) = \{14\} & \rightarrow A(1), B(1) \\
A((123)) &= \{02, 13\}, B((123)) = \{24\} & \rightarrow A(4), B(4) \\
A((132)) &= \{03, 12\}, B((132)) = \{34\} & \rightarrow A(6), B(6) \\
A((124)) &= \{02, 34\}, B((124)) = \{12\} & \\
A((142)) &= \{04, 13\}, B((142)) = \{24\} & \\
A((134)) &= \{03, 24\}, B((134)) = \{13\} & \\
A((143)) &= \{04, 12\}, B((143)) = \{34\} & \\
A((234)) &= \{01, 34\}, B((234)) = \{12\} & \rightarrow A(3), B(3) \\
A((243)) &= \{01, 24\}, B((243)) = \{13\} & \rightarrow A(2), B(2) \\
A((12)(34)) &= \{02, 14\}, B((12)(34)) = \{23\} & \rightarrow A(5), B(5) \\
A((13)(24)) &= \{03, 14\}, B((13)(24)) = \{23\} & \\
A((14)(23)) &= \{04, 23\}, B((14)(24)) = \{14\} &
\end{aligned}
$$

Note that the group action also generates other configurations which are not in the solution, for example $A((124)) = \{02, 34\}$. However, this should not be a problem since if we look into the row of $A((124))$ in matrix M_4, it is exactly the same as row $A((234)) = \{01, 34\}$, thus we still get a solution by choosing any one of $A((234))$ or $A((124))$.

Another way to look into the symmetry of the above solution is through geometry. Treat four pipes as four vertices of a regular tetrahedron. Let vertex 1 be water-in pipe and vertex 4 be water-out pipe, 2 and 3 are connected by Alice while 1 and 4 are connected by Bob. This is the initial configuration $(A(1), B(1))$ as before, and call it *the base construction*. Now rotate or reflect the tetrahedron (see figure 2), then the rotation group A_4 or the symmetric group S_4 of the tetrahedron are exactly the same as we discussed from the first perspective. Also the meaning of equivalence between $A((234))$ and $A((124))$ discussed above can be explained here as we are not making any difference by switching the water-in and water-out pipe since it does not change the side where water comes out.

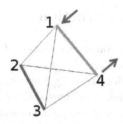

Fig. 2. A geometric explanation of symmetric solutions

4   Group Invariant Framework

The symmetry discussion in previous section can be summarized by the following formula:

$$\boxed{\text{symmetric solution } = \text{ base construction } + \text{ group action}}$$

Now we generalize this idea here.

Let m be an even integer as before. Let t be the number of hoses that Bob uses. Without loss of generality we assume that on input $(1,1)$,

$$A(1) = \{01, 23, \cdots, (2t-2)(2t-1), (2t+1)(2t+2), \cdots, (m-1)m\}$$

$$B(1) = \{12, 34, \cdots, (2t-1)(2t)\}$$

Here $(A(1), B(1))$ is *the base construction*.

Let $G \leq S_m$ be a permutation group. For every $g \in G$, define *the group action* as

$$A(g) = \{0g(1), g(2)g(3), \cdots, g(2t-2)g(2t-1), g(2t+1)g(2t+2), \cdots, g(m-1)g(m)\}$$

$$B(g) = \{g(1)g(2), g(3)g(4), \cdots, g(2t-1)g(2t)\}. \tag{1}$$

Now we can check matrix M_m if the intersection of rows $\{A(g)|g \in G\}$ and columns $\{B(g)|g \in G\}$ forms a permutation submatrix. If so, then we call G an (m,t)-**garden hose permutation group**. However, in some cases, the intersection may contain repeated rows or columns. If we remove those repeated ones and the rest still forms a permutation submatrix, then we call G a **weak** (m,t)-garden hose permutation group. Correspondingly, call the solution $\{A(g), B(g)|g \in G\}$ a (weak) group invariant solution. For example, S_4 and A_4 discussed in section 3 are both $(4,1)$-weak garden hose permutation groups.

Now we are reducing the problem of the garden hose complexity for the equality function to the problem of deciding which group is a garden hose permutation group. If there is an (m,t)-garden hose permutation group G, then the garden hose complexity for the equality function

$$\text{GH}(\text{EQ}_n) \leq \frac{m}{\log|G|} n + O(1). \tag{2}$$

5   Group Invariant Construction

For a given m, we can check all conjugacy classes of the subgroups of symmetric group S_m, to see which one is a (weak) garden hose permutation group. Each of them gives an upper bound for $\mathrm{GH}(\mathrm{EQ}_n)$ according to formula (2). Tables of Marks [PM] shows conjugacy classes of the subgroups of small symmetric groups. Note that to check if a submatrix in M_m is a permutation matrix is much easier than searching a permutation matrix in it. Moreover, lemma 5 helps us check if a group is a gardenhose permutation group in a more efficient way by scanning only one row or column of submatrix generated by that group. Appendix B lists some results by computer programs.

Lemma 5. *G is a garden hose permutation matrix if and only if the intersection of row $A(1)$ and columns $B(g) : g \in G$ has exactly one 1.*

Proof. $\Rightarrow$ By definition.
$\Leftarrow$ Denote by $M = \{A(g)|g \in G\} \times \{B(g)|g \in G\}$. It suffices to show that (1) every row in M has exactly one 1; (2) every column in M has exactly one 1. (1) follows from the observation that the intersection of $A(h_1)$ and $B(h_2)$ in M has the same value as the intersection of $A(g^{-1}h_1)$ and $B(g^{-1}h_2)$ for any $g, h_1, h_2 \in G$. (2) is verified since the column $B(1)$ in M has exactly one 1 by the same observation. $\qquad \square$

5.1   Wreath Product Construction

There is a particular group which stands out in our search results. Denote it by W_2. $W_2(\le S_{10})$ has generators

$$(3,5,10),(2,7,8),(1,2,3)(4,7,10)(5,6,8).$$

For those familiar with wreath product[OOR04], we know that $W_2 \cong C_3 \wr C_3$, the wreath product of two cyclic group of order 3. Computer program tells us that W_2 is a $(10,2)$ garden hose permutation group. Since $|W_2| = 3^4 = 81$, by formula (2) we get $\frac{m}{\log |W_2|} = \frac{10}{\log 81} \approx 1.577$. This ratio is worse than best result by our simulated annealing program, and it is even worse than 1.448 by Margalit and Matsliah[MM12]. However, the reason that why it stands out from others is its special structure: the wreath product, from which there is a way we can generalize, in hope of beating best current result.

A natural question is: is there any large l such that there exists a garden hose permutation group W_l and $W_l \cong \underbrace{C_3 \wr \cdots \wr C_3}_{l}$? We conjecture that this is the case.

Conjecture 1 (3-tree conjecture). For every $l \ge 2$ there exists a group $W_l \le S_{3^l+1}$, $W_l \cong \underbrace{C_3 \wr \cdots \wr C_3}_{l}$ and W_l is a $(3^l + 1, t)$ garden hose permutation group for some t.

One consequence of the 3-tree conjecture is: we can push the upper bound for $GH(EQ_n)$ even further.

Theorem 2. *If conjecture 1 holds, then the garden hose complexity of the equality function:*

$$GH(EQ_n) \leq \frac{2}{\log 3} \cdot n + O(1) \approx 1.262n + O(1).$$

Proof. Since

$$|W_l| = 3^{1+3+\cdots+3^{l-1}} = 3^{\frac{3^l-1}{2}},$$

and $m = 3^l + 1$, thus by formula (2), we have

$$\frac{m}{\log|W_l|} = \frac{3^l + 1}{\log(3^{\frac{3^l-1}{2}})} = \frac{2}{\log 3} \cdot \frac{3^l + 1}{3^l - 1},$$

for large enough l, the above tends to $\frac{2}{\log 3} \approx 1.262$.

We already know that the 3-tree conjecture holds for $l = 2$, actually it also holds for $l = 3$ which gives us better upper bound for $GH(EQ_n)$ unconditionally. And we are ready to prove theorem 1.

Proof (proof of theorem 1). It suffices to provide conjugate representatives for $W_3 = g^{-1}Kg \cong C_3 \wr C_3 \wr C_3$ where
$g = (3, 17, 4, 15)(6, 24, 20, 11)(2, 28, 18, 5, 26, 22, 14, 2)(7, 21, 13, 8, 25, 23, 19, 10, 16)$
and K is generated by:
$(1, 2, 3), (4, 5, 6), (1, 4, 7)(2, 5, 8)(3, 6, 9),$
$(10, 11, 12), (13, 14, 15), (10, 13, 16)(11, 14, 17)(12, 15, 18),$
$(19, 20, 21), (22, 23, 24), (19, 22, 25)(20, 23, 26)(21, 24, 27),$
$(1, 10, 19)(2, 11, 20) \cdots (9, 18, 27)$
Then, by computer programs, one can check it is a $(28, 7)$ garden hose permutation group. Thus $\frac{28}{\log 3^{13}} \approx 1.359$.

6   Further Research

At least two questions arise. First try to prove the 3-tree conjecture. Second, is there any other garden hose permutation group besides W_2 and W_3 such that the structure can be generalized in order to get better upper bound for $GH(EQ_n)$?

The authors would like to thank Mike Saks for useful discussions.

References

[BCF+11] Buhrman, H., Chandran, N., Fehr, S., Gelles, R., Goyal, V., Ostrovsky, R., Schaffner, C.: Position-Based Quantum Cryptography: Impossibility and Constructions. In: Rogaway, P. (ed.) CRYPTO 2011. LNCS, vol. 6841, pp. 429–446. Springer, Heidelberg (2011)

[BFS11]   Buhrman, H., Fehr, S., Schaffner, C.: Position-Based Quantum Cryptog-
           raphy. ERCIM News 2011(85), 16–17 (2011)
[BFSS13]   Buhrman, H., Fehr, S., Schaffner, C., Speelman, F.: The garden-hose
           model. In: ITCS, pp. 145–158 (2013)
[CGMO09]   Chandran, N., Goyal, V., Moriarty, R., Ostrovsky, R.: Position Based Cryp-
           tography. In: Halevi, S. (ed.) CRYPTO 2009. LNCS, vol. 5677, pp. 391–407.
           Springer, Heidelberg (2009)
[MM12]   Margalit, O., Matsliah, A.: Mage - the CDCL SAT solver developed and
           used by IBM for formal verification. Personal Communication (2012),
           http://ibm.co/P7qNpC
[OOR04]   Orellana, R.C., Orrison, M.E., Rockmore, D.N.: Rooted trees and iterated
           wreath products of cyclic groups. Advances in Applied Mathematics 33(3),
           531–547 (2004)
[PM]   Pfeiffer, G., Merkwitz, T.: GAP Data Library "Tables of Marks",
           http://www.gap-system.org/Datalib/tom.html

A   Simulated Annealing Program

This program searches for a large permutation submatrix of the configuration matrix M_m.

We assume m is even, Alice's configuration has only one open pipe, and Bob's hoses covers half of the pipes.

First we find a small permutation submatrix, than it grows larger. In each iteration, we try to add one row or one column. If it fails to grow at some point, we remove one row and/or one column from it, and try again. We give the pseudo-code as below.

```
/*
 * X is a set of rows. Y is a set of columns.
 * The submatrix X * Y is the permutation submatrix we want to find.
 */
X = empty set
Y = empty set

add_a_row :
for x in (all rows \ X) in random order {
    if submatrix {x} * Y is all-zero {
        goto add_a_column
    }
}
/* Failed to add a row. Remove a row and a column. */
Pick a 1 in submatrix X * Y at random.
Denote the location of the 1 by (x1,y1).
Remove x1 from X.
Remove y1 form Y.
goto add_a_row

add_a_column :
for y in (all columns \ Y) in random order {
    if submatrix X * {y} is all-zero and (x,y) is one {
        /* We have a larger one. */
        X = X union {x}
        Y = Y union {y}
        print X * Y is a permutation submatrix
        goto add_a_row
    }
}
```

```
/* Failed to add a row. */
with prob. 1/4 {
    goto add_a_row /* Discard y. */
}
/* with prob. 3/4 */
/* Remove a row and a column. */
Pick a 1 in submatrix X * Y at random.
Denote the location of the 1 by (x1,y1).
Remove x1 from X.
Remove y1 form Y.
goto add_a_column
```

We run the program on a PC. If it has not found a larger permutation submatrix (no "print") in a long time, we kill the program and run it again. The following table shows the search results on $m \leq 12$, where k is the size of the maximum permutation submatrix the program find.

m	4	6	8	10	12
k	6	15	48	144	395
running time	< 1 sec	< 1 sec	10 sec	1 hour	1 day
$m/\log k$	1.547...	1.535...	1.432...	1.394...	1.391...

B List of Some (Weak) Garden Hose Permutation Groups

# of pipes	Base construction	Group action	# of solutions	# of solutions by Simulated Annealing
4		$G \cong A_4, S_4$	6	6
6		$G \cong A_5$	15	15
		$G \cong C_3 \times C_3$	9	
8		$G \cong S_4 \rtimes C_2$	24	48
		$G \cong S_4$	24	
		$G \cong (C_3 \wr C_2) \times C_2$	36	
10		$G \cong C_3 \wr C_3$	81	144
		$G \cong (C_5 \times C_5) \rtimes C_8$	100	
12		$G \cong (C_3 \wr C_3) \times C_3$	243	395
		$G \cong (C_3 \wr C_2) \times (C_3 \wr C_2)$	324	

Finding Robust Minimum Cuts

Barbara Geissmann and Rastislav Šrámek

Institute of Theoretical of Computer Science, ETH Zurich, Zurich, Switzerland
geibarba@student.ethz.ch, rsramek@inf.ethz.ch

Abstract. We study the minimum cut problem in the presence of uncertainty and show how to apply a novel robust optimization approach, which aims to exploit the similarity in subsequent graph measurements or similar graph instances, without posing any assumptions on the way they have been obtained. With experiments we show that the approach works well when compared to other approaches that are also oblivious towards the relationship between the input datasets.

1 Introduction

Dealing with uncertainty is an ever more common problem. We are flooded with data recorded by virtually all modern devices from cars to cellular phones, data about various networks, observations of different phenomena. In order to be able to extract meaningful information from this data, we need to be able to remove or at least identify the noise that is inherently present, whether due to measurement errors or due to systematic influence of unknown factors. In this paper we consider a novel method of robust optimization introduced by Buhmann et al. [1], and apply it to the problem of searching for the global minimum cut in a graph.

Finding the global minimum cut in a graph is a well studied problem with applications ranging from information retrieval [2] to computer vision [3]. The problem is to separate the set of graph vertices V into two non-empty disjoint sets X and $V \setminus X$, such that the sum of the weights of edges that have one endpoint in X and another in $V \setminus X$ is minimized. Since a cut is fully determined by the subset X, we will denote it only by X with the possible caveat that X and $V \setminus X$ denotes the same cut. We are in particular interested in the minimum cut as a measure of network robustness [4]: If the weight of an edge represents the effort needed to cut that particular edge, the minimum cut represents the least effort necessary to disconnect the graph.

Suppose that we are looking for a minimum cut in a graph, for instance one that represents connections between nodes in a sensor network. However, instead of the "true" graph we are only given two snapshots of it from two different points in time, with the same topology, but with different edge-weights. What should we do in order to identify a minimum cut in the "true" underlying graph, or a cut that will be minimum in a third, similar, snapshot?

It is clear that without very precise understanding of the process by which we obtain the graph measurements, we are unable to answer this question with

Q. Gu, P. Hell, and B. Yang (Eds.): AAIM 2014, LNCS 8546, pp. 124–136, 2014.

full confidence and thus any solution will be only an heuristic. Nevertheless, the setting is a realistic and very common one, and we should not give up. For instance, one intuitive course of action would be to average the weights provided by two instances edge by edge and compute a minimum cut on the resulting graph. In this paper we want to show that a different method, also oblivious to the properties of the data generator, might yield better results.

1.1 Approximation Set Optimization

We will introduce the aforementioned robust optimization method of Buhmann et al. [1] in greater detail. We will refer to it as approximation set optimization. Recall that the weight of a graph cut X is the sum of the weights of the edges that have one endpoint in X and another in $V \setminus X$, we will denote it by $w(X)$.

Definition 1 (ρ-Approximate Cut)
Let $\lambda(G)$ denote the weight of a global minimum cut in G. For a parameter $\rho \geq 1$, a ρ-approximate cut X is a cut with weight at most $\rho\lambda(G)$, $w(X) \leq \rho\lambda(G)$.

Definition 2 (ρ-Approximation Set)
A ρ-approximation set of G, denoted by $A_\rho(G)$, is the set of all ρ-approximate cuts in G, $A_\rho(G) = \{X \in V \mid w(X) \leq \rho\lambda(G)\}$.

Let G_1 and G_2 be two weighted graphs with the same topology but different edge-weights. The approximation set optimization method states that we should find a factor ρ, for which the intersection of the ρ-approximation sets $A_\rho(G_1) \cap A_\rho(G_2)$ is the largest, when compared to the expected size of this intersection if the instances were generated at random. We then pick a solution at random from the intersection of the resulting ρ-approximation sets. Formally, we look for ρ^* such that

$$\rho^* = \arg\max_\rho \frac{|A_\rho(G_1) \cap A_\rho(G_2)|}{Es\left(|A_\rho(G_1)|, |A_\rho(G_2)|\right)} , \tag{1}$$

where $Es(|A_\rho(G_1)|, |A_\rho(G_2)|)$ is the expected size of the intersection of the ρ-approximation sets of the given size. We call the value

$$\frac{|A_{\rho^*}(G_1) \cap A_{\rho^*}(G_2)|}{Es\left(|A_{\rho^*}(G_1)|, |A_{\rho^*}(G_2)|\right)} \tag{2}$$

unexpected similarity. It is a measure of similarity of G_1 and G_2, with respect to the optimization problem of looking for the minimum cut.

In order to successfully apply the method, we need to be able to solve five problems: Count the number of ρ-approximate cuts in a graph G, count the number of cuts in the intersection of the approximate sets of two graphs G_1 and G_2, compute the function for the expected intersection Es, find the optimal factor ρ^*, and choose a cut at random from the set of all cuts that are ρ-approximate for the graphs G_1 and G_2 at the same time.

1.2 Related Work

Robust optimization is a widely studied subject. However, in order to be able to derive provably optimal methods, one needs to restrict the scope of inputs, or make other strong assumptions about them. For instance *Stochastic optimization* [5, 6] and *Robust optimization* [7] expect that we know respectively the complete distribution of an instance and the complete set of instances. Various methods of optimization for stable inputs on the other hand suppose that the input cannot change too much [8–10]. In our case, by not assuming anything about the input we lose the ability to apply any of these but greatly increase the scope of problems for which we can hope to achieve good solutions.

The remainder of the paper is structured as follows. In Sect. 2 we show how to count the sizes of ρ-approximation sets of cuts and their intersections, in Sect. 3 we will derive a approximate formula for the expected size of the approximation set intersection on random instances, followed by experimental evaluation of the method in Sect. 4 and concluding remarks in Sect. 5.

2 Algorithms for Counting Small Cuts

For many combinatorial optimization problems, the problem of counting approximate solutions is #P-complete, even if the optimization problem itself is efficiently solvable. The reason for this lies in the possibly exponential number of solutions. For instance, there can be n^{n-2} short spanning trees in a graph with n vertices or 2^{n-2} short s-t paths in a directed acyclic graph [11]. For minimum cuts, however, the possible number of near-optimal cuts is small. Dinits et al. [12] showed that there can be at most $\binom{n}{2} = O(n^2)$ minimum cuts in a graph and Karger [13] showed that the number of ρ-approximate cuts is at most $O(n^{2\rho})$. This makes our life significantly easier, since we can afford to enumerate, not only count the cuts in the approximation sets. Note that calculating the number of cuts shorter than an arbitrary threshold is still #P-complete [14]. This is not surprising, since with rising threshold the problem must turn from easy to difficult, as calculating the maximum graph cut is a NP-complete problem.

There are at least two different algorithms that can compute the ρ-approximation sets of a graph. One is by Nagamochi, Nishimura and Ibaraki [15] and it solves the task deterministically in $O(mn^{2\rho})$ time if m is the number of edges of the graph with n vertices. The other is an adaption of the *recursive contraction algorithm* by Karger and Stein [16], and it finds all ρ-approximate cuts in $O(n^{2\rho} \log^3 n)$ time with high probability.

We will use the approach of Karger and Stein because it is the fastest currently known algorithm. Apart from that it allows us to make an adaptation with which we can directly compute the approximation sets.

2.1 Karger and Stein's Algorithm

The main idea of the recursive contraction algorithm [16], described by RE-CURSIVECONTRACT in Algorithm 1, to find a minimum cut are random edge contractions. It means that edges are repeatedly chosen at random and then contracted, i.e., the two end vertices of an edge are merged into a single vertex. The algorithm starts with two entire copies of the graph. On each of them it performs random edge contractions until the graph has shrunk down to a certain size. That is, routine CONTRACT(G, x) repeatedly performs edge contractions in G until only x vertices remain. Then the graph is copied again and the algorithm continues recursively, again on both graphs, until only two vertices and one connecting edge remain. These two vertices define the cut, since every edge contracted into one of them corresponds to a vertex of the graph, and the weight of the remaining edge corresponds to the cost of this cut. The algorithm keeps track of the found cuts and the best cut is returned. The intuition behind the algorithm is that in the beginning, the probability of contracting an edge from a minimum cut, and thus excluding this cut from the set of possible results, is low. As the algorithm progresses, this chance increases, but this is combated by the increased number of concurrent evaluations.

Algorithm 1 (Recursive Contraction Algorithm).

> RECURSIVECONTRACT(G)
> **if** $|V| \leq 6$ **then**
>> $G \leftarrow$ CONTRACT(G, 2)
>> **return** the cut
> **else**
>> **repeat twice**
>>> $G' \leftarrow$ CONTRACT(G, $\lceil n/\sqrt{2} + 1 \rceil$)
>>> RECURSIVECONTRACT(G')
>> **return** the smaller cut
> **end**

The whole algorithm RECURSIVECONTRACT runs in $O(n^2 \log n)$ time. A contraction of a graph to $\lceil n/\sqrt{2} + 1 \rceil$ vertices needs $O(n^2)$ time and the depth of the recursion is in $O(\log n)$. The probability that the algorithm finds a particular minimum cut is at least $\Omega(1/\log n)$, since the bound of $\lceil n/\sqrt{2} + 1 \rceil$ in the contraction procedure ensures that the probability of not contracting an edge of the minimum cut is always greater than fifty percent. Finally, if we repeat the algorithm $O(\log^2 n)$ times, we will find any particular minimum cut with high probability. For more details we refer to [13] and [16].

The algorithm can be adjusted in such a way that it returns all minimum cuts it finds instead of only one. Since the total number of unique minimum cuts in a graph is bounded from above by $\binom{n}{2}$, we can find every minimum cut with high probability, within the total time complexity of $O(n^2 \log^3 n)$.

Karger and Stein's algorithm can also be modified to find all ρ-approximate cuts [16], by changing the reduction factor from $\lceil n/\sqrt{2} + 1 \rceil$ to $\lceil n/\sqrt[2\rho]{2} + 1 \rceil$ and stopping the contraction when 2ρ vertices remain. In this case, all remaining possible cuts are evaluated. The running time increases to $O(n^{2\rho} \log n)$, whereas the success probability remains the same. Since the number of ρ-approximate cuts is bounded by $\Theta(n^{2\rho})$, we can find all ρ-approximate cuts with high probability by repeating the algorithm $O(\log^2 n)$ times. This gives the overall time of $O(n^{2\rho} \log^3 n)$.

2.2 Approximation Set Optimization Algorithm

Recall that we want to determine ρ that maximizes the unexpected similarity of the two graphs G_1 and G_2 with respect to the minimum cut problem. To this end we first need to compute the ρ-approximation sets of G_1 and G_2 and their intersection. The former is done by APPROXIMATIONSET(G, ρ), which is an adapted version of Karger and Stein's recursive contraction algorithm. Afterwards, we are ready to compute the expected and unexpected similarity, Es and u_sim. By sampling for the best ρ we find the intersection of $A_{\rho^*}(G_1)$ and $A_{\rho^*}(G_2)$ from which we can pick a cut at random, as a solution that generalizes for both instances. The whole process is described in Algorithm 2.

Algorithm 2 (Approximation Set Optimization Algorithm).
>**for** $\rho = 1$ **until** $\rho = \text{MAX}$ **do**
>>$A_\rho(G_1) \leftarrow$ APPROXIMATIONSET(G_1, ρ)
>>$A_\rho(G_2) \leftarrow$ APPROXIMATIONSET(G_2, ρ)
>>intersection $\leftarrow$ $intersect(A_\rho(G_1), A_\rho(G_2))$
>>u_sim $\leftarrow$ |intersection| / $Es(|A_\rho(G_1)|, |A_\rho(G_2)|)$
>>**if** u_sim $>$ max_sim **then**
>>>max_sim $\leftarrow$ u_sim
>>>ρ^*_intersection $\leftarrow$ intersection
>>**end**
>**end**

Now let us discuss some issues of the algorithm in more detail and derive its time complexity. Karger and Stein's version of the algorithm returns the cuts in an implicit way. Since we want to be able to compute the intersection of the approximation sets of two different graphs as well as to choose a cut from the intersection and apply it to a third graph, we need them explicitly. One simple possibility to meet this requirement is to store for each vertex whether it is in the cut or not. The entire cut can then be represented as a bit string of length n. Notice, that this notation is ambiguous, since the inverse of a bit string describes the same cut. We can fix this by allowing only cuts that have the first bit set to zero. By treating the bit strings as numbers, we can sort the cuts in the approximation sets and then build the two intersection in a merging fashion in $O(n^2 \log n)$ time, since the number of cuts in each approximation set is bounded from above by $\Theta(n^{2\rho})$.

Returning the cuts in an explicit manner also implies extra work during the computation of the approximation sets in APPROXIMATIONSET(G, ρ). After every recursion phase the union of the two found approximation sets is returned. To overcome difficulties like different smallest cut weights and duplicates, one possibility is to again sort the cuts. This extra work requires a factor of $O(\log n)$

additional time for the entire approximation set computation. So we end up with an approximation set algorithm that takes $O(n^{2\rho} \log^4 n)$ time. The time to compute u_sim, the unexpected similarity, depends on the complexity of the function Es. We postpone this to Sect. 3.

The last thing we have to look at is the range and step size of the values for ρ in the for loop. To choose a good bound for the largest ρ we want to test is not easy. It depends a lot on the structure of the graphs and the range of their weights, as well as on how different the instances are. Therefore, we cannot avoid sampling in a relatively unbounded area. But we may want to start with rather big steps and refine them as we go on.

3 Expected Intersection Size

Having described an algorithm that counts the size of individual approximation sets and their intersection, we turn to the question of deriving a formula for the expected size of the intersection of the approximation sets. For a more detailed exposure we refer the reader to Chap. 3.2 of the bachelor thesis of Barbara Geissmann [17].

For the expected similarity, we will only consider cuts on complete graphs. Otherwise we would need to track whether each cut X cuts the graph into only two parts, since the Karger-Stein algorithm and its modifications return only such cuts.

We first show that an arbitrary subset of cuts does not necessarily have to form a valid approximation set.

Definition 3 (Crossing Cuts). *Two cuts X and Y cross each other if $X \cap Y \neq \emptyset$, $X - Y \neq \emptyset$, $Y - X \neq \emptyset$, and $V - X - Y \neq \emptyset$.*

Definition 4 (Composed Cuts). *Let X and Y be two cuts that cross each other. Then they must define four further cuts:*

$$Z_1 = X \cap Y \qquad Z_2 = X - Y$$
$$Z_3 = Y - X \qquad Z_4 = V - X - Y . \qquad (3)$$

We call Z_1, Z_2, Z_3, and Z_4 the composed cuts of X and Y.

Theorem 1. *If two cuts X and Y in the approximation set $A_\rho(G)$ cross each other, then at least two of the four composed cuts of X and Y have to be in $A_\rho(G)$ as well.*

Proof. According to the Fig. 1 we denote by a, b, c, and d the sums of the weights of the cut edges between the sets Z_1, Z_2, Z_3, and Z_4, as in Definition 4. Without loss of generality, let us suppose that $a \leq b$, $c \leq d$, $b \leq d$. Then, for cuts X and Y to be in the approximation set, there must be a threshold $t := \rho\lambda(G)$ such that $a + b \leq t$ and $c + d \leq t$. The four composed cuts will have weights $a + c$, $a + d$, $b + c$, and $b + d$. However, it must hold that $a + c \leq t$ because both a and c are at most $t/2$, and also $b + c \leq t$ because we can replace d in $c + d \leq t$ with b which is at most as large. $\qquad \square$

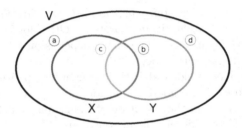

Fig. 1. Crossing cuts

Theorem 1 shows that not every subset of cuts forms a feasible approximation set. Using Theorems 1 and 2 from [1] we can conclude that the expression $|A_\rho(G_1)||A_\rho(G_2)|/|S|$, where S denotes the set of all cuts, is a lower bound on the expected size of the intersection, but not its true value. We see that as soon as we have crossing cuts, we lose freedom in the number of cuts which we can freely choose. We will show that this loss of freedom is substantial enough that by restricting ourselves to approximation sets without crossing cuts we get a approximation of the true expected value.

The number of ways in which we can cut a graph of n vertices m times so that the cuts do not cross is equal to the number of ways we can partition a set of n integers into $m+1$ non-empty subsets. The latter describes the well known Stirling number of the second kind, denoted by $\{{n \atop m+1}\}$, and defined by the explicit formula

$$\begin{Bmatrix} n \\ k \end{Bmatrix} = \frac{1}{k!} \sum_{j=0}^{k} (-1)^{k-j} \binom{k}{j} j^n \ . \tag{4}$$

Observe that in a complete graph, a non-empty cut on n vertices can be chosen in $2^{n-1} - 1$ ways. Furthermore, the smallest approximation set that can contain crossing cuts is of size 4. Such an approximation set would be approached by $\{{n \atop 5}\}$. We conclude that at least if the number of vertices n is large compared to the number of cuts in the approximation set, the loss of freedom to choose 2 additional cuts significantly outweighs the additional flexibility we gained by choosing the second cut in $2^{n-1} - 2$ ways. Note that with increasing number of crossing cuts, the number of composed cuts grows even further.

We now calculate the expected similarity for non-crossing cuts and use it as an approximation for the unexpected similarity when cuts cross. Let $k := |A_\rho(G_1)|$, $l := |A_\rho(G_2)|$, and let $\mathcal{F}_x$ denote all approximation sets that contain exactly x cuts. Then by Lemma 2 of [1] we have

$$Es(k,l) = \frac{1}{|\mathcal{F}_k||\mathcal{F}_l|} \sum_{\substack{F_1 \in \mathcal{F}_k \\ F_2 \in \mathcal{F}_l}} |F_1 \cap F_2| = \frac{1}{|\mathcal{F}_k||\mathcal{F}_l|} \sum_{s \in S} |\{F \in \mathcal{F}_k \mid s \in F\}| \cdot |\{F \in \mathcal{F}_l \mid s \in F\}|$$

$$= \frac{1}{\{{n \atop k+1}\}\{{n \atop l+1}\}} \cdot \frac{\sum_{i=1}^{n-1} \left(\binom{n}{i} \sum_{j=0}^{k-1} \left(\{{i \atop j+1}\}\{{n-i \atop k-j}\} \right) \cdot \sum_{j=0}^{l-1} \left(\{{i \atop l+1}\}\{{n-i \atop l-j}\} \right) \right)}{(2^{k+1} - 2)(2^{l+1} - 2)} \ . \tag{5}$$

The factors $2^{k+1} - 2$ and $2^{l+1} - 2$ prevent from double-counting by choosing the same cuts in a different order.

Deriving a closed formula for Es seems difficult, due to the Stirling numbers of the second kind. We can, however, evaluate the expression algorithmically for every necessary k and l. In order to avoid straight-forward $O(n^7)$ computation, we pre-compute all binomial coefficients from $\binom{n}{1}$ to $\binom{n}{n-1}$ in linear time using the identity $\binom{n}{i+1} = \binom{n}{i} \cdot \frac{(n-i)}{(i+1)}$. Similarly, using the combinatorial identity $\begin{Bmatrix} n \\ k \end{Bmatrix} = k\begin{Bmatrix} n-1 \\ k \end{Bmatrix} + \begin{Bmatrix} n-1 \\ k-1 \end{Bmatrix}$, we can pre-compute all Stirling numbers from $\begin{Bmatrix} 0 \\ 0 \end{Bmatrix}$ to $\begin{Bmatrix} n \\ n \end{Bmatrix}$ in time $O(n^2)$. Using the previously pre-computed values, we can compute all inner summands for different values of l and k in $O(n^3)$ time and space. An evaluation of the formula for two particular values of k and l thus needs only $O(n)$ time and the evaluation for all possible pairs of k and l thus fits into the $O(n^3)$ time necessary for the pre-computations.

4   Experimental Results

In order to evaluate the performance of the approximation set optimization for this problem, we tested it on various sets of input instances and compared the performance to two other algorithms. The first being an algorithm where we average edge weights edge by edge and compute minimum cut on the resulting graph and the second being an algorithm where we increase ρ until the intersection of ρ-approximation sets is non-empty for the first time and we choose the cut from this intersection. We look at this second algorithm because it intuitively seems to be a very good approach.

4.1   Tests

Every test is as follows. Three complete, undirected, weighted graph instances are taken as input, where the first two are used to predict a good solution for a future one. Then, this solution is tested against the third instance. Fig. 2 illustrates all tests done.

4.2   Data

We run experiments on three different kind of graphs to evaluate the quality of the found solution: On graphs constructed with real world data which we expect to be similar, on totally random graphs which we do not expect to be similar at all, and on artificially generated similar random graphs, which all have some small cuts in common. The tests on real world data are based on the historical daily prices between 1999 and 2010 of thirteen different stock indices[1] [18]. The vertices of our graph correspond to individual stock indices and the edges between them correspond to their similarity with respect to the

---

[1] BEL-20, Dow Jones, Hang Seng, Nikkei, AEX, CAC-40, Dax, Eurotop100, FTSE100, JSX, Nasdaq, AS30, RTSIndex, SMI.

Input: Three complete, undirected, weighted graphs, G_1, G_2, and G_3.

Output: Four different results:

- AVERAGE: Add the edge weights of G_1 and G_2 pairwise. Compute a minimum cut for the new formed graph. Apply the solution on G_3.
- FIRSTINTERSECTION: Find the smallest ρ that results in a non-empty intersection of the ρ-approximation sets for G_1 and G_2. Pick a random cut from the intersection and apply it on G_3.
- BESTSIMILARITY: Find ρ^* which maximizes the unexpected similarity of G_1 and G_2. Pick a random cut from the intersection and apply it on G_3.
- OPTIMUM: Compute a minimum cut of G_3.

Fig. 2. Specification of the Experiment

problem of finding a contiguous sub-array of maximum sum[2], as calculated by the approximation set optimization method [1]. Every graph corresponds to one year. For the random graphs, we assign a random weight to every edge. For the artificially made similar graphs we randomly define some cuts to be small and allocate small weights to their edges. To all the other edges we randomly assigned a weight from a larger range.

4.3   Results

Real World Data. Since similarity compares exponentially growing quantities, we took logarithms of the generated edge weights. The results are listed in Fig. 3. In addition to results on all tests, we extracted pairs of instances with higher than median unexpected similarity and tried to use only those to predict results. As perhaps the only unexpected result, this did not seem to improve the specificity. It seems that the differences between various years vary too much (which corresponds to our ability to predict market behaviour, which is, in general, poor).

	sum of all tests (858)	% of opt	sum of all tests with $U \geq \tilde{U}$ (462)	% of opt
Average	65062.20	188.70%	34826.08	187.28%
First Intersection	63682.42	184.70%	34702.42	186.62%
Best Rho	63116.60	183.06%	34702.42	186.62%
Optimum	34478.63	100.00%	18595.24	100.00%

Fig. 3. Stock Market Data (Logarithmised)

---

[2] In other words, finding out when to buy and when to sell in order to maximize profit, if we are only allowed to do each operation once.

Random Graphs. These experiments are mainly done for control purposes. If the data is truly random, we do not expect any algorithm to hold a significant edge, and indeed, the results reflect this. Note that while no algorithm works well here, we are able to realize that this will be so due to low unexpected similarity between instances. Figs. 4 and 5 list the results.

	sum of all tests	% of opt	sum of all tests with $U \geq \tilde{U}$ (260)	% of opt
Average	460763	132.74%	232340	132.91%
First Intersection	459802	132.47%	232688	133.11%
Best Rho	458033	131.96%	232546	133.03%
Optimum	347112	100.00%	174810	100.00%

Fig. 4. Random Graphs of 15 Vertices, Edge Weight Range [0-255]

	sum of all tests (512)	% of opt	sum of all tests with $U \geq \tilde{U}$ (261)	% of opt
Average	1616070	122.16%	820594	121.99%
First Intersection	1612010	121.86%	821715	122.15%
Best Rho	1602646	121.15%	820574	121.99%
Optimum	1322892	100.00%	672683	100.00%

Fig. 5. Random Graphs of 50 Vertices, Edge Weight Range [0-255]

Similar Random Graphs. With these experiments we wanted to verify our expectation that results improve with increasing similarity of graphs, e.g. the larger the expected value of a random cut gets compared to the expected value of a small cut, the better are our results, see Figs. 6, 7, and 8. For fixed small cut cost we get even better results, see Figs. 9, 10, and 11.

	sum of all tests (512)	% of opt	sum of all tests with $U \geq \tilde{U}$ (259)	% of opt
Average	142105	114.43%	65818	108.95%
First Intersection	139536	112.36%	64596	106.93%
Best Rho	139331	112.20%	64596	106.93%
Optimum	124182	100.00%	60410	100.00%

Fig. 6. Similar Graphs with Small Range [0,31] and Big Range [0,255]

	sum of all tests (512)	% of opt	sum of all tests with $U \geq \tilde{U}$ (286)	% of opt
Average	132521	116.03%	72361	113.60%
First Intersection	131285	114.95%	70983	111.44%
Best Rho	128573	112.57%	70983	111.44%
Optimum	114213	100.00%	63697	100.00%

Fig. 7. Similar Graphs with Small Range [0,31] and Big Range [0,127]

	sum of all tests (512)	% of opt	sum of all tests with $U \geq \tilde{U}$ (258)	% of opt
Average	126123	119.87%	61841	116.91%
First Intersection	126831	120.54%	61923	117.06%
Best Rho	123126	117.02%	61581	116.42%
Optimum	105220	100.00%	52897	100.00%

Fig. 8. Similar Graphs with Small Range [0,31] and Big Range [0,63]

	sum of all tests (512)	% of opt	sum of all tests with $U \geq \tilde{U}$ (401)	% of opt
Average	79607	110.85%	57501	104.94%
First Intersection	78311	109.04%	57603	105.13%
Best Rho	77953	108.54%	57573	105.08%
Optimum	71818	100.00%	54792	100.00%

Fig. 9. Similar Graphs with Small Cut Weight 240 and Random Weight Range [0,255]

	sum of all tests (512)	% of opt	sum of all tests with $U \geq \tilde{U}$ (402)	% of opt
Average	155440	111.97%	117571	106.03%
First Intersection	155676	112.14%	117530	106.00%
Best Rho	153880	110.84%	117530	106.00%
Optimum	138828	100.00%	110881	100.00%

Fig. 10. Similar Graphs with Small Cut Weight 500 and Random Weight Range [0,255]

Computation Time. Intuitively, the smaller and the more similar the graph instances are, the faster we should find ρ^*. This is the case because the size of the intersection of the two approximation sets is non-empty even for small values of ρ, and we can stop the search sooner. Our experiments confirmed this. However, even the test cases with the largest graphs and worst similarity values,

	sum of all tests (512)	% of opt	sum of all tests with $U \geq \tilde{U}$ (374)	% of opt
Average	292254	107.90%	217775	104.36%
First Intersection	291094	107.47%	217787	104.36%
Best Rho	288942	106.68%	217685	104.32%
Optimum	270853	100.00%	208679	100.00%

Fig. 11. Similar Graphs with Small Cut Weight 1000 and Random Weight Range [0,255]

i.e., random graphs with fifty vertices, completed in less than a minute on a single core of a usual processor.

5 Conclusion

We showed how to apply approximation set optimization to the problem of looking for a minimum cut in a graph by adapting a known minimum cut algorithm and estimating the expected intersection of two sets of small cuts.

In general, the experimental results reaffirm our expectation that the algorithm is better at generalizing than other simple heuristic algorithms. In addition to this, the unexpected similarity gives us additional information about the usefulness of our result. In some applications this can be a significant benefit. Having information about the quality of the calculated solution may be very important, in particular when the calculated solution is far from optimal.

By the choice of the optimal parameter ρ, our approach selects a set of minimum cuts which are expected to have low weight in the following graph instances. This can be of significant help as it divides the solution space into sets of relevant and irrelevant cuts, for instance, in a network robustness scenario, it separates the cuts that are likely to be critical from those that are not.

References

1. Buhmann, J.M., Mihalák, M., Šrámek, R., Widmayer, P.: Robust optimization in the presence of uncertainty. In: Proceedings of the 4th Conference on Innovations in Theoretical Computer Science, ITCS 2013, pp. 505–514. ACM, New York (2013)
2. Botafogo, R.A.: Cluster Analysis for Hypertext Systems. In: Proceedings of the 16th Annual International ACM SIGIR Conference on Research and Development in Information Retrieval, pp. 116–125. ACM (1993)
3. Boykov, Y., Kolmogorov, V.: An experimental comparison of min-cut/max- flow algorithms for energy minimization in vision. IEEE Transactions on Pattern Analysis and Machine Intelligence 26(9), 1124–1137 (2004)
4. Ramanathan, A., Colbourn, C.J.: Counting almost minimum cutsets with reliability applications. Mathematical Programming 39(3), 253–261 (1987)
5. Schneider, J.J., Kirkpatrick, S.: Stochastic Optimization. Springer (2007)

6. Kall, P., Mayer, J.: Stochastic Linear Programming: Models, Theory, and Computation. Springer (2005)
7. Ben-Tal, A., El Ghaoui, L., Nemirovski, A.: Robust Optimization. Princeton Series in Applied Mathematics. Princeton University Press (October 2009)
8. Bilu, Y., Linial, N.: Are stable instances easy? In: Proceedings of the First Symposium on Innovations in Computer Science (ICS), pp. 332–341 (2010)
9. Bilò, D., Böckenhauer, H.-J., Hromkovič, J., Královič, R., Mömke, T., Widmayer, P., Zych, A.: Reoptimization of steiner trees. In: Gudmundsson, J. (ed.) SWAT 2008. LNCS, vol. 5124, pp. 258–269. Springer, Heidelberg (2008)
10. Mihalák, M., Schöngens, M., Šrámek, R., Widmayer, P.: On the complexity of the metric TSP under stability considerations. In: Černá, I., Gyimóthy, T., Hromkovič, J., Jefferey, K., Královič, R., Vukolić, M., Wolf, S. (eds.) SOFSEM 2011. LNCS, vol. 6543, pp. 382–393. Springer, Heidelberg (2011)
11. Mihalák, M., Šrámek, R.: Counting approximately-shortest paths in directed acyclic graphs (2013), http://arxiv.org/abs/1304.6707
12. Dinits, E.A., Karzanov, A.V., Lomonosov, V.: On the Structure of a Family of Minimal Weighted Cuts in a Graph (1976)
13. Karger, D.R.: Global min-cuts in RNC, and other ramifications of a simple min-out algorithm. In: Proceedings of the Fourth Annual ACM-SIAM Symposium on Discrete Algorithms, pp. 21–30 (1993)
14. Scott Provan, J., Ball, M.O.: The complexity of counting cuts and of computing the probability that a graph is connected. SIAM Journal on Computing 12(4), 777–788 (1983)
15. Nagamochi, H., Nishimura, K., Ibaraki, T.: Computing all small cuts in an undirected network. SIAM Journal on Discrete Mathematics 10(3), 469–481 (1997)
16. Karger, D.R., Stein, C.: A new approach to the minimum cut problem. Journal of the ACM 43(4), 601–640 (1996)
17. Geissmann, B.: Approximation set optimization for minimum cut (August 2012), http://www.100acrewood.org/~rasto/publications/ ThesisBarbaraGeissmann.pdf
18. Market rates online (August 2012), http://www.marketratesonline.com

A Hybrid Genetic Algorithm for Solving the Unsplittable Multicommodity Flow Problem: The Maritime Surveillance Case

Hela Masri[1], Saoussen Krichen[2], and Adel Guitouni[3]

[1] LARODEC Laboratory, Institut Supérieur de Gestion de Tunis,
2000 Bardo, Tunisia
masri_hela@yahoo.fr
[2] LARODEC Laboratory, Faculty of Law, Economics and Management,
University of Jendouba, Avenue de l'UMA, 8189 Jendouba, Tunisia
saoussen.krichen@isg.rnu.tn
[3] Peter B. Gustavson School of Business, University of Victoria, Victoria ,Canada
adelg@uvic.ca

Abstract. Large volume surveillance missions are characterized by the employment and collaboration of several agents processing diverse information sources' inputs in order to ensure a surveillance task. Given the time dependant relevance of the shared information, an efficient global routing policy needs to be set up to optimize information exchange in the backbone of the surveillance network. We propose to model this problem as a single path multicommodity flow problem, where several commodities are to be shared in a capacitated network. The considered objective function is to minimize the overall network congestion. As the problem is NP-Hard, a hybrid genetic approach is proposed as a solution approach. A greedy search procedure based on the nearest neighbor method is transplanted into the genetic algorithm. The empirical validation is done using a simulation environment called Inform Lab. A comparison to a state-of-the-art ant colony system approach is performed based on a real case of maritime surveillance application and some randomly generated instances. The analysis of the results obtained in the two sets was supported by statistical nonparametric Wilcoxon signed-rank tests. The experimental results show that the hybrid genetic algorithm performs consistently well for large sized problems.

1 Introduction

In a surveillance system, information processing is not necessarily performed by a centralized unit. With the advent of social networking, distributed computing and smart sensors, surveillance task has become a network distributed process. A surveillance mission is characterized by the collaboration of several nodes processing diverse information sources' inputs in order to achieve a global goal. These interactions rely on a communication network supporting the information sharing among the dispersed entities. The execution of any surveillance mission requires

Q. Gu, P. Hell, and B. Yang (Eds.): AAIM 2014, LNCS 8546, pp. 137–148, 2014.

a web of heterogeneous communication networks to exchange information and co-ordinate actions. This network is generally composed by mobile and fixed surveillance assets. The stationary nodes represent the network backbone. Given that such network is private and configurable, a centralized global routing algorithm for the backbone can be designed prior to the mission execution to ensure meeting the global goal. The considered routing problem consists of sending various messages from a set of sources to different destinations. Each node in the network can be an information producer (source) or/and an information consumer (destination) or simply a neutral relay node. An arc is characterized by a transmission delay. We assume that each source has a preforcasted bandwidth demand, that represents the transmission rate of this node. Different routing algorithms were proposed in the literature assuming a centralized control of route selection. In the context of multiple pairs of source-destination have to be managed, the problem is defined as a multicommodity flow problem (MCFP) [1] which was largely applied to transportation problems. In telecommunication, a commodity represents a certain demand of a telecommunication traffic between two nodes.

In this paper, the routing objective is to ensure having a load balance over the network, by minimizing the most congested arc traffic value, with the constraint that each communication request can only be communicated on a single path between its source and destination nodes. Although the requirement that the data flow of each request is communicated using a single path may reduce the utilization of network resources, this assumption is prevalent in a number of application contexts especially for real-time applications [10]. For instance a video streaming in a surveillance context requires keeping the traffic intact, that is, without demultiplexing at the source, independent switching at intermediate nodes, and multiplexing at the destination. The solution of this problem consists of defining how to exchange messages from pairs of producers-consumers nodes by generating a single transmission path, such that it minimizes the traffic in the most congested arc.

This routing problem can be modeled as a single path multicommodity flow problem (SMCFP) [3]. The problem was recently studied by Li et al. [10], in the context of an optical switching network, where minimizing network congestion is an important concern in traffic grooming over wavelength division multiplexing (WDM) for a given logical topology. As a solution approach, and given the NP-hardness of the SMCFP, an ant colony system metaheuristic was designed [10]. In this paper, we propose a more effective metaheuristic based in hybrid genetic algorithm (HGA). The k-shortest paths for every source-destination pair is used to initialize the population. To avoid the premature converge of the GA, a greedy local search method is transplanted to push the search process towards unexplored areas. The local search procedure is based in the nearest neighbor search method, integrated during the mutation phase. A series of experiments is conducted based on real application of maritime surveillance as well as some randomly generated large instances. The experimental results prove the efficiency of the HGA.

This paper is organized as follows. Section 2 provides an the literature overview of related routing algorithms. Section 3 presents the problem description and its mathematical programming formulation. Section 4 describes the proposed solution approach. Section 5 provides some experimental results illustrating the efficiency of the proposed method.

2   Literature Review

An efficient routing algorithm should find an optimum path for packet transmission so as to satisfy some quality of services (transmission delay, bandwidth consumption, packet loss) [3,10,11]. Different routing protocols were designed in the literature depending on network architecture and the application context. On one end of the spectrum are the widely used network protocols designed based on engineering heuristic algorithms. As noted by Chiang et al. [6] these algorithms lack the theoretical foundation to analyze how well the network performs globally (e.g., whether the network resources are optimally shared). On the other end of the spectrum, recent progress has put many routing protocols on a mathematical foundation as well using network flow models. In these optimization frameworks, traffic demand is usually assumed to be known a priori. In the standard network flow problem the object is to send a flow through a network from some set of source nodes to a set of destination nodes, in some optimal fashion. For a solution to be feasible it must comply with the mass balance constraints for each node, i.e. the flow into a node must equal the flow out of the node, and the flow on each arc must not exceed the arc's capacity. If multiple pairs of source-destination nodes are managed, the problem is modeled as a multicommodity flow problem (MCFP) [1]. In the MCFP problem several entities, or commodities, share the network. Each commodity has its own mass balance constraints, but the arc capacity constraints are shared.

Different objective functions were considered in the literature, such as, the cost, average delay, maximum congestion, reliability and maximum flow. Assuming that the demand can be split along several paths, the complexity of the MFCP is polynomial. The reader can refer to Ahuja et al. [1], for a comprehensive survey of linear MCFPs and the different solution approaches.

However for some applications, an upper bound on the number of paths used by each commodity needs to be set. This NP-hard problem is called the multicommodity k-splittable flow problem (MCkFP). MCkFP was introduced by Baier et al. [5], they presented approximation algorithms for both single and multicommodity k-splittable flow problems, while considering the maximum flow problem with the maximum budget-constrained. Truffot et al. [15] used branch-and-price to solve the maximum MCkFP.

When $k = 1$, i.e all flow for each commodity must be sent via just one path, such problem is denoted the SMCFP or the unsplittable MCFP, first introduced by Kleinberg [8] and proven to be NP-hard. This assumption of having a single path is necessary in real time applications. It may be also applicable to circuit-oriented technologies such as optical networks employing WDM. In such network, to be able to send data from one access node to another, one needs to establish a single route, also called a light path, between the two nodes and to allocate a free wavelength on all of the links on the path. Bandyopadhyay

[2] studied the SMCFP to model the non-bifurcated traffic grooming problem, having the objective of minimizing the cost of the network by minimizing the number of lightpaths and maximizing the throughput of the network. Decomposition method is considered to computationally solve this problem on large networks. Alternatively, Barnhart et al. [3] proposed an exact method using branch-and-price-and-cut algorithm. Such exact algorithms are feasible only for small instances. More recently, a metaheuristic approach based on ant colony system method was developed [10], while studying two variants of the problem. The first version minimizes the network congestion, in order to solve the problem traffic grooming over WDM. The second version considers the general case of the minimum cost SMCFP.

3   Problem Description

In this paper, we address the optimization of information routing in surveillance network. This problem is about exchanging various messages from a set of sources to different destinations. Each node in the network can be an information provider (source) or/and a destination requiring an information or simply a relay node. These nodes are connected across a web of heterogeneous links. An arc is characterized by a limited capacity c, we assume throughout the paper that all the arcs have the same capacity. Given these statements, the network can be modeled as a directed graph (N, A) where $N = \{v_1, .., v_{|N|}\}$ is the set of nodes and $A = \{e_1, .., e_{|A|}\}$ is the set of arcs. Each arc from a node i to node j is represented by (i, j). The solution of this problem consists of defining how to exchange messages from pairs of producers-consumers nodes by generating a single transmission path, such that it minimizes the traffic in the most congested arc. The SMCFP can be described using a mathematical model [10] as follows:

Notation:

N	the set of nodes $\{v_1, .., v_{|N|}\}$
A	the set of arcs $\{e_1, .., e_{|A|}\}$
K	the set of commodoties
s_k	the source node of commodity k
d_k	the destination node of commodity k
$size_k$	the size of the supply of commodity k
λ_{max}	network congestion that equals the maximum traffic load on network's arcs
x_k	the flow vector of commodity k
x_{ij}^k	a binary decision variable $\begin{cases} 1 \text{ if the entire quantity of} \\ \quad \text{commodity } k \text{ uses arc } (i, j) \\ 0 \text{ otherwise} \end{cases}$

$$min \ \lambda_{max} \tag{1}$$

$s.t$

$$\sum_{k \in K} size_k \ x_{ij}^k \leq \lambda_{max} \qquad (i,j) \in A \tag{2}$$

$$\sum_{j:(i,j) \in A} x_{ij}^k - \sum_{j:(j,i) \in A} x_{ji}^k = \begin{cases} 1 & \text{if } i = s_k \\ -1 & \text{if } i = d_k \\ 0 & \text{otherwise} \end{cases} \quad i \in N \ \ k \in K \tag{3}$$

$$x_{ij}^k \in \{0,1\} \qquad (i,j) \in A \ \ k \in K \tag{4}$$

The objective of SMCFP is to minimize network congestion value λ_{max}. Constraints (2) define the congestion value in the network. Constraints (3) and (4) ensure that each commodity is sent along a single path linking its source s_k to its destination d_k.

4   Solution Approach: A Hybrid Genetic Algorithm (HGA)

The combinatorial structure of the proposed model makes generation of the solution difficult and time consuming. In addition, the problem complexity is NP-hard due to the single path constraint. These reasons justify the computational impracticality of exact algorithms for solving this problem. Therefore, we propose to solve it using a metaheuristic method, based on genetic algorithm (GA). Owing to the distinctive features such as domain independence, robustness and parallel nature, GAs have been proved to be an effective approach for solving optimization problems. The successful application of GAs for solving similar routing problems [14] motivated the choice of the metaheuristic. Starting with an initial population, constructed by a greedy procedure, the individuals evolve to new solutions that approximate better the global optimum. The basic outline of the algorithm is described, and each procedure is briefly explained.

4.1   Chromosome Representation

Better efficiency of GA-based search could be achieved by well defining the chromosome representation and its related operators so as to generate feasible solutions and avoid repair mechanism. Figure 1 depicts the structure of a chromosome. A chromosome is coded as a vector of K substrings where K is the number of commodities. Each substring is composed of the list of indexes i of the used arcs a_i composing the generated path.

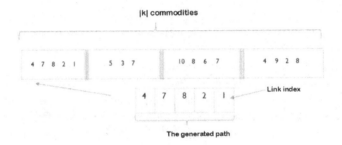

Fig. 1. Chromosome representation

Hybrid genetic algorithm

Initialization

$i = 0$

Set the parameters values: Population size L, propability crossover P_c and the stopping criteria

Generate first population P_0 of size L

Iterative process

  repeat

    $i = i + 1$

    Crossover: generate the set R_i of offspring of size L

    Mutation: apply the nearest neighbour method

    $P_i = P_{i-1} \cup R_i$

    Evaluate the generated offsprings

    Select best L solution from P_i

  until (Stopping criteria is met)

4.2 Generating the First Population

A hybrid approach is used for the initialization of the population. For every source-destination pair the k-shortest paths connecting them are generated using Yen's algorithm [9]. Each gene in a chromosome represents one of the k-shortest paths selected randomly. Thus a single chromosome contains a set of plausible paths for all the source-destination pairs.

4.3 Crossover

The proposed algorithm applies the random key method for the crossover; the offsprings are produced from two parent solutions following these steps:

1) two chromosomes ch_1 and ch_2 are chosen from the current population P_i

2) a probability p is randomly generated

3) if $p < P_c$ then the substring for a new child is chosen from the first chromosome $ch1$ otherwise it is taken from the second chromosome ch_2

4) Repeat 2 and 3 until reaching the last substring.

4.4   Mutation: Nearest Neighbor Method

A random uniform mutation with a probability of $P_m = 1/L$ is used, where L is the population size. After choosing the chromosome s to be modified, a greedy algorithm based on nearest neighbor method is used. One of the paths in the current solution s will be reconstructed using a greedy procedure, while keeping the same paths for the other commodities. The main idea is to search for a new path that has a minimum congestion while considering the changes made in the available capacity in the network. We propose to use a probabilistic path construction strategy so that we start from the source node and move until reaching its corresponding destination. The choice of a neighbor depends on a local information θ_{ij} of an arc (i, j). This value is expressed in terms of the traffic load l_{ij} of the link;

$$\theta_{ij} = \frac{1}{l_{ij}} \tag{5}$$

While constructing a path, and to move from a current node i to another, a probabilistic selection rule is applied. The next node is chosen using the roulette wheel selection procedure of evolutionary computation. The probability distribution is:

$$p_{ij} = \frac{\theta_{ij}}{\sum_{h \in N(i)} \theta_{ih}} \quad \forall j \in N(i) \tag{6}$$

Where $N(i)$ defines the neighborhood of node i.

4.5   Selection Procedure

After generating the offsprings, the obtained solutions are evaluated according their fitness function (λ_{max}). L solutions are selected to be the population of the next iteration using the roulette wheel selection. In this method the chromosomes are selected according to their fitness. Better chromosomes, are having more chances to be selected as parents. It is the most common method for implementing fitness proportionate selection.

5   Experimental Study

We performed many computational experiments for different problem sizes in order to validate the performance of the HGA. The algorithm is coded in Java. A Core2 duo 2GHZ laptop has been used for these experiments. As a simulation environment, we used Inform-Lab simulation testbed [13,7]. This environment enables to execute different algorithms for distributed information fusion and dynamic resource management. An empirical comparison with the Ant Colony System (ACS) method [10] is performed.

5.1   Parameter Tuning

The parameter setting of a metaheuristic may impact the solution quality. We propose to use an automatic procedure, F-Race to determine the best configuration of some main GA parameters. The F-Race approach is proposed by Birattari [4] to automatically configure a metaheuristic. Among a set of candidate configurations, the final configuration returned by the F-Race procedure is detailed in table 1. The same stopping criteria is used for both HGA and ACS, expressed in terms of a maximum number of iterations, number of iterations without improvement and maximum CPU time.

Table 1. Parameter setting

Parameter	Value
Population size	200
P_c	0.6
Nb of iterations	1000
Nb of iterations without improvement	100
Max CPU time(s)	500

5.2   Experimental Results

The experimental design is in two-fold:

- We test the efficiency of the HGA in a real case of routing in a maritime surveillance problem. As some of these instances are fairly of small sizes, CPLEX 12.2 is also implemented to solve them optimally using the (1)-(4) formulation.
- We propose to characterize the proposed algorithm by providing more empirical results showing its performance (quality of solution and CPU time versus problem size) in a larger set of randomly generated instances.

The analysis of the results obtained in the two sets was supported by statistical nonparametric Wilcoxon signed-rank tests, with a 95% confidence level.

1. **Real case of maritime surveillance problem:**
   Inform lab is a testbed supporting the development of two groups of algorithms, which are particularly useful for wide-area surveillance applications: distributed dynamic information fusion and distributed dynamic resource management. It contains different surveillance vignettes with different scenarios (detecting that a boat is sinking, or some entity is smuggling). In a vignette several cooperative platforms are deployed. These agents are cooperating in order to fulfill a mission. The large volume surveillance problem is characterized by the employment of mobile and fixed surveillance assets to a large geographic area in order to identify, assess and track the maximum number of moving, stopped or drifting objects. Platforms include satellites, airborne platforms (e.g., helicopters, marine patrol aircraft and UAVs),

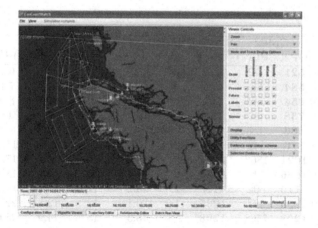

Fig. 2. Inform Lab testbed [13]

seaborne platforms (e.g., coastguard, military and police vessels), stationary and land platforms (e.g., radar stations, land vehicles). Coastal surveillance are good examples of large volume surveillance. During the mission, a set messages has to be shared between these platforms. The proposed HGA is used in order to optimize the messages routing in the backbone of the surveillance network. The real case study supports the mission "is smuggling" in the coastal area of Vancouver Island, contains a network of 50 nodes and 421 arcs. The nodes are fixed assets representing the potential sites in the pacific coast of Vancouver. This networks is heterogeneous composed by both optical and cellular mediums. Considering this network, different scenarios where generated to vary the number of commodities. Traffic estimation is performed prior to the execution of the algorithm to get the sizes of the commodities. A screenshot of the testbed running is depicted in figure 2. The comparison was supported by statistical nonparametric Wilcoxon. We report in table 2 the results of Cplex , HGA and ACS (the best generated solution value *best*, the standard deviation value *Std* and the CPU time) over 30 independent runs of each problem instance. The results of the Wilcoxon test are given in the last column *wilc*, where +, -, and ≈ denote that the HGA performed significantly better than, significantly worse than, or statistically equivalent to the ACS, respectively.

2. **Experiments with randomly generated instances:**
   A set of 15 random large instances is generated. We kept the same empirical design defined by Li et al. [10], in order to have a meaningful comparison. The instance features are summarized in Table 3. The number of nodes varies between 40 to 60, and the number of arcs ranges between 209 to 386. We assume that a communication request exists between each pair of nodes. The reader can refer to [10] for more details about the instances setting. Each Instance is solved with 30 independent runs. We report in table 3 the

Table 2. Experimental results of the surveillance case study

Problem	$|k|$	Cplex		HGA			ACS			Wilc
		OPT	CPU(s)	Best	Std	CPU(s)	Best	Std	CPU(s)	
1	100	**521**	342	**521**	0	9.2	**521**	0	6.4	≈
2	200	**584**	379	**584**	0	10.5	**584**	0	7.1	≈
3	300	**654**	532	**654**	0	12.7	**654**	0	8.9	≈
4	400	**738**	487	**738**	0	13.7	**738**	5.7	11	≈
5	500	**806**	327	811	8.6	14.2	**806**	4.9	13.5	−
6	600	**876**	549	**876**	6.7	16.7	886	5.6	15.3	≈
7	700	**1021**	1863	**1021**	9.4	17.6	1034	10.2	17.8	≈
8	800	**1086**	742	**1086**	10.7	18.4	1099	12.7	18.9	+
9	900	**1119**	802	1121	12.3	15.2	1187	11.5	16.9	+
10	1000	**1235**	637	**1235**	11.7	20.1	1256	12.8	22.3	+
11	1100	**1298**	1568	**1298**	2.8	16.3	1346	14.9	28.6	+
12	1200	**1421**	1084	1426	13.7	21.7	1435	11.6	35.5	+
13	1300	**1527**	893	1537	10	22.7	1566	15.2	43.2	+
14	1400	**1647**	2155	1662	15.4	24.9	1702	20.9	50.4	+
15	1500	**1754**	1428	**1754**	18.2	27.5	1798	17.5	59.6	≈
16	1600	**1884**	1964	1896	17.5	25.8	1996	18.6	64.4	+
17	1700	-	-	**2027**	20.4	29.6	2081	19.4	80.1	+
18	1800	**2178**	2866	2198	23.5	30.1	2254	21.4	112.8	+
19	1900	-	-	**2284**	25.7	31.7	2311	25.9	129.8	+
20	2000	-	-	**2465**	17.8	30.9	2502	24.3	134.8	+
Average	-	1197	1095.1	**1359.7**	**11.5**	**20.47**	**1386.8**	**35.33**	**42.48**	

average of the following measures: the best generated solution value *best*, the standard deviation value *Std*, the CPU time of HGA and ACS and the Wilcoxon test result *Wilc*.

Based on the results of tables 2 and 3, one can notice that:

- The computational difficulty increases significantly with network size and the number of commodities to be routed. From table 2, we can notice that it becomes gradually impractical to solve large instances by Cplex. For instances 17, 18 and 20 no optimal solution is returned when CPLEX terminates. While HGA succeeded in solving these three instances in a an average of 30s.
- Over the two testbeds, The HGA outperformed the ACS in 27 instances (60% of the instances) in terms of solution quality (i.e. *best* values).
- For large instances, the HGA was able to converge to the best solution more rapidly than the ACS method. The computational requirements of the ACS seem to increase rapidly with the problem size.
- Over the different runs, and given the *Std* values, the HGA gives generally a more accurate approximation of the global optimum.
- Based on the Wilcoxon signed-rank tests, HGA performed significantly better than the ACS in 21 instances. While in 12 instances the results from the statistical test are inconclusive. The ACS outperformed the HGA in only two instances.

Table 3. Experimental results of randomly generated instances

Problem	Problem description			HGA			ACS			Wilc
	$\|N\|$	$\|K\|$	$\|A\|$	Best	Std	CPU(s)	Best	Std	CPU(s)	
1	40	1560	209	**3178**	0	35.36	**3178**	0	52.3	$\approx$
2	40	1560	211	**3021**	0	23.7	**3021**	0	45.7	$\approx$
3	40	1560	214	3011	14.2	25.3	**2988**	0	44.3	$\approx$
4	50	2450	321	**2725**	5.8	67.5	2751	21.5	147.8	$\approx$
5	50	2450	322	**2389**	20.5	62.4	2404	28.62	123.7	$\approx$
6	50	2450	324	**2397**	25.7	68.4	2420	0	148.9	+
7	50	2450	328	**2411**	27.6	97.4	2458	48.24	289.3	+
8	55	2970	345	**3452**	38.4	142.3	3679	69.2	350.4	+
9	55	2970	360	**3178**	37.6	136.4	3258	52.7	287.6	+
10	55	2970	350	**3293**	41.2	174.5	3465	46.8	500	+
11	55	2970	357	3326	29.5	168.4	**3316**	28.5	435.5	−
12	60	3540	394	**4238**	47.2	186	4526	58.4	500	+
13	60	3540	396	**4210**	84.2	174	4481	70.5	500	+
14	60	3540	391	**4379**	62.3	194.5	4638	75.3	500	+
15	60	3540	386	**4922**	61.4	245.3	4966	50.2	500	+
Average	-	-	-	**3341.3**	**33.04**	**120.09**	**3435.13**	**39.33**	**295.03**	

6   Conclusion

In this paper, we proposed a new global routing algorithm for optimizing information exchange in large volume surveillance application. The surveillance problem consist in deploying a set of surveillance agents collaborating in order to perform a set of missions in a large geographic operation region. Given the traffic requests in the network backbone and the limited capacity of the links, solving the routing problem consists in finding the best paths between the end-nodes so that we minimize the overall congestion. We proposed to model this problem as a SMCFP. Due to its combinatorial nature, the SMCFP is NP-hard. Therefore, we propose a HGA to solve it. An experimental study is conducted and a comparison with an existing ACS approach is performed, based on the Wilcoxon signed-rank test. The representative results and the comparison show the effectiveness of our algorithm.

References

1. Ahuja, R., Magnanti, T., Orlin, J.: Network Flows: Theory, Algorithms, and Applications. Prentice Hall Inc., Upper Saddle (1993)
2. Bandyopadhyay, S.: Dissemination of Information in Optical Networks: From Technology to Algorithms. Springer, Berlin (2007)
3. Barnhart, C., Hane, C.A., Vance, P.H.: Using branch-and-price-and-cut to solve origin-destination integer multicommodity flow problems. Operations Research 48(2), 318–326 (2000)
4. Birattari, M.: The problem of tuning metaheuristics as seen from a machine learning perspective. PhD thesis, Universite Libre De Bruxelles (2005)

5. Baier, G., Köhler, E., Skutella, M.: On the k-splittable flow problem. In: Möhring, R.H., Raman, R. (eds.) ESA 2002. LNCS, vol. 2461, pp. 101–113. Springer, Heidelberg (2002)
6. Chiang, M., Low, S.H., Calderbank, A.R., Doyle, J.C.: Layering as optimization decomposition: A mathematical theory of network architectures. Proceedings of the IEEE 95(1), 255–312 (2007)
7. Dridi, O., Krichen, S., Guitouni, A.: A multi-objective optimization approach for resource assignment and task scheduling problem: Application to maritime domain awareness. In: IEEE Congress on Evolutionary Computation (CEC), pp. 1–8 (2012)
8. Kleinberg, J.M.: Approximation algorithms for disjoint paths problems, Ph.D. Thesis. MIT, Cambridge, MA (1996)
9. Jiménez, V.M., Marzal, A.: Computing the K Shortest Paths: A New Algorithm and an Experimental Comparison. In: Vitter, J.S., Zaroliagis, C.D. (eds.) WAE 1999. LNCS, vol. 1668, pp. 15–29. Springer, Heidelberg (1999)
10. Li, X.Y., Aneja, Y.P., Baki, F.: An ant colony optimization metaheuristic for single-path multicommodity network flow problems. Journal of the Operational Research Society 61(9), 1340–1355 (2010)
11. Bley, A.: Approximability of unsplittable shortest path routing problems. Networks 54(1), 23–46 (2009)
12. Holmberg, K., Yuan, D.: A Multicommodity Network-Flow Problem with Side Constraints on Paths Solved by Column Generation. INFORMS Journal on Computing 15(1), 42–57 (2003)
13. MacDonald, Dettwiler and Associates Ltd., Inform Lab Wiki
14. Masri, H., Guitouni, A., Krichen, S.: Towards Efficient Information Exchange in Fusion Networks. In: Rutkowski, L., Korytkowski, M., Scherer, R., Tadeusiewicz, R., Zadeh, L.A., Zurada, J.M. (eds.) ICAISC 2013, Part II. LNCS, vol. 7895, pp. 535–546. Springer, Heidelberg (2013)
15. Truffot, J., Duhamel, C., Mahey, P.: Using branch-and-price to solve multicommodity k-splittable flow problems. In: Proceedings of the International Network Optimization Conference (INOC) (2005)

Multiple Sink Location Problems in Dynamic Path Networks

Yuya Higashikawa[1], Mordecai J. Golin[2], and Naoki Katoh[1,*]

[1] Department of Architecture and Architectural Engineering,
Kyoto University, Japan
{as.higashikawa,naoki}@archi.kyoto-u.ac.jp
[2] Department of Computer Science and Engineering,
The Hong Kong University of Science and Technology, Hong Kong
golin@cs.ust.hk

Abstract. This paper considers the k-sink location problem in dynamic path networks. In our model, a dynamic path network consists of an undirected path with positive edge lengths, uniform edge capacity, and positive vertex supplies. Here, each vertex supply corresponds to a set of evacuees. Then, the problem requires to find the optimal location of k sinks in a given path so that each evacuee is sent to one of k sinks. Let x denote a k-sink location. Under the optimal evacuation for a given x, there exists a $(k-1)$-dimensional vector d, called $(k-1)$-divider, such that each component represents the boundary dividing all evacuees between adjacent two sinks into two groups, i.e., all supplies in one group evacuate to the left sink and all supplies in the other group evacuate to the right sink. Therefore, the goal is to find x and d which minimize the maximum cost or the total cost, which are denoted by the minimax problem and the minisum problem, respectively. We study the k-sink location problem in dynamic path networks with continuous model, and prove that the minimax problem can be solved in $O(kn \log n)$ time and the minisum problem can be solved in $O(kn^2)$ time, where n is the number of vertices in the given network.

Keywords: sink location, dynamic network, evacuation planning.

1 Introduction

The Tohoku-Pacific Ocean Earthquake happened in Japan on March 11, 2011, and many people failed to evacuate and lost their lives due to severe attack by tsunamis. From the viewpoint of disaster prevention from city planning and evacuation planning, it has now become extremely important to establish effective evacuation planning systems against large scale disasters. In particular, arrangements of tsunami evacuation buildings in large Japanese cities near the coast has become an urgent issue. To determine appropriate tsunami evacuation buildings, we need to consider where evacuation buildings are assigned and how

* Supported by JSPS Grant-in-Aid for Scientific Research(A)(25240004).

Q. Gu, P. Hell, and B. Yang (Eds.): AAIM 2014, LNCS 8546, pp. 149–161, 2014.

to partition a large area into small regions so that one evacuation building is designated in each region. This produces several theoretical issues to be considered. Among them, this paper focuses on the location problem of multiple evacuation buildings assuming that we fix the region such that all evacuees in the region are planned to evacuate to one of these buildings. In this paper, we consider the simplest case for which the region consists of a single road.

In order to represent the evacuation, we consider the *dynamic* setting in graph networks, which was first introduced by Ford et al. [3]. In a graph network under the dynamic setting, each vertex is given supply and each edge is given length and capacity which limits the rate of the flow into the edge per unit time. We call such networks under the dynamic setting *dynamic networks*. Dynamic networks can be considered in discrete and continuous models. In discrete model, each input value is given as an integer. Then each supply can be regarded as a set of evacuees, and edge capacity is defined as the maximum number of evacuees who can enter an edge per unit time. On the other hand, in continuous model, each input value is given as a real number. Then each supply can be regarded as fluid, and edge capacity is defined as the maximum amount of supply which can enter an edge per unit time. In either model, we assume that all supply at a vertex is sent to the same sink. *The k-sink location problem in dynamic networks* is defined as the problem which requires to find the optimal location of k sinks in a given network so that all supply of each vertex is sent to one of k sinks in the shortest time.

For the 1-sink location problem in dynamic networks, the following two criteria can be naturally considered: *maximum cost criterion* and *total cost criterion* (in static networks, these criteria correspond to the center problem and the median problem in facility location, respectively). If a sink location x is given in a dynamic network with discrete model, the cost of x for an evacuee is defined as the minimum time required to send him/her to x (by taking into account the congestion). Then two criteria are defined as the maximum of cost of x for all evacuees and the sum of cost of x for all evacuees, respectively. Now let us turn to continuous model. In continuous model, we define the *unit* as the infinitesimally small portion of supply, then the cost is defined on each unit. If a sink location x is given in a dynamic network with continuous model, the cost of x for a unit is defined as the minimum time required to send the unit to x. Also two criteria are defined as the maximum of cost of x for all units and the sum of cost of x for all units, respectively. Definitions for k-sink location problem are given later. Then, *the minimax* (resp. *minisum) k-sink location problem in dynamic networks* requires to find a k-sink location in a given dynamic network which minimizes the maximum (resp. total) cost. Mamada et al. [7] studied the minimax 1-sink location problem in dynamic tree networks with discrete model assuming that the sink must be located at a vertex, and proposed an $O(n \log^2 n)$ time algorithm. Higashikawa et al. [4] also studied the same problem as [7] assuming that edge capacity is uniform and the sink can be located at any point in the network, and proposed an $O(n \log n)$ time algorithm. On the other hand, to the authors' knowledge, no one has studied the minisum sink location problem in dynamic networks.

In this paper, we study the k-sink location problem in a dynamic path network with continuous model assuming that edge capacity is uniform and the sink can be located at any point in the network, and prove that the minimax problem can be solved in $O(kn \log n)$ time and the minisum problem can be solved in $O(kn^2)$ time. This paper is the first one which studies the minisum sink location problem in dynamic networks and also the minimax k-sink location problem in dynamic networks.

2   Minimax k-sink Location Problem

2.1   Definitions

Let $P = (V, E)$ be an undirected path where $V = \{v_0, v_1, \ldots, v_n\}$ and $E = \{e_1, e_2, \ldots, e_n\}$ such that v_{i-1} and v_i are endpoints of e_i for $1 \leq i \leq n$. Let $\mathcal{N} = (P, l, w, c, \tau)$ be a dynamic network with the underlying graph being a path P, l is a function that associates each edge e_i with positive length l_i, w is also a function that associates each vertex v_i with positive weight w_i representing the amount of supply at v_i, c is a positive constant representing the amount of supply which can enter an edge per unit time, and τ is also a constant representing the time required by flow for traversing the unit distance. We call such networks with path structures *dynamic path networks*. In the following, we use the notation P to denote the set of all points $p \in P$. Also, for a vertex $v_i \in P$ with $0 \leq i \leq n$, we abuse the notation v_i to denote the distance from v_0 to v_i, and for a point $p \in P$, we abuse the notation p to denote the distance from v_0 to p. Then, we can regard P as embedded on a real line such that $v_0 = 0$. For two points $p, q \in P$ with $p < q$, $[p, q]$ (resp. $[p, q)$, $(p, q]$ and (p, q)) denote the part of P which consists of all points $x \in P$ such that $p \leq x \leq q$ (resp. $p \leq x < q$, $p < x \leq q$ and $p < x < q$).

Suppose that k sinks are located at points $x_1, x_2, \ldots, x_k \in P$ such that $x_1 \leq x_2 \leq \ldots \leq x_k$, respectively. Note that each sink can be located at any point in P. In this paper, we assume that if we place a sink at a vertex, all supply of the vertex can finish the evacuation in no time. So, without loss of generality, we assume $k \leq n + 1$ (otherwise, at least one sink can be located at each vertex). Let $\boldsymbol{x} = (x_1, x_2, \ldots, x_k)$ which is a k-dimensional vector, called k-*sink location*. Let us consider the optimal evacuation for a given $\boldsymbol{x}$. In this paper, we assume that all units of a vertex are sent to the same sink. We call a directed path along which all units of a vertex are sent to a sink *evacuation path*. Then, any two evacuation paths never cross each other in an optimal evacuation (otherwise, we can realize the better or equivalent evacuation by exchanging the two destinations of crossing evacuation paths). Suppose that there exists only one vertex v_j in $[x_i, x_{i+1}]$ and all units of the vertex are sent to x_i, then x_{i+1} can be moved to v_{j+1} without increasing the cost of any unit. Therefore, if we optimally locate k sinks with $k \geq 2$, there exist at least two vertices in $[x_i, x_{i+1}]$ for any i with $1 \leq i \leq k-1$, i.e., there exist two vertices v_j and v_{j+1} with $0 \leq j \leq n-1$ in $[x_i, x_{i+1}]$ such that all supplies on $[x_i, v_j]$ are sent to x_i and all supplies on $[v_{j+1}, x_{i+1}]$ are sent to x_{i+1}. We call such a vertex v_j *dividing vertex*. For an integer i with $1 \leq i \leq k-1$ with $k \geq 2$, let d_i be an index of the dividing vertex in $[x_i, x_{i+1})$. By the above discussion, $d_{i-1} + 1 \leq d_i$ holds for $1 \leq i \leq k$ where $d_0 = -1$ and $d_k = n$. Let $\boldsymbol{d} = (d_1, d_2, \ldots, d_{k-1})$ which is a $(k-1)$-dimensional vector, called $(k-1)$-divider. For a given $\boldsymbol{d}$, let $P_i(\boldsymbol{d}) = [v_{d_{i-1}+1}, v_{d_i}]$ for $1 \leq i \leq k$ where $d_0 = -1$ and $d_k = n$, then we need only consider $\boldsymbol{x}$ such that x_i is given on $P_i(\boldsymbol{d})$ for $1 \leq i \leq k$. For given $\boldsymbol{x}$ and $\boldsymbol{d}$, and also for an integer i with $1 \leq i \leq k$, let $\Theta_i(\boldsymbol{x}, \boldsymbol{d})$ denote the minimum time required to send all supplies on $P_i(\boldsymbol{d})$ to x_i. Letting $\Theta(\boldsymbol{x}, \boldsymbol{d}) = \max\{\Theta_i(\boldsymbol{x}, \boldsymbol{d}) \mid 1 \leq i \leq k\}$, the minimax k-sink location problem is defined as follows:

$$Q_{\text{minimax}} : \text{minimize } \{\Theta(\boldsymbol{x}, \boldsymbol{d}) \mid \boldsymbol{x} \in P^k \text{ and } \boldsymbol{d} \in \{0, 1, \ldots, n\}^{k-1}\}. \tag{1}$$

In the following, for a l-dimensional vector $\boldsymbol{y} = (y_1, y_2, \ldots, y_l)$ and a value z, we use the notation $(\boldsymbol{y}, z)$ to denote a $(l+1)$-dimensional vector $(y_1, y_2, \ldots, y_l, z)$.

2.2   Recursive Formulation

We now consider a subproblem of the above mentioned problem: for some integers i, j and p with $0 \leq i < j \leq n$ and $1 \leq p \leq k$, the p-sink location problem in $[v_i, v_j]$. For $[v_i, v_j]$, let $\boldsymbol{x}^*(p, i, j)$ denote the optimal p-sink location and $\boldsymbol{d}^*(p, i, j)$ denote the optimal $(p-1)$-divider. Note that $\boldsymbol{x}^*(p, i, j)$ is a p-dimensional vector and $\boldsymbol{d}^*(p, i, j)$ is also a $(p-1)$-dimensional vector, so $\boldsymbol{d}^*(p, i, j)$ is not defined for $p = 1$. Also, let $\text{OPT}(p, i, j)$ denote the minimum time required to send all supplies on $[v_i, v_j]$ divided by $\boldsymbol{d}^*(p, i, j)$ to $\boldsymbol{x}^*(p, i, j)$. Note that if $p \geq j - i + 1$ holds, the optimal sink location is trivial, i.e., $\text{OPT}(p, i, j) = 0$.

Next, we show the recursive formula of $\text{OPT}(p, i, j)$. For integers i, j and p with $0 \leq i < j \leq n$ and $1 \leq p \leq k - 1$, let us consider the optimal $(p+1)$-sink location and p-divider for $[v_i, v_j]$, i.e., $\boldsymbol{x}^*(p+1, i, j)$ and $\boldsymbol{d}^*(p+1, i, j)$. Since any two evacuation paths never cross each other in an optimal evacuation, there exists an integer h with $i \leq h \leq j - 1$ such that all supplies on $[v_{h+1}, x_j]$ are sent to the rightmost sink and all supplies on $[x_i, v_h]$ are sent to the other k sinks. Thus, we have the following recursion:

$$\text{OPT}(p+1, i, j) = \min_{i \leq h \leq j-1} \max\{\text{OPT}(p, i, h), \text{OPT}(1, h+1, j)\}. \tag{2}$$

Here, let d be an integer which minimizes the maximum of $\text{OPT}(p, i, h)$ and $\text{OPT}(1, h+1, j)$ on $i \leq h \leq j - 1$:

$$d = \underset{i \leq h \leq j-1}{\text{argmin}} \max\{\text{OPT}(p, i, h), \text{OPT}(1, h+1, j)\}. \tag{3}$$

Then, $\boldsymbol{x}^*(p+1, i, j)$ and $\boldsymbol{d}^*(p+1, i, j)$ can be represented by using d as follows:

$$\boldsymbol{x}^*(p+1, i, j) = (\boldsymbol{x}^*(p, i, d), \boldsymbol{x}^*(1, d+1, j)), \tag{4}$$
$$\boldsymbol{d}^*(p+1, i, j) = (\boldsymbol{d}^*(p, i, d), d). \tag{5}$$

2.3 Properties

In this section, we show several key properties of our problem. Here, for integers p and i with $2 \le p \le k$ and $1 \le i \le n$, let $f_{p,i}(t)$ denote a function defined on $\{t \in \mathbb{Z} \mid 0 \le t \le i-1\}$:

$$f_{p,i}(t) = \max\{\mathsf{OPT}(p-1,0,t), \mathsf{OPT}(1,t+1,i)\}. \qquad (6)$$

Note that for fixed p and i, $\mathsf{OPT}(p-1,0,t)$ is monotonically increasing in t and $\mathsf{OPT}(1,t+1,i)$ is monotonically decreasing in t. Thus, we have the following claim.

Claim 1. *For any integers p and i with $2 \le p \le k$ and $1 \le i \le n$, function $f_{p,i}(t)$ is unimodal in t on $0 \le t \le i-1$.*

Let $d_{p,i}$ be an integer which minimizes $f_{p,i}(t)$ for $0 \le t \le i-1$:

$$d_{p,i} = \operatorname*{argmin}_{0 \le t \le i-1} f_{p,i}(t). \qquad (7)$$

By Claim 1, there uniquely exists $d_{p,i}$. Then, by (4) and (5), we have

$$\boldsymbol{x}^*(p,0,i) = (\boldsymbol{x}^*(p-1,0,d_{p,i}), \boldsymbol{x}^*(1,d_{p,i}+1,i)), \qquad (8)$$
$$\boldsymbol{d}^*(p,0,i) = (\boldsymbol{d}^*(p-1,0,d_{p,i}), d_{p,i}). \qquad (9)$$

We also have the following claim.

Claim 2. *For any integers p and i with $2 \le p \le k$ and $2 \le i \le n$, the following inequality holds:*

$$d_{p,i-1} \le d_{p,i}. \qquad (10)$$

Now, for fixed integers i and j with $0 \le i < j \le n$, let us consider how to compute $\boldsymbol{x}^*(1,i,j)$ and $\mathsf{OPT}(1,i,j)$. Suppose that a sink is located at a point x in $[v_i, v_j]$. Let $\Theta_{i,j}(x)$ denote the minimum time required to send all supplies on $[v_i, v_j]$ to x. Here, let $\Theta^L_{i,j}(x)$ (resp. $\Theta^R_{i,j}(x)$) denote the minimum time required to send all supplies on $[v_i, x]$ (resp. $(x, v_j]$) to x. Then, $\Theta_{i,j}(x)$ is the maximum of $\Theta^L_{i,j}(x)$ and $\Theta^R_{i,j}(x)$, i.e.,

$$\Theta_{i,j}(x) = \max\{\Theta^L_{i,j}(x), \Theta^R_{i,j}(x)\}. \qquad (11)$$

For discrete model, Kamiyama et al. [6] showed that $\Theta^L_{i,j}(x)$ and $\Theta^R_{i,j}(x)$ are expressed as follows:

$$\Theta^L_{i,j}(x) = \max_l \left\{ \tau(x - v_l) + \left\lceil \frac{\sum_{i \le h \le l} w_h}{c} \right\rceil - 1 \;\middle|\; v_l \in [v_i, x) \right\},$$
$$\Theta^R_{i,j}(x) = \max_l \left\{ \tau(v_l - x) + \left\lceil \frac{\sum_{l \le h \le j} w_h}{c} \right\rceil - 1 \;\middle|\; v_l \in (x, v_j] \right\}.$$

From these, we can immediately develop the formulae for continuous model as follows:

$$\Theta_{i,j}^L(x) = \max_l \left\{ \tau(x - v_l) + \frac{\sum_{i \leq h \leq l} w_h}{c} \;\middle|\; v_l \in [v_i, x) \right\},\tag{12}$$

$$\Theta_{i,j}^R(x) = \max_l \left\{ \tau(v_l - x) + \frac{\sum_{l \leq h \leq j} w_h}{c} \;\middle|\; v_l \in (x, v_j] \right\}.\tag{13}$$

Note that $\Theta_{i,j}^L(x)$ (resp. $\Theta_{i,j}^R(x)$) is a piecewise linear monotone increasing (resp. decreasing) function of x. Therefore, function $\Theta_{i,j}(x)$ is unimodal in x. By the properties shown in [2] and [5], we immediately have the following two claims.

Claim 3. *For any integers i and j with $0 \leq i < j \leq n$ and a point $x \in [v_i, v_j]$,*
(i) if $\Theta_{i,j}^L(x) \leq \Theta_{i,j}^R(x)$ holds, $x^*(1, i, j) \geq x$ holds, and
(ii) if $\Theta_{i,j}^L(x) \geq \Theta_{i,j}^R(x)$ holds, $x^*(1, i, j) \leq x$ holds.

Claim 4. *For given integers i and j with $0 \leq i < j \leq n$, suppose that for the interval $[v_l, v_{l+1}]$ with $i \leq l \leq j - 1$, $\Theta_{i,j}^L(v_l) \leq \Theta_{i,j}^R(v_l)$ and $\Theta_{i,j}^L(v_{l+1}) \geq \Theta_{i,j}^R(v_{l+1})$ hold, and let α^* denote the solution to an equation for α: $\Theta_{i,j}^R(v_l) - \alpha\tau(v_{l+1} - v_l) = \Theta_{i,j}^L(v_{l+1}) - (1 - \alpha)\tau(v_{l+1} - v_l)$. Then,*
(i) if $0 \leq \alpha^* \leq 1$ holds, $x^*(1, i, j)$ is a point dividing the interval $[v_l, v_{l+1}]$ with the ratio of α^* to $1 - \alpha^*$ and $\mathsf{OPT}(1, i, j) = \Theta_{i,j}^R(v_l) - \alpha^*\tau(v_{l+1} - v_l)$ holds,
(ii) if $\alpha^* < 0$ holds, $x^*(1, i, j) = v_l$ and $\mathsf{OPT}(1, i, j) = \Theta_{i,j}^R(v_l)$ hold, and
(iii) if $\alpha^* > 1$ holds, $x^*(1, i, j) = v_{l+1}$ and $\mathsf{OPT}(1, i, j) = \Theta_{i,j}^L(v_{l+1})$ hold.

We also have the following claim.

Claim 5. *For any integers i and j with $0 \leq i < j \leq n$, the following inequality holds:*

$$x^*(1, i, j) \leq x^*(1, i + 1, j).\tag{14}$$

Also, for any integers i and j with $0 \leq i \leq j \leq n - 1$, the following inequality holds:

$$x^*(1, i, j) \leq x^*(1, i, j + 1).\tag{15}$$

2.4   Algorithm

The algorithm basically computes $\mathsf{OPT}(1, 0, 1)$, ..., $\mathsf{OPT}(1, 0, n)$, $\mathsf{OPT}(2, 0, 1)$, ..., $\mathsf{OPT}(2, 0, n)$, ..., $\mathsf{OPT}(k, 0, 1)$, ..., $\mathsf{OPT}(k, 0, n)$ in this order. For some integers p and i with $2 \leq p \leq k$ and $2 \leq i \leq n$, let us consider how to obtain $\mathsf{OPT}(p, 0, i)$. Actually, in order to obtain $\mathsf{OPT}(p, 0, i)$, the algorithm needs $\mathsf{OPT}(p - 1, 0, l)$ for $l = 1, 2, \ldots, n$ and $\mathsf{OPT}(p, 0, i - 1)$. Suppose that the algorithm has already obtained $\mathsf{OPT}(p-1, 0, l)$ for $l = 1, 2, \ldots, n$ and $\mathsf{OPT}(p, 0, i-1)$. By (2), (6) and (7), we have

$$\mathsf{OPT}(p, 0, i) = f_{p,i}(d_{p,i}) = \max\{\mathsf{OPT}(p - 1, 0, d_{p,i}), \mathsf{OPT}(1, d_{p,i} + 1, i)\}.\tag{16}$$

Here, we assumed that $\mathsf{OPT}(p-1,0,d_{p,i})$ has already been obtained. Thus, in order to obtain $\mathsf{OPT}(p,0,i)$, we only need to compute $\mathsf{OPT}(1,d_{p,i}+1,i)$. Recall that by (7), $d_{p,i}$ is the unique point which minimizes function $f_{p,i}(t)$. Now, the algorithm knows where $d_{p,i-1}$ exists, and by Claim 2, $d_{p,i-1} \leq d_{p,i}$ holds. So the algorithm starts to compute $f_{p,i}(t)$ for $t = d_{p,i-1}$, and continues to compute in ascending order of t, as will be shown below. Note that function $f_{p,i}(t)$ is unimodal in t by Claim 1, which implies that $f_{p,i}(t)$ is strictly decreasing until $t = d_{p,i}$. Thus, if the algorithm reaches the first integer $t^* \geq d_{p,i-1}+1$ such that $f_{p,i}(t^*-1) \leq f_{p,i}(t^*)$, it outputs t^*-1 as $d_{p,i}$. Then, the algorithm also outputs $f_{p,i}(t^*-1)$ as $\mathsf{OPT}(p,0,i)$.

Computation of $f_{p,i}(t)$ for $t \geq d_{p,i-1}$: As above mentioned, the algorithm first computes $f_{p,i}(t)$ with $t = d_{p,i-1}$ which is defined as follows:

$$f_{p,i}(d_{p,i-1}) = \max\{\mathsf{OPT}(p-1,0,d_{p,i-1}), \mathsf{OPT}(1,d_{p,i-1}+1,i)\}. \tag{17}$$

Since the algorithm has already obtained $\mathsf{OPT}(p-1,0,d_{p,i-1})$, we only need to compute $\mathsf{OPT}(1,d_{p,i-1}+1,i)$. To do this, we actually need to find $x^*(1,d_{p,i-1}+1,i)$. By (15) in Claim 5, $x^*(1,d_{p,i-1}+1,i-1) \leq x^*(1,d_{p,i-1}+1,i)$ holds. On the other hand, the algorithm has already obtained $\mathsf{OPT}(p,0,i-1)$ as follows:

$$\mathsf{OPT}(p,0,i-1) = \max\{\mathsf{OPT}(p-1,0,d_{p,i-1}), \mathsf{OPT}(1,d_{p,i-1}+1,i-1)\}, \tag{18}$$

which implies that $x^*(1,d_{p,i-1}+1,i-1)$ has been obtained. Suppose that there exists $x^*(1,d_{p,i-1}+1,i-1) \in [v_l, v_{l+1}]$ with $d_{p,i-1}+1 \leq l \leq i-2$. By Claim 3, for any interval $[v_h, v_{h+1}]$ with $d_{p,i-1}+1 \leq h \leq i-1$, there exists $x^*(1,d_{p,i-1}+1,i)$ in $[v_h, v_{h+1}]$ if $\Theta^L_{d_{p,i-1}+1,i}(v_h) \geq \Theta^R_{d_{p,i-1}+1,i}(v_h)$ and $\Theta^L_{d_{p,i-1}+1,i}(v_{h+1}) \leq \Theta^R_{d_{p,i-1}+1,i}(v_{h+1})$ hold. Therefore, if we maintain the data structure (which will be explained in the next subsection) so that we can compute these values, the algorithm can test if there exists $x^*(1,d_{p,i-1}+1,i)$ in $[v_h, v_{h+1}]$ or not. Then, the algorithm starts to test if there exists $x^*(1,d_{p,i-1}+1,i) \in [v_h, v_{h+1}]$ for $h = l$, and continues to test in ascending order of h. If an interval $[v_{l^*}, v_{l^*+1}]$ where $x^*(1,d_{p,i-1}+1,i)$ exists is found, then $x^*(1,d_{p,i-1}+1,i)$ and $\mathsf{OPT}(1,d_{p,i-1}+1,i)$ can be computed in $O(1)$ time by Claim 4. The computation of $f_{p,i}(t)$ for $t \geq d_{p,i-1}+1$ can be treated in the similar manner as above.

2.5   Data Structure

For the computation mentioned in Section 2.4, the algorithm maintains a data structure $D(i,j)$ for integers i and j with $0 \leq i < j \leq n$ so that $\Theta^L_{i,j}(v_s)$ and $\Theta^R_{i,j}(v_s)$ can be efficiently computed for any integer s with $i \leq s \leq j$. This data structure is based on that in [5]. Basically, $D(i,j)$ consists of two binary heaps $T_L(i,j)$ and $T_R(i,j)$, and two values $os_L(i)$ and $os_R(j)$ explained below. In order to compute $\Theta^L_{i,j}(v_s)$ (resp. $\Theta^R_{i,j}(v_s)$), the algorithm uses $T_L(i,j)$ and $os_L(i)$ (resp. $T_R(i,j)$ and $os_R(j)$). Here, we explain $T_L(i,j)$ and $os_L(i)$ in detail ($T_R(i,j)$ and $os_R(j)$ can be constructed in a symmetric manner).

$T_L(i, j)$ is a binary heap with $i - j + 1$ leaves $i, i + 1, \ldots, j$ corresponding to vertices $v_i, v_{i+1}, \ldots, v_j$ and internal nodes such that each internal node has pointers to left and right children. For a node ν in $T_L(i, j)$, let $\kappa_{i,j}^L(\nu)$ (resp. $\kappa_{i,j}^R(\nu)$) denote the left (resp. right) child of ν, and $i_{\min}(\nu)$ (resp. $i_{\max}(\nu)$) denote the index of a minimum (resp. maximum) leaf of a subtree rooted at ν, which are stored at ν. Note that for a leaf l, $i_{\min}(l) = i_{\max}(l) = l$ holds. Then, each node (including leaf) ν in $T_L(i, j)$ also stores

$$value(\nu) = \max_l \left\{ -v_l \tau + \frac{\sum_{0 \leq h \leq l} w_h}{c} \,\middle|\, i_{\min}(\nu) \leq l \leq i_{\max}(\nu) \right\}, \qquad (19)$$

and the corresponding index of the leaf that attains the maximum.

On the other hand, the value $os_L(i)$ is the offset value defined as

$$os_L(i) = \frac{\sum_{0 \leq h \leq i-1} w_h}{c}. \qquad (20)$$

Here, for an integer s with $i \leq s \leq j$, let $Path_{i,j}(s)$ denote the path in $T_L(i, j)$ from a leaf s to the root. Then, by (12), $\Theta_{i,j}^L(v_s)$ can be represented as follows:

$$\Theta_{i,j}^L(v_s) = v_s \tau + \max\{value(\kappa_{i,j}^L(\nu)) \mid \nu \in Path_{i,j}(s)\} - os_L(i), \qquad (21)$$

which can be computed in $O(\log n)$ time by following $Path_{i,j}(s)$. Since $\Theta_{i,j}^R(v_s)$ can be also computed in $O(\log n)$ by using $T_R(i, j)$ and $os_R(j)$, we have the following claim.

Claim 6. *For any integers i, j and s with $0 \leq i < j \leq n$ and $i \leq s \leq j$, $\Theta_{i,j}^L(v_s)$ and $\Theta_{i,j}^R(v_s)$ can be computed in $O(\log n)$ time once $D(i, j)$ has been obtained.*

Note that $T_L(i, j)$ can be updated to $T_L(i+1, j)$ or $T_L(i, j+1)$ in $O(\log n)$ time and $os_L(i)$ can be updated to $os_L(i+1)$ in $O(1)$ time. In general, we have the following claim.

Claim 7. (i) *For any integers i and j with $0 \leq i < j \leq n$, $D(i, j)$ can be updated to $D(i+1, j)$ in $O(\log n)$ time.*
(ii) *For any integers i and j with $0 \leq i < j \leq n - 1$, $D(i, j)$ can be updated to $D(i, j+1)$ in $O(\log n)$ time.*

2.6   Time Complexity

As mentioned in Section 2.4, in order to obtain $\mathsf{OPT}(p, 0, i)$ for fixed p and i, the algorithm computes $f_{p,i}(d_{p,i-1}), \ldots, f_{p,i}(d_{p,i}), f_{p,i}(d_{p,i}+1)$. Thus, in order to obtain $\mathsf{OPT}(p, 0, i)$ for fixed p and all $i = 1, 2, \ldots, n$, the algorithm does such $O(n)$ computations. And through these computations, $O(n)$ intervals are tested in total (see Section 2.4). In order to test if there exists $\boldsymbol{x}^*(1, i, j)$ in some interval $[v_h, v_{h+1}]$ or not, the algorithm needs to confirm that $\Theta_{i,j}^L(v_h) \geq \Theta_{i,j}^R(v_h)$ and $\Theta_{i,j}^L(v_{h+1}) \leq \Theta_{i,j}^R(v_{h+1})$ hold by Claim 3, which takes $O(\log n)$ time by Claim 6. Thus, such $O(n)$ computations take $O(n \log n)$ time in total.

On the other hand, let us consider the total time required to update the data structure. For fixed p and i, when $\mathsf{OPT}(p, 0, i-1)$ is obtained, the algorithm maintains $D(d_{p,i-1} + 1, i - 1)$ and first updates to $D(d_{p,i-1} + 1, i)$. After repeatedly updating, the algorithm maintains $D(d_{p,i} + 1, i)$ when $\mathsf{OPT}(p, 0, i)$ is obtained. Thus, in order to obtain $\mathsf{OPT}(p, 0, i)$, the algorithm updates the data structure $d_{p,i} - d_{p,i-1} + 1$ times, and so, for fixed p and all $i = 1, 2, \ldots, n$, the algorithm does $O(n)$ times by Claim 2, which takes $O(n \log n)$ time by Claim 7. Therefore, $\mathsf{OPT}(p, 0, i)$ for all $i = 1, 2, \ldots, n$ and $p = 1, 2, \ldots, k$ can be obtained in $O(kn \log n)$ time.

Theorem 1. *The minimax k-sink location problem in a dynamic path network with uniform capacity can be solved in $O(kn \log n)$ time.*

3   Minisum k-sink Location Problem

In this section, an input graph of this problem is a dynamic path network defined in Section 2. As a preliminary step, let us consider the minisum 1-sink location problem.

3.1   Properties of the Minisum 1-sink Location Problem

Suppose that a sink is located at a point $x \in P$ where P is the input path with $n+1$ vertices. In continuous model, the cost is defined on each infinitesimal unit of supply, i.e., the cost of x for a unit is defined as the minimum time required to send the unit to x. Let $sum(x)$ denote the total cost of x, i.e., the sum of cost of x for all units on P. Here, let $sum_L(x)$ (resp. $sum_R(x)$) denote the sum of cost of x for all units on $[v_0, x]$ (resp. $(x, v_n]$). Then, $sum(x)$ is the maximum of $sum_L(x)$ and $sum_R(x)$, i.e.,

$$sum(x) = sum_L(x) + sum_R(x). \tag{22}$$

Without loss of generality, we assume $sum_L(v_0) = 0$ and $sum_R(v_n) = 0$. Now, suppose that x is located in an open interval (v_h, v_{h+1}) with $0 \le h \le n-1$, then let us explain how function $sum_L(x)$ is determined.

Case 1: For every integer i with $1 \le i \le h$, $\tau(v_i - v_{i-1}) > w_i/c$ holds. In this case, the first unit of each vertex on $[v_0, v_h]$ can reach x after leaving the original vertex without being blocked due to the existence of other units at an intermediate vertex. For an integer i with $0 \le i \le h$, let $sum^i(x)$ denote the sum of cost of x for all units of v_i. Here, suppose that there are α units at v_i with sufficiently large α, i.e., the size of each unit is equal to w_i/α, and these units continuously reach x. Then by (12), the l-th unit finishes reaching x at time $\tau(x - v_i) + l \cdot (w_i/\alpha)/c$. Therefore, by taking α to the infinity, $sum^i(x)$ can be represented as follows:

$$sum^i(x) = \lim_{\alpha \to \infty} \sum_{l=1}^{\alpha} \frac{w_i}{\alpha} \left(\tau(x - v_i) + l \cdot \frac{w_i}{\alpha} \cdot \frac{1}{c} \right) = w_i \tau(x - v_i) + \frac{w_i^2}{2c}, \tag{23}$$

and then, $sum_L(x)$ is represented as follows:

$$sum_L(x) = \sum_{0 \le i \le h} sum^i(x) = \sum_{0 \le i \le h} \left(w_i \tau(x - v_i) + \frac{w_i^2}{2c} \right). \qquad (24)$$

Case 2: There exists an integer j with $1 \le j \le h$ such that $\tau(v_j - v_{j-1}) \le w_j/c$ holds. First, we set $\rho_i = v_i$ and $\sigma_i = w_i$ for $0 \le i \le h$. Suppose that j is the minimum integer such that $1 \le j \le h$ and $\tau(\rho_j - \rho_{j-1}) \le \sigma_j/c$. In this case, the first unit of ρ_{j-1} must catch up with the last unit of ρ_j before the last unit of ρ_j leaves ρ_j, then shifting the supply corresponding to σ_{j-1} from ρ_{j-1} to ρ_j does not change the cost of x for any unit. We update ρ_i and σ_i for $0 \le i \le h - 1$ as follows:

$$\begin{aligned} \rho_i &\leftarrow \rho_i \quad \text{and} \quad \sigma_i \leftarrow \sigma_i & \text{for } 0 \le i \le j - 2, \\ \rho_{j-1} &\leftarrow \rho_j \quad \text{and} \quad \sigma_{j-1} \leftarrow \sigma_{j-1} + \sigma_j, \\ \rho_i &\leftarrow \rho_{i+1} \quad \text{and} \quad \sigma_i \leftarrow \sigma_{i+1} & \text{for } j \le i \le h - 1, \end{aligned} \qquad (25)$$

and delete ρ_h and σ_h. As long as there exist two vertices denoted by ρ_{j-1} and ρ_j such that $\tau(\rho_j - \rho_{j-1}) \le \sigma_j/c$ holds, we repeatedly update or delete ρ_i and σ_i in the similar manner. Suppose that $\rho_0, \dots, \rho_{h^*}$ eventually remain such that $\tau(\rho_i - \rho_{i-1}) > \sigma_i/c$ holds for any i with $1 \le i \le h^*$ or $h^* = 0$. Then by (22), $sum_L(x)$ is represented as follows:

$$sum_L(x) = \sum_{0 \le i \le h^*} \left(\sigma_i \tau(x - \rho_i) + \frac{\sigma_i^2}{2c} \right). \qquad (26)$$

We can compute $sum_R(x)$ in the similar manner as $sum_L(x)$. Thus, for an open interval (v_j, v_{j+1}) with $0 \le j \le n - 1$, function $sum(x)$ is linear in x with slope $\tau(\sum_{0 \le i \le j} w_i - \sum_{j+1 \le i \le n} w_i)$. Now let us consider an open interval (v_j, v_{j+1}) with $0 \le j \le n - 1$ such that $\sum_{0 \le i \le j} w_i - \sum_{j+1 \le i \le n} w_i \ge 0$ holds. Then, we can see that for any two points $p, q \in (v_j, v_{j+1})$ with $p < q$, $sum(p) \le sum(q)$ holds. We will show that for sufficiently small $\epsilon > 0$, $sum(v_j) \le sum(v_j + \epsilon)$ holds. We confirm $sum_R(v_j) = sum_R(v_j + \epsilon) + \left(\sum_{j+1 \le i \le n} w_i \right) \cdot \tau\epsilon$ and $sum_L(v_j + \epsilon) \ge sum_L(v_j) + \left(\sum_{0 \le i \le j} w_i \right) \cdot \tau\epsilon$. From these and the assumption of $\sum_{0 \le i \le j} w_i - \sum_{j+1 \le i \le n} w_i \ge 0$, we can derive $sum(v_j) \le sum(v_j + \epsilon)$. In general, we have the following claim.

Claim 8. (i) *For an open interval* (v_j, v_{j+1}) *with* $0 \le j \le n - 1$ *such that* $\sum_{0 \le i \le j} w_i - \sum_{j+1 \le i \le n} w_i \ge 0$, $sum(v_j) \le sum(p)$ *holds where* $p \in (v_j, v_{j+1})$. (ii) *For an open interval* (v_j, v_{j+1}) *with* $0 \le j \le n - 1$ *such that* $\sum_{0 \le i \le j} w_i - \sum_{j+1 \le i \le n} w_i < 0$, $sum(v_{j+1}) < sum(p)$ *holds where* $p \in (v_j, v_{j+1})$.

Let x^* denote the optimal sink location which minimizes $sum(x)$. Then, Claim 8 implies that x^* is located at some vertex.

Claim 9. *There exists* x^* *at a vertex.*

3.2 Algorithm and Time Complexity for the Minisum 1-sink Location Problem

We propose the algorithm which can solve the minisum 1-sink location problem in a dynamic path network. Basically, the algorithm first computes $sum_L(v_i)$ for $1 \leq i \leq n$ in ascending order of i, and next $sum_R(v_i)$ for $0 \leq i \leq n-1$ in descending order of i. After computing all these values, $sum(v_i)$ can be computed and evaluated for $0 \leq i \leq n$ in $O(n)$ time. Then, by Claim 9, the optimal sink location x^* is at a vertex which minimizes $sum(v_i)$ for $0 \leq i \leq n$. Below, we show how to compute $sum_L(v_i)$ (computation of $sum_R(v_i)$ can be treated in the similar manner).

Now, suppose that for some integer j with $1 \leq j \leq n-1$, $sum_L(v_j)$ has been already computed as $sum_L(v_j) = \sum_{0 \leq i \leq h(j)} \left(\sigma_i \tau(v_j - \rho_i) + \sigma_i^2/2c \right)$, where $h(j)$ is a non-negative integer, and ρ_i and σ_i is obtained for $0 \leq i \leq h(j)$ in the same manner as mentioned in Case 2, Section 3.1. Let $W_{j-1} = \sum_{0 \leq i \leq j-1} w_i = \sum_{0 \leq i \leq h(j)} \sigma_i$ and suppose that W_{j-1} has also been computed. We then show how to compute $sum_L(v_{j+1})$. The algorithm newly sets

$$sum' = sum_L(v_j), \quad \text{and} \quad W' = W_{j-1}. \tag{27}$$

Next, the algorithm tests if $\tau(v_j - \rho_i) \leq w_j/c$ for $0 \leq i \leq h(j)$ in descending order. If so, it updates sum' and W' as follows:

$$sum' \leftarrow sum' - \left(\sigma_i \tau(v_j - \rho_i) + \frac{\sigma_i^2}{2c} \right), \quad \text{and} \quad W' \leftarrow W' - \sigma_i, \tag{28}$$

and deletes ρ_i. If the maximum integer m such that $\tau(v_j - \rho_m) > w_j/c$ is found or $\tau(v_j - \rho_0) \leq w_j/c$ is obtained, the algorithm stops testing. In the former case, after the algorithm tests $h(j) - m + 1$ times, $\rho_0, \ldots, \rho_m$ remain. Then, after computing W_j as $W_j = W_{j-1} + w_j$, by (26), $sum_L(v_{j+1})$ can be computed as

$$sum_L(v_{j+1}) = sum' + W'\tau(v_{j+1} - v_j) + \\ \left((W_j - W')\tau(v_{j+1} - v_j) + \frac{(W_j - W')^2}{2c} \right). \tag{29}$$

For the next recursive step, the algorithm eventually sets

$$h(j+1) = m + 1, \quad \rho_{m+1} = v_j, \quad \text{and} \quad \sigma_{m+1} = W_j - W'. \tag{30}$$

Since the algorithm tests $h(j) - m + 1 = h(j) - h(j+1) + 2$ times to compute $sum_L(v_{j+1})$, it needs to test $\sum_{1 \leq i \leq n-1}(h(i) - h(i+1) + 2)$ times to compute $sum_L(v_i)$ for $2 \leq i \leq n$. By $h(1) = 0$, we have $\sum_{1 \leq i \leq n-1}(h(i) - h(i+1) + 2) = -h(n) + 2(n-1) = O(n)$.

Lemma 1. *The minisum 1-sink location problem in a dynamic path network with uniform capacity can be solved in $O(n)$ time.*

3.3   Extension to the Minisum k-sink Location Problem

For a subproblem, that is, the p-sink location problem in $[v_i, v_j]$ with $0 \leq i < j \leq n$ and $1 \leq p \leq k$, let $\mathsf{OPT}(p, i, j)$ denote the optimal cost (which is defined in the same manner as mentioned in Section 2.2). Then, for integers p and i with $2 \leq p \leq k$ and $1 \leq i \leq n$, $\mathsf{OPT}(p, 0, i)$ can be recursively represented as follows:

$$\mathsf{OPT}(p, 0, i) = \min_{0 \leq t \leq i-1} \{\mathsf{OPT}(p - 1, 0, t) + \mathsf{OPT}(1, t + 1, i)\}, \qquad (31)$$

and let $d_{p,i}$ be an integer which minimizes $\mathsf{OPT}(p - 1, 0, t) + \mathsf{OPT}(1, t + 1, i)$ for $0 \leq t \leq i - 1$. Then, we can show that $d_{p,i-1} \leq d_{p,i}$ holds for any integers p and i with $2 \leq p \leq k$ and $2 \leq i \leq n$ as with Claim 2 for the minimax problem (details are omitted). Thus, in the similar manner as mentioned in Section 2.4, after solving 1-sink location problem in the subgraph $O(kn)$ times, we can obtain the solution for the k-sink location problem. By Lemma 1, it takes $O(kn^2)$ time in total. Recall that in the minimax problem, each 1-sink problem can be solved in $O(\log n)$ time by using the unimodality of the objective function and the special data structure. On the other hand, in the minisum problem, the objective function for 1-sink problem is not unimodal, so there is the difference between time bounds for two problems.

Theorem 2. *The minisum k-sink location problem in a dynamic path network with uniform capacity can be solved in $O(kn^2)$ time.*

4   Conclusion

In this paper, we prove that the minimax k-sink location problem can be solved in $O(kn \log n)$ time and the minisum k-sink location problem can be solved in $O(kn^2)$ time. On the other hand, we leave as an open problem to reduce the time bound to $O(kn)$ for the minimax problem or the minisum problem, and extend the solvable networks into dynamic path networks with general capacities or more general networks (e.g., trees).

References

1. Chen, D., Chen, R.: A relaxation-based algorithm for solving the conditional p-center problem. Operations Research Letters 38(3), 215–217 (2010)
2. Cheng, S.W., Higashikawa, Y., Katoh, N., Ni, G., Su, B., Xu, Y.: Minimax Regret 1-Sink Location Problems in Dynamic Path Networks. In: Chan, T.-H.H., Lau, L.C., Trevisan, L. (eds.) TAMC 2013. LNCS, vol. 7876, pp. 121–132. Springer, Heidelberg (2013)
3. Ford Jr., L.R., Fulkerson, D.R.: Constructing maximal dynamic flows from static flows. Operations Research 6, 419–433 (1958)
4. Higashikawa, Y., Golin, M.J., Katoh, N.: Minimax Regret Sink Location Problem in Dynamic Tree Networks with Uniform Capacity. In: Pal, S.P., Sadakane, K. (eds.) WALCOM 2014. LNCS, vol. 8344, pp. 125–137. Springer, Heidelberg (2014)

5. Higashikawa, Y., Augustine, J., Cheng, S.W., Golin, M.J., Katoh, N., Ni, G., Su, B., Xu, Y.: Minimax Regret 1-Sink Location Problem in Dynamic Path Networks. Theoretical Computer Science (2014), doi:10.1016/j.tcs.2014.02.010

6. Kamiyama, N., Katoh, N., Takizawa, A.: An efficient algorithm for evacuation problem in dynamic network flows with uniform arc capacity. IEICE Transactions 89-D(8), 2372–2379 (2006)

7. Mamada, S., Uno, T., Makino, K., Fujishige, S.: An $O(n \log^2 n)$ Algorithm for the Optimal Sink Location Problem in Dynamic Tree Networks. Discrete Applied Mathematics 154(16), 2387–2401 (2006)

Narrowing the Complexity Gap
for Colouring (C_s, P_t)-Free Graphs

Shenwei Huang[1], Matthew Johnson[2], and Daniël Paulusma[2,*]

[1] School of Computing Science, Simon Fraser University
Burnaby B.C., V5A 1S6, Canada
shenweih@sfu.ca
[2] School of Engineering and Computing Sciences, Durham University,
Science Laboratories, South Road, Durham DH1 3LE, United Kingdom
{matthew.johnson2,daniel.paulusma}@durham.ac.uk

Abstract. Let k be a positive integer. The k-COLOURING problem is to decide whether a graph has a k-colouring. The k-PRECOLOURING EXTENSION problem is to decide whether a colouring of a subset of a graph's vertex set can be extended to a k-colouring of the whole graph. A k-list assignment of a graph is an allocation of a list — a subset of $\{1, \ldots, k\}$ — to each vertex, and the LIST k-COLOURING problem asks whether the graph has a k-colouring in which each vertex is coloured with a colour from its list. We prove a number of new complexity results for these three decision problems when restricted to graphs that do not contain a cycle on s vertices or a path on t vertices as induced subgraphs (for fixed positive integers s and t).

1 Introduction

It is well-known deciding whether a graph can be coloured with at most k colours is NP-complete even if $k = 3$ [18], and so the problem has been studied for special graph classes; see the surveys of Randerath and Schiermeyer [21] and Tuza [23], and the very recent survey of Golovach, Johnson, Paulusma and Song [8]. In this paper, we consider the computational complexity of several graph colouring problems for graph classes defined in terms of forbidden induced subgraphs. We introduce some notation and terminology before stating our results.

Terminology. Let $G = (V, E)$ be a graph. A *colouring* of G is a mapping $c : V \to \{1, 2, \ldots\}$ such that $c(u) \neq c(v)$ whenever $uv \in E$. We call $c(u)$ the *colour* of u. A k-*colouring* of G is a colouring with $1 \le c(u) \le k$ for all $u \in V$. We study the following decision problem:

k-COLOURING
Instance: A graph G.
Question: Is G k-colourable?

---

* Author supported by EPSRC (EP/G043434/1).

Q. Gu, P. Hell, and B. Yang (Eds.): AAIM 2014, LNCS 8546, pp. 162–173, 2014.

A *k-precolouring* of $G = (V, E)$ is a mapping $c_W : W \to \{1, 2, \ldots k\}$ for some subset $W \subseteq V$. A *k-colouring* c is an *extension* of c_W if $c(v) = c_W(v)$ for each $v \in W$. Another decision problem:

k-PRECOLOURING EXTENSION
Instance: A graph G and a *k*-precolouring c_W of G.
Question: Can c_W be extended to a *k*-colouring of G?

A *list assignment* of a graph $G = (V, E)$ is a function L that assigns a list $L(u)$ of *admissible* colours to each $u \in V$. If $L(u) \subseteq \{1, \ldots, k\}$ for each $u \in V$, then L is also called a *k-list assignment*. A colouring c *respects* L if $c(u) \in L(u)$ for all $u \in V$. Here is our next decision problem:

LIST *k*-COLOURING
Instance: A graph G and a *k*-list assignment L for G.
Question: Is there a colouring of G that respects L?

Note that *k*-COLOURING can be viewed as a special case of *k*-PRECOLOURING EXTENSION which is, in turn, a special case of LIST *k*-COLOURING.

Let G be a graph and $\{H_1, \ldots, H_p\}$ be a set of graphs. We say that G is $(H_1, \ldots, H_p)$-*free* if G has no induced subgraph isomorphic to a graph in $\{H_1, \ldots, H_p\}$; if $p = 1$, we write H_1-free instead of (H_1)-free. We denote the cycle, complete graph and path, each on r vertices, by C_r, K_r and P_r, respectively. The *complement* of a graph $G = (V, E)$, denoted by $\overline{G}$, has vertex set V and an edge between two distinct vertices if and only if these vertices are not adjacent in G. The disjoint union of two graphs G and H is denoted $G + H$, and the disjoint union of r copies of G is denoted rG.

Our Results. Several papers [4,9,13] have considered the computational complexity of the three decision problems defined above when restricted to (C_s, P_t)-free graphs. In this paper, we continue this investigation. Our first contribution is to state the following theorem that provides a complete summary of our current knowledge. In Section 5, we prove the theorem by providing references for results that demonstrate or imply each case. The cases marked with an asterisk are new results presented in this paper. We use p-time to mean polynomial-time throughout the paper.

Theorem 1. *Let* k, s, t *be three positive integers. The following statements hold for* (C_s, P_t)*-free graphs.*

(i) LIST *k*-COLOURING *is* NP-*complete if*

    1.* $k \geq 4$, $s = 3$ *and* $t \geq 8$ 2.* $k \geq 4$, $s \geq 5$ *and* $t \geq 6$.

    LIST *k*-COLOURING *is p-time solvable if*

    3. $k \leq 2$, $s \geq 3$ *and* $t \geq 1$ 7. $k \geq 4$, $s = 3$ *and* $t \leq 6$
    4. $k = 3$, $s = 3$ *and* $t \leq 6$ 8. $k \geq 4$, $s = 4$ *and* $t \geq 1$
    5. $k = 3$, $s = 4$ *and* $t \geq 1$ 9. $k \geq 4$, $s \geq 5$ *and* $t \leq 5$.
    6. $k = 3$, $s \geq 5$ *and* $t \leq 6$

(ii) k-PRECOLOURING EXTENSION *is* NP-*complete if*

1. $k = 4$, $s = 3$ *and* $t \geq 10$
2. $k = 4$, $s = 5$ *and* $t \geq 7$
3. $k = 4$, $s = 6$ *and* $t \geq 7$
4.* $k = 4$, $s = 7$ *and* $t \geq 8$

5. $k = 4$, $s \geq 8$ *and* $t \geq 7$
6. $k \geq 5$, $s = 3$ *and* $t \geq 10$
7.* $k \geq 5$, $s \geq 5$ *and* $t \geq 6$.

k-PRECOLOURING EXTENSION *is p-time solvable if*

8. $k \leq 2$, $s \geq 3$ *and* $t \geq 1$
9. $k = 3$, $s = 3$ *and* $t \leq 6$
10. $k = 3$, $s = 4$ *and* $t \geq 1$
11. $k = 3$, $s \geq 5$ *and* $t \leq 6$

12. $k \geq 4$, $s = 3$ *and* $t \leq 6$
13. $k \geq 4$, $s = 4$ *and* $t \geq 1$
14. $k \geq 4$, $s \geq 5$ *and* $t \leq 5$.

(iii) k-COLOURING *is* NP-*complete if*

1.* $k = 4$, $s = 3$ *and* $t \geq 39$
2. $k = 4$, $s = 5$ *and* $t \geq 7$
3. $k = 4$, $s = 6$ *and* $t \geq 7$
4. $k = 4$, $s = 7$ *and* $t \geq 9$

5. $k = 4$, $s \geq 8$ *and* $t \geq 7$
6. $k \geq 5$, $s = 5$ *and* $t \geq 7$
7. $k \geq 5$, $s \geq 6$ *and* $t \geq 6$.

k-COLOURING *is p-time solvable if*

8. $k \leq 2$, $s \geq 3$ *and* $t \geq 1$
9. $k = 3$, $s = 3$ *and* $t \leq 7$
10. $k = 3$, $s = 4$ *and* $t \geq 1$
11. $k = 3$, $s \geq 5$ *and* $t \leq 7$
12. $k = 4$, $s = 3$ *and* $t \leq 6$
13. $k = 4$, $s = 4$ *and* $t \geq 1$

14. $k = 4$, $s = 5$ *and* $t \leq 6$
15. $k = 4$, $s \geq 6$ *and* $t \leq 5$
16. $k \geq 5$, $s = 3$ *and* $t \leq k + 2$
17. $k \geq 5$, $s = 4$ *and* $t \geq 1$
18. $k \geq 5$, $s \geq 5$ *and* $t \leq 5$.

We describe the rest of the paper.

In Section 2, we consider LIST k-COLOURING restricted to (C_s, P_t)-free graphs and prove two results. We first show that LIST 4-COLOURING is NP-complete for $(C_5, C_6, K_4, \overline{P_1 + 2P_2}, \overline{P_1 + P_4}, P_6)$-free graphs, thus strengthening the NP-completeness result of LIST 4-COLOURING for P_6-free graphs [10]. (We observe that $\overline{P_1 + 2P_2}$ is also known as the 5-vertex wheel and $\overline{P_1 + P_4}$ is sometimes called the gem or the 5-vertex fan.) We also show that LIST 4-COLOURING is NP-complete for P_8-free bipartite graphs.

In Section 3, we show that for all $k \geq 4$, k-PRECOLOURING EXTENSION is NP-complete for P_{10}-free bipartite graphs extending a result of Kratochvíl [17] who showed that 5-PRECOLOURING EXTENSION is NP-complete for P_{13}-free bipartite graphs. We also prove that 4-PRECOLOURING EXTENSION is NP-complete for $(C_5, C_6, C_7, C_8, P_8)$-free graphs and that for all $k \geq 5$, k-PRECOLOURING EXTENSION is NP-complete for (C_s, P_t)-free graphs if $s \geq 5$ and $t \geq 6$.

In Section 4, we show that 4-COLOURING is NP-complete for (C_3, P_{39})-free graphs improving a result of Golovach et al. [9] who showed that 4-COLOURING is NP-complete for (C_3, P_{164})-free graphs.

In Section 5, we prove Theorem 1 by combining a number of previously known results with our new results, and in Section 6 we summarize the open cases and pose a number of related open problems.

Related Work. In this paper, we focus on (C_s, P_t)-free graphs. We comment that this can be seen as a natural continuation of investigations into the complexity of k-COLOURING and LIST k-COLOURING for P_r-free graphs (see [8]). The sharpest results are the following. Hoàng et al. [14] proved that, for all $k \geq 1$, LIST k-COLOURING is p-time solvable on P_5-free graphs. Huang [15] proved that 4-COLOURING is NP-complete for P_7-free graphs and that 5-COLOURING is NP-complete for P_6-free graphs. Recently, Chudnovsky, Maceli and Zhong [5,6] announced a p-time algorithm for solving 3-COLOURING on P_7-free graphs. Broersma et al. [3] proved that LIST 3-COLOURING is p-time solvable for P_6-free graphs. Golovach, Paulusma and Song [10] proved that LIST 4-COLOURING is NP-complete for P_6-free graphs. These results lead to the following table (in which the open cases are denoted by "?").

Table 1. The complexity of k-COLOURING, k-PRECOLOURING EXTENSION and LIST k-COLOURING for P_r-free graphs

	k-COLOURING				k-PRECOLOURING EXTENSION				LIST k-COLOURING			
	$k=3$	$k=4$	$k=5$	$k \geq 6$	$k=3$	$k=4$	$k=5$	$k \geq 6$	$k=3$	$k=4$	$k=5$	$k \geq 6$
$r \leq 5$	P	P	P	P	P	P	P	P	P	P	P	P
$r=6$	P	?	NP-c	NP-c	P	?	NP-c	NP-c	P	NP-c	NP-c	NP-c
$r=7$	P	NP-c	NP-c	NP-c	?	NP-c	NP-c	NP-c	?	NP-c	NP-c	NP-c
$r \geq 8$	?	NP-c	NP-c	NP-c	?	NP-c	NP-c	NP-c	?	NP-c	NP-c	NP-c

2 New Results for List Colouring

We start by proving that LIST 4-COLOURING is NP-complete for the class of $(C_5, C_6, K_4, \overline{P_1 + 2P_2}, \overline{P_1 + P_4})$-free graphs. This result will follow from a closer analysis of the hardness reduction for LIST 4-COLOURING for P_6-free graphs [10], which is from the problem NOT-ALL-EQUAL 3-SAT with positive literals only. This problem was shown to be NP-complete by Schaefer [22], and is defined as follows. The input I consists of a set $X = \{x_1, x_2, \ldots, x_n\}$ of variables, and a set $C = \{D_1, D_2, \ldots, D_m\}$ of 3-literal clauses over X in which all literals are positive. The question is whether there exists a truth assignment for X such that each D_i contains at least one true literal and at least one false literal. We may assume without loss of generality (see, for example, [10]) that each D_i contains either two or three literals and that each literal occurs in at most three different clauses. Given such an instance, Golovach et al. [10] define the following graph J_I and 4-list assignment L.

- a-type and b-type vertices: for each clause D_j, there are two *clause components* D_j and D'_j each isomorphic to P_5. Considered along the paths the vertices in D_j are $a_{j,1}, b_{j,1}, a_{j,2}, b_{j,2}, a_{j,3}$ with lists of admissible colours $\{2,4\}, \{3,4\}, \{2,3,4\}, \{3,4\}, \{2,3\}$, respectively, and the vertices in D'_j are $a'_{j,1}, b'_{j,1}, a'_{j,2}, b'_{j,2}, a'_{j,3}$ with lists of admissible colours $\{1,4\}, \{3,4\}, \{1,3,4\}, \{3,4\}, \{1,3\}$, respectively.

- x-type vertices: for each variable x_i, there is a vertex x_i with list of admissible colours $\{1, 2\}$.
- For every clause D_j with variables $x_{i_1}, x_{i_2}, x_{i_3}$, there are edges $a_{j,h}x_{i_h}$ and $a'_{j,h}x_{i_h}$ for $h = 1, 2, 3$.
- There is an edge from every x-type vertex to every b-type vertex.

See Figure 1 for an example of the graph J_I. In this figure, D_j is a clause with ordered variables $x_{i_1}, x_{i_2}, x_{i_3}$. The thick edges indicate the connection between these vertices and the a-type vertices of the two copies of the clause gadget. Indices from the labels of the clause gadget vertices have been omitted to increase visibility.

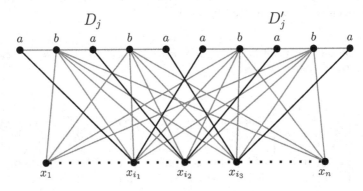

Fig. 1. An example of a graph J_I, as shown in [10]. Only the clause $D_j = \{x_{i_1}, x_{i_2}, x_{i_3}\}$ is displayed

The following two lemmas are known.

Lemma 1 ([10]). *The graph J_I has a colouring that respects L if and only if I has a satisfying truth assignment in which each clause contains at least one true and at least one false literal.*

Lemma 2 ([10]). *The graph J_I is P_6-free.*

We are now ready to prove our main result.

Theorem 2. *The* LIST 4-COLOURING *problem is* NP-*complete for the class of* $(C_5, C_6, K_4, \overline{P_1 + 2P_2}, \overline{P_1 + P_4}, P_6)$-*free graphs.*

Proof. Lemma 1 shows that the LIST 4-COLOURING problem is NP-hard for the class of graphs J_I, where $I = (X, \mathcal{C})$ is an instance of NOT-ALL-EQUAL 3-SAT with positive literals only, in which every clause contains either two or three literals and in which each literal occurs in at most three different clauses. Lemma 2 shows that each J_I is P_6-free. As the LIST 4-COLOURING problem is readily seen to be in NP, it remains to prove that each J_I is $(C_5, C_6, K_4, \overline{P_1 + 2P_2}, \overline{P_1 + P_4})$-free. For contradiction, assume that some J_I has an induced subgraph H isomorphic to a graph in $\{C_5, C_6, K_4, \overline{P_1 + 2P_2}, \overline{P_1 + P_4}\}$.

First suppose that $H \in \{C_5, C_6\}$. The total number of x-type and b-type vertices can be at most 3, as otherwise H contains an induced C_4 or a vertex of degree at least 3, which is not possible. Because $|V(H)| \geq 5$ and the subgraph of H induced by its b-type and x-type vertices is connected, H must contain at least two adjacent a-type vertices. This is not possible.

Now suppose that $H = K_4$. Because the b-type and x-type vertices induce a bipartite graph, H must contain an a-type vertex. Every a-type vertex has degree at most 3. If it has degree 3, then it has two non-adjacent neighbours (which are of b-type). Hence, this is not possible.

Finally suppose that $H \in \{\overline{P_1 + 2P_2}, \overline{P_1 + P_4}\}$. Let u be the vertex that has degree 4 in H. Then u cannot be of a-type, because no a-type vertex has more than three neighbours in J_I. Suppose u is of b-type. Then every other vertex of H is either of a-type or of x-type. Because vertices of the same type are not adjacent, H must contain two a-type vertices and two x-type vertices. Then an a-type vertex is adjacent to two x-type vertices. This is not possible. Suppose u is of x-type. Then every other vertex of H is either of a-type or of b-type. Because vertices of the same type are non-adjacent, H must contain two a-type vertices and two b-type vertices. However, then u is adjacent to two a-type vertices in the same clause-component. This is not possible.   □

Our second hardness result is also based on the hardness reduction of LIST 4-COLOURING for P_6-free graphs. Let J_I be defined as before. We subdivide every edge between an a-type vertex and an x-type vertex and give each new vertex the list $\{1, 2\}$ (we say that these new vertices are of c-type). This results in a new graph J'_I with list assignment L' which extends the original list assignment L for J_I.

Lemma 3. *The graph J'_I is P_8-free and bipartite.*

Proof. The graph J'_I is readily seen to be bipartite. Below we prove that J'_I P_8-free (but not P_7-free).

Let P be an induced path in J'_I. If P contains no x-type vertex, then P contains vertices of at most one clause-component together with at most two c-type vertices. This means that $|V(P)| \leq 7$. If P contains no b-type vertex, then P can contain at most one x-type vertex (as any two x-type vertices can only be connected by a path that uses at least one b-type vertex). Consequently, P can have at most two a-type vertices and at most two c-type vertices. Hence, $|V(P)| \leq 5$ in this case. From now on assume that P contains at least one b-type vertex and at least one x-type vertex. Also note that P can contain in total at most three vertices of b-type and x-type.

First suppose that P contains exactly three vertices of b-type and x-type. Then these vertices form a 3-vertex subpath in P of types b, x, b or x, b, x. In both cases we can extend both ends of the subpath only by an a-type vertex and an adjacent c-type vertex, which means that $|V(P)| \leq 7$. Now suppose that P contains exactly two vertices of b-type and x-type. Because these vertices are of different type, they are adjacent and we can extend both ends of the corresponding 2-vertex subpath of P only by an a-type vertex and an adjacent c-type vertex. This means that $|V(P)| \leq 6$. This completes our proof.   □

The following lemma can be proven by exactly the same arguments that were used to prove Lemma 1.

Lemma 4. *The graph J_I' has a colouring that respects L' if and only if I has a satisfying truth assignment in which each clause contains at least one true and at least one false literal.*

Lemmas 3 and 4 imply the last result of this section.

Theorem 3. LIST 4-COLOURING *is* NP-*complete for P_8-free bipartite graphs.*

3   New Results for Precolouring Extension

In this section we give three results on the k-PRECOLOURING EXTENSION problem.

Let $k \geq 4$. Consider the bipartite graph J_I' with its list assignment L' from Section 2. The list of admissible colours $L'(u)$ of each vertex u is a subset of $\{1, 2, 3, 4\}$. We add $k - |L'(u)|$ pendant vertices to u and precolour these vertices with different colours from $\{1, \ldots, k\} \setminus L'(u)$. This results in a graph J_I'' with a k-precolouring c_W, where W is the set of all the new pendant vertices.

Lemma 5. *The graph J_I'' is P_{10}-free and bipartite.*

Proof. Because J_I' is P_8-free and bipartite by Lemma 3, and moreover, we only added pendant vertices, J_I'' is P_{10}-free and bipartite.   □

The following lemma can be proven by exactly the same arguments that were used to prove Lemma 1.

Lemma 6. *The graph J_I'' has a k-colouring that is an extension of c_W if and only if I has a satisfying truth assignment in which each clause contains at least one true and at least one false literal.*

Lemmas 5 and 6 imply the first result of this section.

Theorem 4. *For all $k \geq 4$, k-PRECOLOURING EXTENSION is* NP-*complete for the class of P_{10}-free bipartite graphs.*

Here is our second result.

Theorem 5. *The 4-PRECOLOURING EXTENSION problem is* NP-*complete for the class of $(C_5, C_6, C_7, C_8, P_8)$-free graphs.*

Proof. Let J_I be the instance with list assignment L as constructed in Section 2. Instead of considering lists, we introduce new vertices, which we precolour (we do not precolour any old vertices). For each clause D_j we add five new vertices, s_j, t_j, $u_{j,1}$, $u_{j,2}$, $u_{j,3}$. We add edges $a_{j,1}s_j$, $a_{j,3}t_j$ and $a_{j,h}u_{j,h}$ for $h = 1, \ldots 3$. We precolour s_j, t_j, $u_{j,1}$, $u_{j,2}$, $u_{j,3}$ by colours 3, 4, 1, 1, 1, respectively. For each clause D'_j we add five new vertices, s'_j, t'_j, $u'_{j,1}$, $u'_{j,2}$, $u'_{j,3}$. We add edges $a'_{j,1}s'_j$, $a'_{j,3}t'_j$ and $a'_{j,h}u'_{j,h}$ for $h = 1, \ldots 3$. We precolour s_j, t_j, $u_{j,1}$, $u_{j,2}$, $u_{j,3}$ by colours 3, 4, 2, 2, 2, respectively. Finally, we add two new vertices c_1, c_2, which we make adjacent to all x-type vertices, and two new vertices y_1, y_2, which we make adjacent to all b-type vertices. We colour c_1, c_2, y_1, y_2 with colours 3, 4, 1, 2, respectively. This results in a new graph J_I^*. Because y_1, y_2 can be viewed as x-type vertices and c_1, c_2 as b-type vertices, because every other new vertex is a pendant vertex and because J_I is (C_5, C_6, P_6)-free (by Theorem 2), we find that J_I^* is $(C_5, C_6, C_7, C_8, P_8)$-free. Moreover, our precolouring forces the lists $L(v)$ upon every vertex v of J_I. Hence, J_I^* has a 4-colouring extending this precolouring if and only if J_I has a colouring that respects L. By Lemma 1 the latter is true if and only if I has a satisfying truth assignment in which each clause contains at least one true and at least one false literal. □

Broersma et al. [3] showed that 5-PRECOLOURING EXTENSION for P_6-free graphs is NP-complete. It can be shown that the gadget constructed in their NP-hardness reduction is C_s-free for all $s \geq 5$. By adding $k-5$ dominating vertices, precoloured with colours $6, \ldots, k$, to each vertex in their gadget, we can extend their result from $k = 5$ to $k \geq 5$. This leads to the following theorem.

Theorem 6. *For all $k \geq 5$, k-PRECOLOURING EXTENSION is NP-complete for (C_s, P_t)-free graphs if $s \geq 5$ and $t \geq 6$.*

4   New Results for Colouring

In this section, we prove that 4-COLOURING is NP-complete for (C_3, P_{39})-free graphs. We do this by modifying the graph J_I'' from Section 3 when $k = 4$.

First we review a well-known piece of graph theory. The *Mycielski construction* of a graph $G = (V, E)$ is the new graph G' constructed from G by adding a new vertex v' for each $v \in V$ that is adjacent to every neighbour of v in G, followed by adding a further new vertex u adjacent to every new vertex v'. By repeating this construction from K_2, a sequence of graphs $M_2, M_3, \ldots$ is obtained. Here, $M_2 = K_2$, $M_3 = C_5$ and M_4 is the well-known Grötzsch graph. Mycielski [19] showed that every M_k is C_3-free and has chromatic number k. Moreover, any proper subgraph of M_k is $(k-1)$-colourable (see for example [1]).

We focus on M_5. For any pair of adjacent vertices p and r, $M_5 - pr$ is 4-colourable and, in every 4-colouring, p and r are coloured alike (else a 4-colouring of M_5 has been found). We let M_{pq} be the graph obtained from $M_5 - pr$ by adding a new vertex q and making it adjacent to r only. Note that M_{pq} is 4-colourable and that, in any 4-colouring of M_{pq}, the vertices p and q must have different colours.

Let G be a graph with $e = xy \in E(G)$. The *M-identification* of e in G is the following operation: delete the edge $e = xy$ and add a copy of M_{pq} between x and y by identifying $p \in M_{pq}$ and $q \in M_{pq}$ with x and y, respectively. We denote this copy of M_{pq} by M_e.

We are now ready to explain how we modify the graph J_I''. Recall that $k = 4$. First we take a complete graph on four new vertices $t_1, \ldots, t_4$. We perform an M-identification of every edge $t_i t_j$. Recall that we had defined a precolouring W for a subset $W \subseteq V(J_I'')$. We add an edge between a vertex t_i and a vertex $u \in W$ if and only if $c_W(u) \neq i$. This results in a new graph J_I'''.

In the next three lemmas we show three properties of J_I'''. The proof of the third lemma has been omitted due to page restrictions.

Lemma 7. *The graph J_I''' is 4-colourable if and only if I has a satisfying truth assignment in which each clause contains at least one true and at least one false literal.*

Proof. We claim that J_I''' is 4-colourable if and only if J_I'' has a 4-colouring that is an extension of c_W. This follows by construction and from the fact that p and q have different colours in any 4-colouring of M_{pq}. In order to prove the lemma it remains to apply Lemma 6.   □

Lemma 8. *The graph J_I''' is C_3-free.*

Proof. The graph J_I''' is C_3-free because of the following three reasons. Firstly, M_{pq} is C_3-free. Secondly, we applied an M-identification for every edge $t_i t_j$. So, the vertices $t_1, \ldots, t_4$ form an independent set of J_I'''. Thirdly, the neighbours of $t_1, \ldots, t_4$ in J_I'' are all in W, and W is an independent set of J_I'', and thus of J_I'''.   □

Lemma 9. *The graph J_I''' is P_{39}-free.*

The main result of this section now follows from Lemmas 7–9.

Theorem 7. 4-COLOURING *is* NP-*complete for* (C_3, P_{39})-*free graphs.*

5   Proof of Theorem 1

To prove Theorem 1 we need first to discuss some additional results. Kobler and Rotics [16] showed that for any constants p and k, LIST k-COLOURING is p-time solvable on any class of graphs that have clique-width at most p, assuming that a p-expression is given. Oum [20] showed that a $(8^p - 1)$-expression for any n-vertex graph with clique-width at most p can be found in $O(n^3)$ time. Combining these two results leads to the following theorem.

Theorem 8. *Let $\mathcal{G}$ be a graph class of bounded clique-width. For all $k \geq 1$, LIST k-COLOURING can be solved in p-time on $\mathcal{G}$.*

We also need the following result due to Gravier, Hoáng and Maffray [11] who slightly improved upon a bound of Gyárfás [12] who showed that every (K_s, P_t)-free graph can be coloured with at most $(t-1)^{s-2}$ colours.

Theorem 9 ([11]). *Let $s, t \geq 1$ be two integers. Then every (K_s, P_t)-free graph can be coloured with at most $(t-2)^{s-2}$ colours.*

We now prove Theorem 1 by considering each case. For each we either refer back to an earlier result, or give a reference; the results quoted can clearly be seen to imply the statements of the theorem.

We first consider the intractable cases of LIST k-COLOURING and note that (i).1 follows from Theorem 3, and Theorem 2 implies that LIST 4-COLOURING is NP-complete for the class of (C_5, C_6, P_6)-free graphs which proves (i).2.

Now the tractable cases. Erdös, Rubin and Taylor [7] and Vizing [24] observed that 2-LIST COLOURING is p-time solvable on general graphs implying (i).3. Broersma et al. [3] showed that LIST 3-COLOURING is p-time solvable for P_6-free graphs from which we can infer (i).4 and (i).6. Golovach et al. [9] proved that for all $k, r, s, t \geq 1$, LIST k-COLOURING can be solved in linear time for $(K_{r,s}, P_t)$-free graphs. By taking $r = s = 2$, we obtain (i).5 and (i).8. The class of (C_3, P_6)-free graphs was shown to have bounded clique-width by Brandstädt, Klembt and Mahfud [2]; using Theorem 8 we see that LIST k-COLOURING is p-time solvable on (C_3, P_6)-free graphs for all $k \geq 1$ demonstrating (i).7. Hoàng, Kamiński, Lozin, Sawada, and Shu [14] proved that for all $k \geq 1$, LIST k-COLOURING is p-time solvable on P_5-free graphs proving (i).9.

We now consider k-PRECOLOURING EXTENSION. The tractable cases all follow from the results on LIST k-COLOURING just discussed. So we are left to consider the NP-complete cases. Theorem 4 implies (ii).1 and (ii).6. Theorems 5 and 6 imply (ii).4 and (ii).7 And (ii).2, (ii).3 and (ii).5 follow immediately from corresponding results for k-COLOURING proved by Hell and Huang [13].

Finally, we consider k-COLOURING; first the NP-complete cases. Theorem 7 gives us (iii).1. Golovach, Paulusma and Song [9] proved that for all $s \geq 5$, there exists a constant $t(s)$ such that 4-COLOURING is NP-complete for $(C_5, \ldots, C_s, P_{t(s)})$-free graphs. In particular, they showed that 4-COLOURING is NP-complete for (C_5, P_{23})-free graphs, and this result has been strengthened by Hell and Huang [13] who proved all the other NP-completeness subcases.

Chudnovsky, Maceli and Zhong [5,6] announced that 3-COLOURING is p-time solvable on P_7-free graphs, and Chudnovsky, Maceli, Stacho and Zhong [4] announced that 4-COLOURING is p-time solvable for (C_5, P_6)-free graphs. Theorem 9 gives us (iii).16. All other tractable cases follow from the corresponding tractable cases for LIST k-COLOURING.   □

6 Open Problems

From Theorem 1, we see that the following cases are open in the classification of the complexity of graph colouring problems for (C_s, P_t)-free graphs:

(i) For LIST k-COLOURING the following cases are open:

- $k = 3$, $s = 3$ and $t \geq 7$
- $k = 3$, $s \geq 5$ and $t \geq 7$
- $k \geq 4$, $s = 3$ and $t = 7$.

(ii) For k-PRECOLOURING EXTENSION the following cases are open:

- $k = 3$, $s = 3$ and $t \geq 7$
- $k = 3$, $s \geq 5$ and $t \geq 7$
- $k = 4$, $s = 3$ and $7 \leq t \leq 9$
- $k = 4$, $s \geq 5$ and $t = 6$
- $k = 4$, $s = 7$ and $t = 7$
- $k \geq 5$, $s = 3$ and $7 \leq t \leq 9$

(iii) For k-COLOURING the following cases are open:

- $k = 3$, $s = 3$ and $t \geq 8$
- $k = 3$, $s \geq 5$ and $t \geq 8$
- $k = 4$, $s = 3$ and $7 \leq t \leq 38$
- $k = 4$, $s \geq 6$ and $t = 6$
- $k = 4$, $s = 7$ and $7 \leq t \leq 8$
- $k \geq 5$, $s = 3$ and $t \geq k + 3$
- $k \geq 5$, $s = 5$ and $t = 6$.

Besides solving these missing cases (and the missing cases from Table 1) we pose the following problems specifically. First, does there exist a graph H and an integer $k \geq 3$ such that LIST k-COLOURING is NP-complete and k-COLOURING is p-time solvable for H-free graphs? Theorem 1 shows that if we forbid two induced subgraphs then the complexity of these two problems *can* be different: take $k = 4$, $H_1 = C_5$ and $H_2 = P_6$. Second, is LIST 4-COLOURING NP-complete for P_7-free bipartite graphs? This is the only missing case of LIST 4-COLOURING for P_t-free bipartite graphs due to Theorems 1 and 3.

References

1. Bondy, J.A., Murty, U.S.R.: Graph Theory. Springer Graduate Texts in Mathematics vol. 244 (2008)
2. Brandstädt, A., Klembt, T., Mahfud, S.: P_6- and triangle-free graphs revisited: structure and bounded clique-width. Discrete Mathematics & Theoretical Computer Science 8, 173–188 (2006)
3. Broersma, H.J., Fomin, F.V., Golovach, P.A., Paulusma, D.: Three complexity results on coloring P_k-free graphs. European Journal of Combinatorics 34, 609–619 (2013)
4. Chudnovsky, M., Maceli, P., Stacho, J., Zhong, M.: Four-coloring graphs with no induced six-vertex path, and no C_5, personal communication
5. Chudnovsky, M., Maceli, P., Zhong, M.: Three-coloring graphs with no induced six-edge path I: the triangle-free case (in preparation)
6. Chudnovsky, M., Maceli, P., Zhong, M.: Three-coloring graphs with no induced six-edge path II: using a triangle (in preparation)
7. Erdős, P., Rubin, A.L., Taylor, H.: Choosability in graphs. In: Proceedings of the West Coast Conference on Combinatorics, Graph Theory and Computing (Humboldt State Univ., Arcata, Calif., 1979), Congress. Numer., XXVI, Winnipeg, Man., Utilitas Math, pp. 125–157 (1980)
8. Golovach, P.A., Johnson, M., Paulusma, D., Song, J.: A survey on the computational complexity of colouring graphs with forbidden subgraphs (manuscript)
9. Golovach, P.A., Paulusma, D., Song, J.: Coloring graphs without short cycles and long induced paths. Discrete Applied Mathematics 167, 107–120 (2014)

10. Golovach, P.A., Paulusma, D., Song, J.: Closing complexity gaps for coloring problems on H-free graphs. Information and Computation (to appear)
11. Gravier, S., Hoàng, C.T., Maffray, F.: Coloring the hypergraph of maximal cliques of a graph with no long path. Discrete Mathematics 272, 285–290 (2003)
12. Gyárfás, A.: Problems from the world surrounding perfect graphs. Zastosowania Matematyki Applicationes Mathematicae XIX(3-4), 413–441 (1987)
13. Hell, P., Huang, S.: Complexity of coloring graphs without paths and cycles. In: Pardo, A., Viola, A. (eds.) LATIN 2014. LNCS, vol. 8392, pp. 538–549. Springer, Heidelberg (2014)
14. Hoàng, C.T., Kamiński, M., Lozin, V., Sawada, J., Shu, X.: Deciding k-colorability of P_5-free graphs in p-time. Algorithmica 57, 74–81 (2010)
15. Huang, S.: Improved complexity results on k-coloring P_t-free graphs. In: Chatterjee, K., Sgall, J. (eds.) MFCS 2013. LNCS, vol. 8087, pp. 551–558. Springer, Heidelberg (2013)
16. Kobler, D., Rotics, U.: Edge dominating set and colorings on graphs with fixed clique-width. Discrete Applied Mathematics 126, 197–221 (2003)
17. Kratochvíl, J.: Precoloring extension with fixed color bound. Acta Mathematica Universitatis Comenianae 62, 139–153 (1993)
18. Lovász, L.: Coverings and coloring of hypergraphs. In: Proc. 4th Southeastern Conference on Combinatorics, Graph Theory, and Computing, Utilitas Math., pp. 3–12 (1973)
19. Mycielski, J.: Sur le coloriage des graphes. Colloq. Math. 3, 161–162 (1955)
20. Oum, S.-I.: Approximating rank-width and clique-width quickly. ACM Transactions on Algorithms 5 (2008)
21. Randerath, B., Schiermeyer, I.: Vertex colouring and forbidden subgraphs - a survey. Graphs Combin. 20, 1–40 (2004)
22. Schaefer, T.J.: The complexity of satisfiability problems. In: Proc. STOC 1978, pp. 216–226 (1978)
23. Tuza, Z.: Graph colorings with local restrictions - a survey. Discuss. Math. Graph Theory 17, 161–228 (1997)
24. Vizing, V.G.: Coloring the vertices of a graph in prescribed colors, in Diskret. Analiz., no. 29. Metody Diskret. Anal. v. Teorii Kodov i Shem 101, 3–10 (1976)

New Lower Bounds on Broadcast Function

Hayk Grigoryan and Hovhannes A. Harutyunyan

Department of Computer Science and Software Engineering,
Concordia University, Montreal, Quebec, H3G 1M8, Canada
h_grig@encs.concordia.ca, haruty@cse.concordia.ca

Abstract. This paper studies the broadcast function $B(n)$. We consider the possible vertex degrees and possible connections between vertices of different degrees in graphs with $b(G) = \lceil \log_2 n \rceil$. Using this, we present improved lower bounds on $B(n)$ when $n = 2^k - 2^p$ and $n = 2^k - 2^p + 1$ $(3 \leq p < k)$. Also, we prove that $B(24) \geq 36$ for graphs with maximum vertex degree at most 4.

Keywords: Broadcasting, minimum broadcast graphs, broadcast function, lower bounds on broadcast function.

1 Introduction

Broadcasting is the process of distributing a message from a node, called the *originator*, to all other nodes of a communication network. Broadcasting is accomplished by placing series of calls over the communication channels of the network and takes place in discrete time units, sometimes called rounds. Each call involves only two nodes (one sender and one receiver), requires one time unit, and each node participates in at most one call at each time unit.

A network can be modeled as a connected graph $G = (V, E)$, where V is the set of all nodes and E is the set of all communication lines. The *broadcast time* $b(v, G)$ or just $b(v)$ of a vertex v in a connected graph G is defined as the minimum time required to inform all the vertices of G from originator v. The broadcast time $b(G)$ of a graph G is defined as $b(G) = max\{b(v) \mid v \in V\}$.

The set of calls used to distribute the message from originator v to all other vertices is called a *broadcast scheme* for vertex v. The broadcast scheme for v is a spanning tree rooted at v where all the communication lines are labeled with the transmission time. Each communication line is used exactly once and the message is always transmitted from a parent to a child.

For any graph G on n vertices, $b(G) \geq \lceil \log n \rceil$ (all logarithms in this paper are base two), since after each time unit the number of informed vertices can at most double. A graph G with $b(G) = \lceil \log n \rceil$, is called a *broadcast graph (bg)*. A broadcast graph with the minimum possible number of edges is called a *minimum broadcast graph (mbg)*. An *mbg* has very important practical implications; it represents the cheapest possible architecture to build a network, in which broadcasting can be accomplished in theoretically minimum possible

Q. Gu, P. Hell, and B. Yang (Eds.): AAIM 2014, LNCS 8546, pp. 174–184, 2014.
© Springer International Publishing Switzerland 2014

time. The *broadcast function* $B(n)$ is defined as the number of edges in an *mbg* on n vertices.

$B(n)$ is known only for very few values of n, in particular for all $n \leq 32$ except for $n = 23, 24$ and 25. The values of $B(n)$ for $1 \leq n \leq 15$ are presented in [6], also, $B(17) = 22$ [17], $B(18) = 23, B(19) = 25$ [3],[19], $B(20) = 26, B(21) = 28, B(22) = 31$ [16], $B(23) = 33$ or 34 [4],[16], $B(26) = 42$ [18],[20], $B(27) = 44, B(28) = 48, B(29) = 52$ [18], $B(30) = 60$ [3],[13], $B(31) = 65$ [3]. $B(n)$ is also known for $n = 2^k$, $B(2^k) = k2^{k-1}$ [6], for $n = 2^k - 2$, $B(2^k - 2) = (k-1)(2^{k-1}-1)$ [5],[13], $B(58) = 121$, $B(59) = 124$, $B(60) = 130$, $B(61) = 136$ [18], $B(61) = 136$, $B(63) = 162$ [15], $B(127) = 389$ [8].

Since mbg's seem to be extremely difficult to find, a long sequence of papers presented techniques to construct broadcast graphs and to obtain upper bounds on $B(n)$. Most techniques combine several known mbg's and bg's on smaller sizes to create new ones of a larger size (see e.g. [1],[2],[3],[5],[7],[8],[9],[11],[12],[13]).

However, it is extremely difficult to prove a lower bound on $B(n)$ that matches the obtained upper bound from a broadcast graph construction. When $n = 2^k$ or $n = 2^k - 2$, a simple lower bound based on the minimum vertex degree matches the known upper bounds, thus the known mbg's are for these cases are k-regular and $k - 1$-regular graphs respectively. However, when $n < 2^k - 2$, mbg's are not regular graphs, hence the simple lower bounds based only on minimum vertex degree are not very helpful. For small n, an *mbg* can be found by exhaustive case analysis, but when n becomes large, the number of possible graphs grows exponentially and this technique is no longer useful.

Similar to the approach taken in [16],[18] in this paper we consider the possible vertex degrees in any broadcast graph on n vertices and possible connections between vertices of different degrees.

Our first observation is that in a broadcast graph on $n = 2^k - x$ vertices where $1 \leq x \leq 2^{k-1}$, the minimum vertex degree must be at least $k - \lfloor \log x \rfloor$. A broadcast tree rooted at some vertex v of a smaller degree will contain at most $n = 2^k - x - 1$ vertices. This number is smaller than the total number of vertices. It means that not all vertices will be able to receive the broadcast message by the time k from originator v. From this observation it follows that

$$B(2^k - x) \geq \frac{2^k - x}{2} \cdot (k - \lfloor \log x \rfloor).$$

The fact that a given graph is a broadcast graph determines not only the minimum possible vertex degree in it, but also the possible connections between vertices of different degrees. By making more accurate observations the above mentioned bound can be improved. This approach was used in [16] to obtain lower bounds on $B(n)$ when $n = 2^k - 3, n = 2^k - 4, n = 2^k - 5$ and $n = 2^k - 6$.

The following bounds are presented:

$$B(2^k - 3) \geq \left\lceil \frac{2^k - 3}{2} \cdot (k - 2 + \frac{3k - 5}{k^2 - k - 1}) \right\rceil,$$

$$B(2^k - 4) \geq \left\lceil \frac{2^k - 4}{2} \cdot (k - 2 + \frac{4}{2k + 1}) \right\rceil,$$

$$B(2^k - 5) \geq \left\lceil \frac{2^k - 5}{2} \cdot (k - 2 + \frac{2}{2k - 1}) \right\rceil,$$

$$B(2^k - 6) \geq \left\lceil \frac{2^k - 6}{2} \cdot (k - 2 + \frac{1}{k}) \right\rceil.$$

The same approach is also used in [15] to get a lower bound on $B(n)$ when $n = 2^k - 1$.

$$B(2^k - 1) \geq \left\lceil \frac{2^k - 1}{2} \cdot (k - 1 + \frac{1}{k + 1}) \right\rceil.$$

We find this method of getting lower bounds on $B(n)$ promising and we will use it to find lower bounds on $B(n)$ when $n = 2^k - 2^p$ and $n = 2^k - 2^p + 1$ ($3 \leq p < k$). The main difficulty in the above approach is that when x increases, the lower bound on the minimum degree presented above decreases and then the number of possibilities of different relations between vertices of different degree increases as well and it becomes more and more difficult to deal with them and derive an improved lower bound on $B(2^k - x)$.

One of the motivations for looking on these two particular forms of n is that the smallest values for which $B(n)$ is not known are $n = 23, n = 24$ and $n = 25$. The latter two have a form $n = 2^k - 7$ and $n = 2^k - 8$ respectively. Where are known broadcast graphs on 24 and 25 vertices having 36 and 40 edges respectively [3] but whether these graphs are mbg's or not is not known. Tight lower bounds on $B(24)$ and $B(25)$ may help to address this problem.

2   Lower Bound on $B(2^k - 7)$

As mentioned above, there are tight lower bounds on $B(n)$ when $n = 2^k - 1, 2^k - 3, 2^k - 4, 2^k - 5, 2^k - 6$. We continue on this line and in this section we present a new lower bound on $B(n)$ when $n = 2^k - 7$. In the following section, we generalize the presented result for $n = 2^k - 2^p + 1$. Thus, the proof in this section will help to follow the proof of Section 3.

In our approach, we extend the technique presented by Sacle in [18].

Theorem 1. $B(2^k - 7) \geq \frac{2^k - 7}{2}((k - 3) + \frac{5k - 11}{(k + 1)(k - 2)})$.

Proof. Recall that in a broadcast graph on $n = 2^k - 7$ vertices, the minimum possible vertex degree is $k - 3$. Let us look at the broadcast tree rooted at a vertex u of degree $k - 3$. We observe that u must have at least one neighbour of degree at least k, at least two neighbours of degree at least $k - 1$ and at least three neighbours of degree at least $k - 2$. We also observe that a vertex cannot have all neighbours of degree $k - 3$. In other words each vertex in the graph must have at least one neighbour of degree at least $k - 2$. Let v_i denote the number of vertices of degree i. We can write the following inequalities:

$$\sum_{i \geq k} (i - 1)v_i \geq v_{k-3},$$

$$\sum_{i \geq k-1} (i - 1)v_i \geq 2v_{k-3},$$

$$\sum_{i \geq k-2} (i - 1)v_i \geq 3v_{k-3}.$$

For the number of edges in the graph, denoted by m, we will have

$$2m = \sum_{i \geq k-3} iv_i = n + \sum_{i \geq k-3} (i - 1)v_i.$$

This implies that

$$\sum_{i \geq k-3} (i - 1)v_i = 2m - n.$$

After substituting this in the above three inequalities we will get

$$2m - n - (k - 4)v_{k-3} - (k - 3)v_{k-2} - (k - 2)v_{k-1} \geq v_{k-3},$$

$$2m - n - (k - 4)v_{k-3} - (k - 3)v_{k-2} \geq 2v_{k-3},$$

$$2m - n - (k - 4)v_{k-3} \geq 3v_{k-3}.$$

After rearranging the terms we will have

$$2m - n \geq (k - 3)v_{k-3} + (k - 3)v_{k-2} + (k - 2)v_{k-1},$$

$$2m - n \geq (k - 2)v_{k-3} + (k - 3)v_{k-2},$$

$$2m - n \geq (k - 1)v_{k-3}.$$

After subtracting v_{k-1} and v_{k-3} from the right hand sides of the first and the second inequalities respectively, we will get

$$2m - n \geq (k - 3)(v_{k-3} + v_{k-2} + v_{k-1}),$$

$$2m - n \geq (k - 3)(v_{k-3} + v_{k-2}),$$

$$2m - n \geq (k - 1)v_{k-3}.$$

It follows that

$$v_{k-3} + v_{k-2} + v_{k-1} \leq \frac{2m - n}{k - 3},$$

$$v_{k-3} + v_{k-2} \leq \frac{2m - n}{k - 3},$$

$$v_{k-3} \leq \frac{2m-n}{k-1}.$$

Alternatively, for the number of edges we also have the following expression

$$2m \geq nk - (v_{k-1} + 2v_{k-2} + 3v_{k-3}) =$$

$$= nk - (v_{k-1} + v_{k-2} + v_{k-3}) - (v_{k-2} + v_{k-3}) - v_{k-3}.$$

By substituting the above 3 inequalities in this inequality we get

$$2m \geq nk - (2m-n)(\frac{2}{k-3} + \frac{1}{k-1}).$$

From which it follows that

$$m \geq \frac{n}{2} \cdot \frac{k + (\frac{2}{k-3} + \frac{1}{k-1})}{1 + (\frac{2}{k-3} + \frac{1}{k-1})} = \frac{n}{2} \cdot \frac{k + (\frac{2}{k-3} + \frac{1}{k-1})}{1 + (\frac{2}{k-3} + \frac{1}{k-1})}.$$

This gives the following lower bound on $B(2^k - 7)$

$$B(2^k - 7) \geq \frac{n}{2} \cdot \frac{k + (\frac{1}{k-1} + \frac{2}{k-3})}{1 + (\frac{1}{k-1} + \frac{2}{k-3})} = \frac{2^k - 7}{2}((k-3) + \frac{5k - 11}{(k+1)(k-2)}).$$

3   Lower Bound on $B(2^k - 2^p + 1)$

In this section we obtain a new lower bound on $B(n)$ where $n = 2^k - 2^p + 1$ based on the degree sequence restrictions of any broadcast graph on $2^k - 2^p + 1$ vertices.

Theorem 2. $B(2^k - 2^p + 1) \geq \frac{2^k - 2^p + 1}{2}((k-p) + \frac{k(2p-1)-(p^2+p-1)}{k(k-1)-(p-1)})$.

Proof. We observe that in an *mbg* on $2^k - 2^p + 1$ vertices, each vertex of degree $k - p$ must have at least one neighbour of degree at least k, two neighbours of degree at least $k - 1$, three neighbours of degree at least $k - 2$, ... , p neighbours of degree at least $k - p + 1$. After noticing that a vertex cannot have all its neighbours of degree $k - p$ we are getting the following inequalities

$$\sum_{i \geq k}(i - 1)v_i \geq v_{k-p},$$

$$\sum_{i \geq k-1}(i - 1)v_i \geq 2v_{k-p},$$

$$\sum_{i \geq k-2}(i - 1)v_i \geq 3v_{k-p},$$

$$\cdots$$

$$\sum_{i \geq k-p+1}(i - 1)v_i \geq pv_{k-p}.$$

For the number of edges in the graph, denoted by m we will have

$$2m = \sum_{i \geq k-p} iv_i = n + \sum_{i \geq k-p} (i-1)v_i.$$

This implies that

$$\sum_{i \geq k-p} (i-1)v_i = 2m - n.$$

After substituting this in the above p inequalities and reversing their order we will get

$$2m - n - (k-p-1)v_{k-p} \geq pv_{k-p},$$

$$2m - n - (k-p-1)v_{k-p} - (k-p)v_{k-p+1} \geq (p-1)v_{k-p},$$

$$2m - n - (k-p-1)v_{k-p} - (k-p)v_{k-p+1} - (k-p+1)v_{k-p+2} \geq (p-2)v_{k-p},$$

$$\dots$$

$$2m - n - \sum_{j=0}^{i}(k-p-1+j)v_{k-p+j} \geq (p-i)v_{k-p},$$

$$\dots$$

$$2m - n - \sum_{j=0}^{p-1}(k-p-1+j)v_{k-p+j} \geq v_{k-p}.$$

After rearranging the terms we will have

$$2m - n \geq (k-1)v_{k-p},$$

$$2m - n \geq (k-2)v_{k-p} + (k-p)v_{k-p+1},$$

$$2m - n \geq (k-3)v_{k-p} + (k-p)v_{k-p+1} + (k-p+1)v_{k-p+2},$$

$$\dots$$

$$2m - n \geq (k-p)v_{k-p} + \sum_{j=1}^{p-1}(k-p-1+j)v_{k-p+j}.$$

By replacing all the $k-2$, $k-3$, ..., $k-p+1$ coefficients on the right side of these inequalities with $k-p$, which is the smallest one, we will get

$$2m - n \geq (k-1)v_{k-p},$$

$$2m - n \geq (k-p)(v_{k-p} + v_{k-p+1}),$$

$$2m - n \geq (k-p)(v_{k-p} + v_{k-p+1} + v_{k-p+2}),$$

$$\dots$$

$$2m - n \geq (k-p)(v_{k-p} + v_{k-p+1} + v_{k-p+2} + \dots + v_{k-1}).$$

Alternatively, we have the following trivial inequality

$$2m \geq nk - (v_{k-1} + 2v_{k-2} + 3v_{k-3} + \ldots + pv_{k-p}) =$$

$$= nk - (v_{k-1} + \ldots + v_{k-p}) - (v_{k-2} + \ldots + v_{k-p}) - (v_{k-3} + \ldots + v_{k-p}) - \ldots - v_{k-p}.$$

By substituting the terms in parenthesis with their upper bounds from the previous set of inequalities we will get

$$2m \geq nk - (2m - n)(\frac{1}{k-1} + \frac{p-1}{k-p}).$$

It follows that

$$B(2^k - 2^p + 1) \geq m \geq \frac{n}{2} \cdot \frac{k + (\frac{1}{k-1} + \frac{p-1}{k-p})}{1 + (\frac{1}{k-1} + \frac{p-1}{k-p})} =$$

$$= \frac{2^k - 2^p + 1}{2}((k - p) + \frac{k(2p-1) - (p^2 + p - 1)}{k(k-1) - (p-1)}).$$

Note that by plugging $p = 3$ in Theorem 2 we get the lower bound from Theorem 1.

4   Lower Bound on $B(2^k - 2^p)$

In this section we present a new lower bound on $B(2^k - 2^p)$.

Theorem 3. $B(2^k - 2^p) \geq \frac{2^k - 2^p}{2}((k - p) + \frac{k(2p-2) - (p^2 + p - 2)}{k(k-2) - (p-2)})$.

Proof. Most of the proof is omitted due to its similarity to the proof for $B(2^k - 2^p - 1)$. We observe that in an *mbg* on $2^k - 2^p$ vertices, each vertex of degree $k - p$ must have at least two neighbours of degree at least $k - 1$, three neighbours of degree at least $k - 2$, ... , p neighbours of degree at least $k - p + 1$. See Fig. 1. This will give the following inequalities:

$$\sum_{i \geq k-1} (i - 1)v_i \geq 2v_{k-p},$$

$$\sum_{i \geq k-2} (i - 1)v_i \geq 3v_{k-p},$$

$$\ldots$$

$$\sum_{i \geq k-p+1} (i - 1)v_i \geq pv_{k-p}.$$

Note that the first inequality from Theorem 2 is missing here. The reason is that in a broadcast graph on $2^k - 2^p + 1$ vertices, unlike for the $2^k - 2^p$ case, a vertex of degree $k - p$ can have a neghbour of degree $k - 1$. Using these inequalities in a similar way as in Theorem 2, we were able to prove the presented lower bound on $B(2^k - 2^p)$.

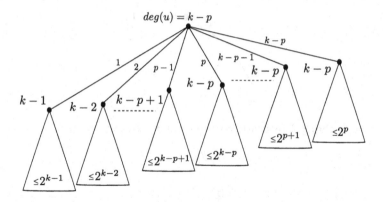

Fig. 1. Broadcast tree rooted at vertex u of degree $k - p$. The neighbours of u are sorted in the decreasing order of their degree and labeled $1, 2, ..., k - p$. The number in each triangle is the maximum possible number of vertices in that subtree. The number on each vertex is the minimum required degreed of that vertex.

5   About the Value of $B(24)$

A broadcast graph on 24 vertices and 36 edges was constructed by Bermond et al. [3]. This gives $B(24) \leq 36$.

We will prove the $B(24) \geq 36$ inequality for graphs G with $\Delta(G) \leq 4$, i.e. for graphs with maximum vertex degree at most 4. It will follow that, if it exists, a broadcast graph on 24 vertices and less than 36 edges, must have at least one vertex of degree at least 5.

Let v_i denote the number of vertices of degree i, and α_{ij} denote the number of all edges between vertices of degree i and j. By our definition $\alpha_{ij} = \alpha_{ji}$.

Lemma 1. *For a broadcast graph G on 24 vertices, $\delta(G) \geq 2$, i.e. $v_1 = 0$.*

Proof. Let G be a broadcast graph on 24 vertices, i.e $b(G) = \lceil \log 24 \rceil = 5$. Let u be the broadcast originator and $deg(u) = 1$. In the first round, it will inform its only neighbor v. In the remaining 4 rounds, v will be able to inform at most $16 = 2^4$ vertices. It follows that in 5 rounds u can inform only at most 17 vertices, as shown in Fig. 2.

Theorem 4. *A broadcast graph G on 24 vertices and $\Delta(G) \leq 4$, must have at least 36 edges.*

Proof. From Lemma 1, it follows that $v_1 = 0$. By counting the number of edges adjacent to vertices of degree 4, we will have

$$4v_4 = \alpha_{42} + \alpha_{43} + 2\alpha_{44}.$$

We observe that the broadcast tree rooted at a vertex u of degree 2 must have a form shown in Figure 3, otherwise in 5 rounds it will not be possible

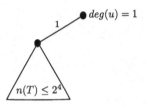

Fig. 2. Broadcast tree rooted at a vertex of degree 1

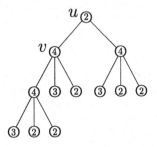

Fig. 3. Subtree of a broadcast tree rooted at a vertex of degree 2

to inform 24 vertices. The number next on each vertex indicates the minimal possible degree for that vertex. For example, a vertex with label 3 may actually have degree 4.

From the figure we observe that a vertex of degree 2 must have both its neighbours of degree 4. Therefore,

$$\alpha_{42} = 2v_2.$$

From the fact that a vertex of degree 4 cannot have all its neighbours having degree 2, it follows that it has at least one adjacent is edge going to vertex of degree 3 or 4. Also we note that a vertex of degree 2 must have a neighbour v (left child in Figure 3) of degree 4 having at least 2 edges going to a vertex of degree 3 or 4. That is v cannot have three neghbours of degree 2. Vertex v can be shared between at most 2 vertices of degree 2. It follows that there are at least $\left\lceil \frac{v_2}{2} \right\rceil$ such vertices "v", i.e. vertices of degree 4 having at least 2 edges going to a vertex of degree 3 or 4. Thus, we can claim that that

$$\alpha_{43} + \alpha_{44} \geq v_4 + \left\lceil \frac{v_2}{2} \right\rceil.$$

From the observation that an edge between vertices of degree 4 in Figure 3 can be shared among at most 4 vertices of degree 2 we have that

$$\alpha_{44} \geq \left\lceil \frac{v_2}{4} \right\rceil.$$

Finally, by using the expressions above we will have

$$4v_4 = \alpha_{42} + \alpha_{43} + 2\alpha_{44} = \alpha_{42} + (\alpha_{43} + \alpha_{44}) + \alpha_{44} \geq$$

$$\geq 2v_2 + v_4 + \left\lceil \frac{v_2}{2} \right\rceil + \left\lceil \frac{v_2}{4} \right\rceil \geq (2 + \frac{1}{2} + \frac{1}{4})v_2 + v_4 = \frac{11}{4}v_2 + v_4.$$

It follows that

$$v_4 \geq \frac{11}{12}v_2.$$

To prove that $b(24) \geq 36 = \frac{24 \cdot 3}{2}$, we must show that in any broadcast graph of on 24 vertices, the average vertex degree is at least 3. In our case, this means that in any broadcast graph G with $\Delta(G) = 4, |G| = 24$, we must show that $v_4 \geq v_2$. From $v_4 \geq \frac{11}{12}v_2$ it almost always follows that $v_4 \geq v_2$. The only pair of values for which it is not the case is $v_2 = 12, v_4 = 11$, but this would mean that $v_3 = 24 - v_2 - v_4 = 1$. We observe that this is not possible, since in any graph the number of vertices of odd degree must be even.

6   Summary

In [7], it was shown that $B(n) \geq \frac{n}{2}(\lfloor \log n \rfloor - \log(1 + 2^{\lfloor \log n \rfloor} - n))$. Let k be the index of the leftmost 0 bit in the binary representation $(\alpha_{p-1}\alpha_{p-2}...\alpha_1\alpha_0)$ of $n - 1$. In [14], the following bound was obtained $B(n) \geq \frac{n}{2}(p - k - 1)$. This bound was later improved in [10] to $B(n) \geq \frac{n}{2}(p - k - 1 + \beta)$ where $\beta = 0$ if $k = 0$ or if $\alpha_0 = \alpha_1 = ... = \alpha_{k-1} = 0$, otherwise $\beta = 1$. For cases $n = 2^p - 2^k + 1$ and $n = 2^p - 2^k$, these bounds give $B(n) \geq \frac{n}{2}(p - k)$. It follows that the bounds from Theorems 2 and 3 are obviously better.

Note that the best known upper bounds on $B(n)$ for both $n = 2^p - 2^k + 1$ and $n = 2^p - 2^k$ is $B(n) \leq \frac{2^p - 2^k}{2}(p - \frac{k+1}{2})$ [9]. So, still there is a gap between the best known lower and upper bounds.

References

1. Bermond, J.C., Fraigniaud, P., Peters, J.G.: Antepenultimate broadcasting. Networks 26(3), 125–137 (1995)
2. Bermond, J.C., Hell, P., Liestman, A.L., Peters, J.G.: Broadcasting in bounded degree graphs. SIAM Journal on Discrete Mathematics 5(1), 10–24 (1992)
3. Bermond, J.C., Hell, P., Liestman, A.L., Peters, J.G.: Sparse broadcast graphs. Discrete Applied Mathematics 36(2), 97–130 (1992)
4. Changhong, L., Kemin, Z.: The broadcast function value B(23) is 33 or 34. Acta Mathematicae Applicatae Sinica (English Series) 16(3), 329–331 (2000)
5. Dinneen, M.J., Ventura, J.A., Wilson, M.C., Zakeri, G.: Compound constructions of broadcast networks. Discrete Applied Mathematics 93(2), 205–232 (1999)
6. Farley, A.M., Hedetniemi, S., Mitchell, S., Proskurowski, A.: Minimum broadcast graphs. Discrete Mathematics 25, 189–193 (1979)
7. Gargano, L., Vaccaro, U.: On the construction of minimal broadcast networks. Networks 19(6), 673–689 (1989)

8. Harutyunyan, H.A.: An efficient vertex addition method for broadcast networks. Internet Mathematics 5(3), 211–225 (2008)
9. Harutyunyan, H.A., Liestman, A.L.: More broadcast graphs. Discrete Applied Mathematics 98(1-2), 81–102 (1999)
10. Harutyunyan, H.A., Liestman, A.L.: Improved upper and lower bounds for k-broadcasting. Networks 37(2), 94–101 (2001)
11. Harutyunyan, H.A., Liestman, A.L.: Upper bounds on the broadcast function using minimum dominating sets. Discrete Mathematics 312(20), 2992–2996 (2012)
12. Hedetniemi, S.M., Hedetniemi, S.T., Liestman, A.L.: A survey of gossiping and broadcasting in communication networks. Networks 18(4), 319–349 (1988)
13. Khachatrian, L.H., Harutounian, O.S.: Construction of new classes of minimal broadcast networks. In: Conference on Coding Theory, Dilijan, Armenia, pp. 69–77 (1990)
14. Konig, J.C., Lazard, E.: Minimum k-broadcast graphs. Discrete Applied Mathematics 53(1-3), 199–209 (1994)
15. Labahn, R.: A minimum broadcast graph on 63 vertices. Discrete Applied Mathematics 53(1-3), 247–250 (1994)
16. Maheo, M., Sacle, J.F.: Some minimum broadcast graphs. Discrete Applied Mathematics 53(1-3), 285 (1994)
17. Mitchell, S., Hedetniemi, S.: A census of minimum broadcast graphs. Journal of Combinatorics, Information and System Sciences 5, 141–151 (1980)
18. Sacle, J.F.: Lower bounds for the size in four families of minimum broadcast graphs. Discrete Mathematics 150(1), 359–369 (1996)
19. Xiao, J., Wang, X.: A research on minimum broadcast graphs. Chinese Journal of Computers 11, 99–105 (1988)
20. Zhou, J., Zhang, K.: A minimum broadcast graph on 26 vertices. Applied Mathematics Letters 14(8), 1023–1026 (2001)

Efficient Memoization for Approximate Function Evaluation over Sequence Arguments

Tamal Biswas and Kenneth W. Regan

Department of CSE
University at Buffalo
Amherst, NY 14260 USA
{tamaltan,regan}@buffalo.edu

Abstract. This paper proposes strategies for maintaining a database of computational results of functions f on sequence arguments $\boldsymbol{x}$, where $\boldsymbol{x}$ is sorted in non-decreasing order and $f(\boldsymbol{x})$ has greatest dependence on the first few terms of $\boldsymbol{x}$. This scenario applies also to symmetric functions f, where the partial derivatives approach zero as the corresponding component value increases. The goal is to pre-compute exact values $f(\boldsymbol{u})$ on a tight enough net of sequence arguments, so that given any other sequence $\boldsymbol{x}$, a neighboring sequence $\boldsymbol{u}$ in the net giving a close approximation can be efficiently found. Our scheme avoids pre-computing the more-numerous partial-derivative values. It employs a new data structure that combines ideas of a trie and an array implementation of a heap, representing grid values compactly in the array, yet still allowing access by a single index lookup rather than pointer jumping. We demonstrate good size/approximation performance in a natural application.

Keywords: Data structures, memoization, sequences, metrics, topology, machine learning, cloud computing.

1 Introduction

In many computational tasks, we need to evaluate a function on many different arguments, in applications such as aggregation where we can tolerate approximation. Evaluations $f(x)$ may be expensive enough to demand *memoization* of pre-computed values, creating a fine enough grid of argument-value pairs to enable approximating $f(x)$ via one or more neighboring pairs $(u, f(u))$. In this paper we limit ourselves to external memoization, here meaning building the complete grid in advance. Such applications of grids in high-dimensional real spaces $\mathbb{R}^\ell$ are well known (see history in [1]). What distinguishes this paper is a different kind of space in which the arguments are homogeneous *sequences* not just arbitrary vectors, where f and the space obey certain large-scale structural properties.

To explain our setting and ideas, consider first the natural grid strategy in $\mathbb{R}^\ell$ of employing the Taylor expansion to approximate $f(x)$ via a nearby gridpoint u:

$$f(x) = f(u) + \sum_{i=1}^{\ell}(x_i - u_i)\frac{\partial f}{\partial x_i}(u) + \frac{1}{2}\sum_{i,j}(x_i - u_i)(x_j - u_j)\frac{\partial^2 f}{\partial u_i \partial u_j} + \cdots$$

Q. Gu, P. Hell, and B. Yang (Eds.): AAIM 2014, LNCS 8546, pp. 185–196, 2014.

If we make the grid fine enough, and assume that f is reasonably smooth, we can ignore the terms with second and higher partials, since $(x_i - u_i)(x_j - u_j)$ is quadratically small in the grid size. Doing this still requires knowledge of the gradient of f on the grid-points, however. Memoizing—that is, precomputing and storing—all the partials on the gridpoints might be ℓ times as expensive as memoizing the values $f(u)$. Hence, depending on the application, one may employ an approximation to the gradient or a recursive estimation of the partials.

In our setting, we are given a different kind of structure with a little more knowledge. Here we need to compute functions f on sequence arguments $\boldsymbol{x} = (x_1, x_2, x_3, \dots)$ under the following circumstances.

(a) The sequence entries and function values belong to $[0, 1]$.
(b) For all $i < j$ and $\boldsymbol{x}$, $\partial f / \partial x_i > \partial f / \partial x_j$ at $\boldsymbol{x}$.
(c) The sequences are non-decreasing, and for all i, $\partial f / \partial x_i$ becomes small as x_i approaches 1.
(d) While exact computation of $f(\boldsymbol{x})$ is expensive, moderate precision suffices, especially when there is no bias in the approximations.

Part (b) says that the initial terms have the highest influence on the result, while (c) together with (b) implies that as the sequence approaches its ceiling, terms lose their influence. Part (c) also allows us to assume all sequences have the same length ℓ, using 1.0 values as trailing padding if needed. Applications obeying (d) include calculation of means and percentiles and other aggregate statistics, as are typical for streaming algorithms [2], and various tasks in curve fitting, machine learning [3], complex function evaluation, and Monte Carlo simulations (see [4]).

For further intuition, note that this setting applies to any function f that is symmetric, that is whose value is independent of the order of the arguments. Such a function really depends on the values of the arguments in a ranking structure. We may without loss of generality restrict the arguments to be sequenced by rank. Then (b) says that the first-ranked arguments matter most, while (c) says that elements with not only poor rank but also poor underlying scores have negligible marginal influence.

The goal is to build a data structure U of arguments and values $f(\boldsymbol{u})$ with these properties:

1. Coding: For all argument sequences $\boldsymbol{x}$ there are $\boldsymbol{u} \in U$ such that $|f(\boldsymbol{u}) - f(\boldsymbol{x})| < \epsilon$.
2. Size: $|U|$ is not too large, as a function of ϵ.
3. Efficiency: A good neighbor or small set of neighbors $\boldsymbol{u}$ can be found in time proportional to the length of the sequence, with only $O(1)$ further computation needed to retrieve the value $f(\boldsymbol{u})$.
4. Only a "black box" memo table of values $f(\boldsymbol{u})$ is needed, with the remainder of the approximation algorithm staying (essentially) independent of f.

Our main contribution is the construction of a family $\mathcal{G}$ of grid structures U and a simple memoizing algorithm A that is parameterized by a weighting function $wt(\cdots)$ used in lieu of the gradient. Among functions we consider are $wt(\cdots) = 0$, which means ignoring the gradient, and

$$wt(i, \boldsymbol{x}, -) = \frac{1}{i}(1 - x_i). \tag{1}$$

The intuition for this is that the first i elements have size no larger than x_i, which is to say equal or higher rank and influence on $f(x)$ to x_i. If they were all equal, then each would have a share $1/i$ of their total influence. The multiplier $(1 - x_i)$ aims to moderate the influence as the argument component x_i itself increases. It is also the simplest multiplier that goes to zero as x_i goes to 1. Our algorithm A uses weights in a balancing strategy that reflects the other sequence elements x_j for $j < i$.

We report success on a fairly general range of functions f that arise from a natural application in which probabilities are estimated. Some of the f represent salient basic mathematical problems in their own right. Key features of our data structure, algorithm, and overall strategy are:

- The grid is not regular but "warped": it starts fine but becomes coarse as i increases, eventually padding with nonce 1 values.
- The grid has an efficient compact mapping to an array, like a "warped" array implementation of a heap, so that values $f(u)$ can be looked up by one index rather than pointer jumping.
- The actual gradient of f is replaced by the universal substitute (1) and various other weighting method described later, which reflects properties (b) and (c) above.
- The algorithm to find a good grid neighbor u to the given argument x uses weights $w_i = wt(i, \cdots)$ to balance rounding.

This seems like a "cookbook" approach, but our point is that we have more structure to work with, before the steps of the recipe that have f as a particular ingredient need to be acted on. In our application there is no connection between the weights w_i and the functions f except for the axiomatic properties (b) and (c) and lack of unusual pathology in f. We demonstrate performance that is almost ten times faster than without memoization, and with four-place accuracy from a grid that starts with only two-place fineness. First we describe the application.

2 Estimating Probabilities and Means

Although our application comes from the chess decision-making model of [5], the present computational task is simple and general and can be described without reference to chess. The main model-design parameter is a real function h, and varying h implicitly defines the functions f treated in this paper. The argument is a non-*increasing* sequence of ℓ-many real numbers a_i beginning with $a_1 = 1$. The goal is to fit ℓ-many probability values p_i according to the equations

$$\frac{h(p_i)}{h(p_1)} = a_i; \qquad 0 \le p_i \le 1, \qquad \sum_{i=1}^{\ell} p_i = 1.$$

In modeling our grid, we transform $x_i = 1 - a_i$ so that the most influential values x_i as those closest to 0. The function value $f(x)$ is just the first probability, p_1. This is hence a fairly broad setting of curve-fitting. Our application further involves data with myriad sequences x, for which we need to compute sums $M = \sum_{x \in S} f(x)$ over sampled subsets S of the data, which when divided by $|S|$ become means. When S is moderately large, it suffices to have good approximation for f provided it is unbiased. Again this is highly typical in computational applications.

When h is the identity function id, we can solve for f in closed form:

$$f_{id}(\boldsymbol{a}) = \frac{1}{\sum_i a_i}, \quad \text{so} \quad f_{id}(\boldsymbol{x}) = \frac{1}{\ell - \sum_i x_i}.$$

For other functions h, however, we know only iterative techniques to compute $f(\boldsymbol{x}) = p_1$, and from p_1 the other probabilities, to desired precision. These iterations are expensive, so we seek to save them. When h is the logarithm function—more precisely the function $h(p) = 1/\log(1/p)$—we obtain the equations:

$$\frac{\log(1/p_1)}{\log(1/p_i)} = a_i, \quad \text{so} \quad p_i = p_1^{1/a_i}.$$

Viewing $b_i = 1/a_i$ as a general non-decreasing sequence, not necessarily beginning at 1, defines the following mathematical problem: Given y and real numbers $b_1, \ldots, b_\ell$, find a real number p such that

$$y = p^{b_1} + p^{b_2} + \cdots + p^{b_\ell}.$$

When $\ell = 1$ so there is just one number b, then $p = y^{1/b}$, so this is just the problem of root-finding. Hence we think of this problem as computing "vectorized roots" $\sqrt[b]{y}$. Above we have the case $y = 1$ and also $b_1 = 1$.

To restore some of the intuition from chess, the x_i values are the perceived differences of various chess moves m_i in a chess position from the optimal move m_1. These are obtained from the differences $\Delta(m_1, m_i)$ given by analysis from a strong computer chess program, after factoring in model parameters representing the skill of a particular chess player P. The p_i are estimates for the probabilities that P will choose the respective moves, and in particular p_1 is the probability of finding the move the computer thinks is best. The value M becomes the expected number of agreements with the computer on the set S of moves. We imagine that for testing a small set of a few hundred moves one might pay the time for exact computation, but for *training* the model on sets of many tens of thousands, using sequences of values from 10–20 different levels of analysis on each move, explains the need for memoization. The better moves have x_i close to zero. If there are k such moves, then p_1 will be order-of $1/k$, and so a $(k+1)$st good move will have influence only at most about $1/(k+1)$. Moves with x_i close to 1 are "blunders," and the exact value of a blunder matters little in the phenomenon of player choice that we are modeling. Hence the axiomatic properties in the Introduction are fulfilled in an intuitive sense based on chess, but our point here is that they flow originally from a quite general mathematical system of equations.

3   The Tapered Grid Data Structures

When working with sequences, it is natural first to think of a *trie* data structure, that is a tree whose root branches to the possible first sequence elements (often restricted to those actually used), each of those nodes to possible second elements, and so on. When presented with a new sequence x, upon replacing entries x_i by ones in the domain if needed, one follows tree links until reaching a node for which x is known to be the unique sequence through it, whereupon $f(x)$ can be stored at that node, or the node is deep enough to store a good approximation. The main advantage of tries is flexibility to add new sequences and values dynamically and compactly, but this comes with high use of main memory and cache misses with pointer jumps.

We sacrifice the dynamism to store pre-computed values. This brings, however, the need to cover all possible next elements, not just those present in the dynamically built data. Since chess positions can have 50 or more legal moves we regard $\ell = 50$, so to store the result for all possible x with 2-decimal-place accuracy would involve powers of 101 that bust the information capacity of the universe. Hence our first task is to finalize the ordered vector s such that $u_i \in s$ for all $u \in U$ where i belong to $[1, \ell]$. Keeping in mind the constraint that $u_i \leq u_j \iff i \leq j$, the grid that we construct is one we call a self-similar tapered tree.

Definition 1. *A self-similar tapered tree is a tree where every node has a branching label b, which is also its number of children. Its children have branches labeled $b, b - 1, \ldots, 1$ going from left to right.*

One more definition is useful in this context:

Definition 2. *The branching factor of the grid is the maximum number of children any node can have at any depth. The branching factor of any depth is maximum number of children any node can have at that particular depth.*

The branching factor of the root is the same as that of the grid, which is the length $|s|$ of s in our case. The choice of elements in s and its length are implementation dependent, though we usually start with compact branches and later space out the gaps as the argument entries approach the ceiling 1.0.

Because the grid grows exponentially in size with increasing depth, we need to use at least one of the two compromises:

- Truncate: cut off the *depth* of the grid at some depth $d_0 \ll \ell$.
- Warp: reduce the *branching factor* geometrically as the depth of the grid d increases.

To demonstrate the ideas, we develop two schemes. The first idea is utilized by both of our schemes. For best exposition we implement the first before introducing the second. After choosing a fixed-spacing vector s, we parameterize our first scheme by the maximum depth d and a fixed branching factor B at every depth. Our *reduced-branching* scheme replaces B by a specifier G of a rounded geometric progression to define the family $\mathcal{G}$ of grids $U = U(d, G)$, in which the omission of "ℓ" as a parameter is deliberate.

For our first scheme, once we finalize the branching factor b, the depth d and the spacing vector s, we can generate all the sequences, where the root always contains the value 0.0. Of all the sequences, $(0.0, 0.0, 0.0, \ldots)$ is the infimum (in which all moves have equal value) and $(0.0, 1.0) \equiv (0.0, 1.0, 1.0, 1.0, \ldots)$ is the supremum. All other monotone sequences are totally ordered between them by prefix lexicographical order.

We evaluate $f(\mathbf{u})$ for each of the sequences in the same order and store the output in a file, array or any sequential data structure that supports random access.

For our second implementation, we add the idea of "warping". The grid is built in such a way that, at every subsequent depth, the branching factor gets reduced by half. If the branching factor for the root is $2^n + 1$, then the maximum depth possible for such a grid is $n + 2$. The leaf nodes will contain values u_0 and $u_{|s|-1}$.

It is easy to generate all the sequences in increasing order, evaluate the function and just store the evaluations in file. The real complication is calculating the location/index where the evaluation for the given vector is stored in the file.

Lemma 1. *For the grid with fixed branching factor B at every depth, given any vector $\mathbf{u} = (u_1, u_2, \ldots, u_d)$ where depth $d > 0$, we can find the index of $\mathbf{u}$ in $O(d)$ time.*

Proof. For performing the indexing, we first need to create a table. The table holds $V_{i,b}$, which is the number of nodes at any depth i for any core branch b. Core branches are branches generated from the root, while i and b index the corresponding row and column of the table respectively. The constructed table has d rows and B columns. The entries for the table for the first implementation can be generated using equation(2).

$$V_{i,b} = \begin{cases} 0 & \text{if } i = 1 \\ 1 & \text{if } i = 2 \\ \sum_{j=b}^{B} V_{i-1,j} & \text{otherwise} \end{cases} \tag{2}$$

or

$$V_{i,b} = \begin{cases} 0 & \text{if } i = 1 \\ 1 & \text{if } i = 2 \\ V_{i-1,b} + V_{i,b+1} & \text{if } (b < B) \wedge (i > 2) \\ V_{i-1,b} & \text{otherwise,} \end{cases}$$

The creation of the table, which is a one-time event, takes $O(Bd)$ time. Once the table is generated, for calculating the index for the vector $\mathbf{u} = (u_1, \ldots, u_d)$, we first calculate the position of u_i where $i \in \{1, 2, \ldots, d\}$ in the table. We can define a mapping function $h : u_i \rightarrow m_i$ where m_i is the corresponding column in the grid for u_i. Then the index would be:

$$index = \sum_{i=1}^{d-1} \sum_{j=h(u_i)}^{h(u_{i+1})-1} V_{d+1-i,j}. \tag{3}$$

Calculating the index requires summing $O(Bd)$ terms. By generating a second table that stores the partial sum $\sum_{j=0}^{b} V_{i,j}$ for every b and i, we can perform the inner sum operation of equation (3)in constant time. This makes the whole indexing take $O(d)$ time.

The fixed branching factor algorithm is well suited for most of the application. But in cases where the precision of evaluating f is most crucial for the initial values of the vector $\mathbf{u}$, we use the "warping" concept. A warped grid of depth d has initial branching factor $2^d + 1$ and at each subsequent depth, the branching factor gets reduced by half.

Lemma 2. *For a grid with exponentially reducing branching factor B, given any vector $u = (u_1, u_2, \ldots, u_d)$ and depth $d > 0$, we can find the index of the sequence in $O(d)$ time.*

Proof. As with the grid for fixed branching, for preforming indexing we first need to generate a table. The constructed table has d rows and B columns. The entries for the table for the second implementation can be generated using equation (4).

$$
V_{i,b} = \begin{cases}
0 & \text{if } i = 1 \\
1 & \text{if } i = 2 \\
V_{i-1,b} + V_{i,b+2^{i-2}} & \text{if } (b + 2^{i-2} \leq B) \wedge (i > 2) \\
V_{i-1,b} & \text{otherwise,}
\end{cases}
\tag{4}
$$

where i ranges from 1 to d, and $V_{i,b}$ is the number of nodes at depth i and for core branch index b. The generation of the table requires $O(Bd)$ time.

The indexing for this scheme is different than that of our first scheme. If the mapping function is $g : (u_i, k) \rightarrow m_{i,k}$ we can evaluate $m_{i,k}$ using equation (5).

$$
m_{i,k} = \begin{cases}
0 & \text{if } (h(u_i) - 1) \bmod 2^{k-1} \neq 0 \\
B - (B-1)/2^{k-1} + (h(u_i) - 1)/2^{k-1}) & \text{otherwise,}
\end{cases}
\tag{5}
$$

Here the function h is the mapping function used for fixed branched grid and k represents the depth for which the index is sought. while $m_{i,k} = 0$ indicates that the corresponding u_i is not present for the particular depth. This information can be stored for future lookup. The modified indexing function for this reduced branched grid is:

$$
index = \sum_{i=1}^{d-1} \sum_{j=g(u_i,i)}^{g(u_{i+1},i)-1} V_{d+1-i,j}
\tag{6}
$$

Again, by the same procedure as described for fixed branching, we can calculate the corresponding index for any vector in $O(d)$ time.

We can further specialize the scheme by reducing the branches only at specific depths. This gives us better control on the overall size and precision of the grid.

Lemma 3. *For a grid with initial branching factor B with a binary vector r of size d indicating reduction in branches at any depth, given any vector $u = (u_1, u_2, \ldots, u_d)$ and depth $d > 0$, we can find the index of the sequence in $O(d)$ time.*

Proof. Like the earlier two approaches, we first need to construct a table for lookup. The table needs to be constructed recursively from depth $j = d$ to 1 due to the variable nature of r. The intermediate rows in the table will be rewritten for each j. The constructed table has d rows and B columns. The first two values in r are 0 which indicates no reduction in branches while the rest of the values can be either 0 or 1 indicating no reduction and reduction in branches respectively. The entries for the table for this implementation can be generated using equation (7) for any intermediate depth i and final depth j where $R(i,j) = \sum_{k=d-j+1}^{d-j+i} r_k$.

$$V_{i,b,j} = \begin{cases} 0 & \text{if } i = 1 \\ 1 & \text{if } i = 2 \\ V_{i-1,b,j} + V_{i,b+2R(i,j)} & \text{if } (b + 2R(i,j) \leq B) \wedge (i > 2) \\ V_{i-1,b,j} & \text{otherwise,} \end{cases} \tag{7}$$

where i ranges from 1 to j, and $V_{i,b,j}$ is the number of nodes at intermediate depth i and for branch index b for the purpose of generating row j of the table. Once the table is created we can ignore the j parameter for indexing purpose. Due to the recursive nature of creation, the generation of the table requires $O(Bd^2)$ time in comparison to $O(Bd)$ time required for earlier two implementation. After the construction of the table the mapping function $g : (u_i, k) \to m_{i,k}$ can be evaluated using equation (8) where $R(k) = \sum_{j=1}^{k+1} r_j$.

$$m_{i,k} = \begin{cases} 0 & \text{if } (h(u_i) - 1) \bmod 2^{R(k)} \neq 0 \\ B - B/2^{R(k)} + (h(u_i) - 1)/2^{R(k)}) & \text{otherwise,} \end{cases} \tag{8}$$

The indexing scheme is the same as *reduced-branching* implementation.

4 Interpolation Algorithm

Although the members of U are totally ordered lexicographically, the values $f(u)$ are generally *not* monotone in this ordering, and this makes the neighborhood topology play havoc when given a non-grid sequence x. For instance, if the spacing starts $0.00, 0.02, 0.04, \ldots$ in the first few index places $i \geq 2$, then the sequence

$$x = 0.00, 0.01, 0.20, 0.40, 0.60, 0.80, 1.0 \ldots$$

has as its *lower* neighbor the sequence $0.00, 0.00, 1.0, 1.0 \ldots$, which generally gives much *higher* p_1, and its *upper* neighbor the sequence $0.00, 0.02, 0.02, 0.02, 0.02 \ldots$ in which all moves are close to equal and p_1 has nearly its minimum possible value, which is $1/\ell$. Thus attempting to use the neighbors in the *trie* structure of the sequences is bad. Instead, given (any) x, we define its *component-wise bounds* x^+ and x^-. In this case, assuming the spacing continues $\ldots, 0.20, 0.35, 0.50, 0.75, 1.0$, we get:

$$x^+ = 0.00, 0.02, 0.20, 0.50, 0.75, 1.0, 1.0 \ldots$$
$$x^- = 0.00, 0.00, 0.20, 0.35, 0.50, 0.75, 1.0 \ldots$$

Since these are members of U, it is plausible to output some weighted average of $f(x^+)$ and $f(x^-)$. We mix this idea with interpolating to find a better argument u between x^+ and x^-. For this purpose we use the spacing vector s, which contains all possible points in ascending order, ending with the maximum, 1.0. We also let $wt(i, x, j, s)$ stand for a weighting function which can be the aforementioned $(1 - x_i)/i$ gradient or might reflect s and its index j as well.

Our procedure INTERPOLATE$(x, s, d; u)$ creates u of depth d by first truncating x to length d (or padding x with extra 1.0 values if its length is $< d$), initializing j and "credit" c to 0, and then executing the following loop for $i = 1$ to d:

1. while $(s[j + 1] < x[i])$ do $j := j + 1$; //so $x[i] \leq s[j + 1]$
2. let $a = (s[j] + s[j + 1])/2$;
3. if $a > x[i] + c$ then do $u[i] := s[j]$ and $c := c + wt(i, x, j, s) * (x[i] - s[j])$;
4. else do $u[i] = s[j + 1]$; $j := j + 1$; and $c := c - wt(i, x, j, s) * (s[j + 1] - x[i])$;

If we round down in step 3, then higher c may influence us to round up in the next iteration. Once we round up, however, the increment in j makes $s[j + 1]$ the new floor, so we can never round down to a lower value. The $x[i]$ values may stay below this floor—then the while-loop in step 1 does nothing, and negative adjustments in c will prevent further rounding up unless $x[i']$ itself increases for some $i' > i$. The algorithm runs in $O(d)$ time since j never backtracks. A modified version of this interpolation algorithm can be used for reduced and selective branching implementation where at any depth s may have half the elements that its earlier depth in case of reduction of branches at that depth. In that case, we need to replace j by $j/2$. An additional check is required to make sure that for every i, $u[i] \geq u[i - 1]$.

5   Experimental Results

We implemented all of the schemes in C++ with highest optimization on a shared Red Hat Enterprise Linux Server release 5.7 (Tikanga) 32GB 64-bit system configured for non-interactive, CPU-intensive and long-running processes. Our main tests employed the vector-root function from Section 2 as f. The core branches used for the fixed branched grid are shown in table 1. For this scheme, we set the branching factor to 16 and the depth to 15. We generated the evaluation for every possible vector using the core-branch values, and stored the evaluations in a file. The generated file had a total of $77,558,760$ entries, starting from the evaluation for a vector of all 0's to the entry for the length-2 vector $(0, 1)$.

Table 1. Core branches used for fixed-branch grid

Index	1	2	3	4	5	6	7	8	9	10	11	14	13	14	15	16
Value	0.0	0.05	0.10	0.15	0.20	0.25	0.30	0.35	0.40	0.45	0.50	0.60	0.70	0.80	0.90	1.0

For testing the performance of our scheme, we used 8,000 random vectors x. The distribution was controlled by a parameter b governing the expected time to hit the 1.0 ceiling—in chess terms it expressed how many reasonable (non-blunder) moves to expect in a position. This was done by initializing $p = 0$, making $x_1 = 0$ the first entry, and for each iteration, generating a uniformly distributed random number r between p and 1. Then we select the next element of the vector as $p + \frac{r-p}{b}$ and update p to the same value. We have tested our implementation for various b values ranging from 3 to 6.

For each of those 8,000 vectors, we first calculated the exact vector-root using Newton's method and then calculated the closest neighbor of that vector using various interpolation algorithm (refer Section 4) and found the corresponding index for that vector. Finally we accessed the file to fetch the evaluation at that location. For interpolation, we used various weighting and mapping policies. The details of the various interpolation schemes are as follows:

1. The *simple nearest neighbor* (NN) strategy follows the algorithm INTERPOLATE with $wt(\cdots) = 0$. Each value in the input vector was hence matched to the nearest grid point, subject to the monoticity requirement.
2. The *universal gradient* (UG) strategy sets the weight to $(1 - x_i)/i$, per equation (1).
3. The *gridpoint weight* (GW) strategy sets the weight to $wt(i, \boldsymbol{x}, j, \boldsymbol{s}) = (1 - s_j)(1 - x_i)$, where s_j is the nearest gridpoint to which x_i is mapped.
4. The *simple lower bound* (LB) policy maps every element of the vector to the nearest grid value which is smaller or equal to x_i . The generated vector is the pointwise lower bound for $\boldsymbol{x}$.
5. The *simple upper bound* (UB) policy maps every element of the vector to the nearest grid value which is equal or greater than x_i. The generated vector is the pointwise upper bound for $\boldsymbol{x}$.

The last two furnish comparisons to show that cavalier interpolation produces significantly worse results.

Table 2. Comparison between interpolation algorithms

$b =$	3	4	5	6
NN	0.0010	0.0007	0.0004	-0.0003
UG	0.0009	0.0007	0.0002	-0.0007
GW	0.0008	0.0003	-0.0008	-0.0023
LB	-0.2284	-0.2711	-0.2993	-0.3209
UB	0.1871	0.1487	0.1221	0.1011

Table 3. Comparison between interpolation algorithms for selective branching

$b =$	3	4	5	6
NN	0.0004	0.0007	0.0009	0.0005
UG	0.0004	0.0006	0.0008	0.0004
GW	0.0003	0.0004	0.0005	0.0002
LB	-0.0131	-0.0142	-0.0150	-0.0154
UB	0.0158	0.0172	0.0173	0.0163

Table 2 presents the average deviation from the exact vector-root computation for the various strategies and values of b. The results show that each of the first three schemes gives good approximation, but there is no clear winner, rather the best performance depends on the distribution of data.

Figure 1 shows a scatter plot of the deviation from the true values of the vector root function where b was set to 5 and the UG strategy was used.

We ran similar tests for reduced-branching implementation. The branching factor of the grid was set to 257, where each branch was equally spaced between 0 and 1, and the depth of the grid was 10. For interpolation we used 'NN' strategy. We tested the

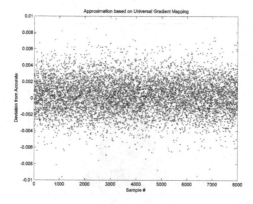

Fig. 1. Performance analysis for the fixed-branch grid

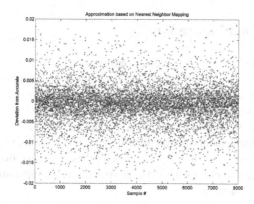

Fig. 2. Approximation performance analysis for the reduced-branched grid

implementation with 8,000 iterations. Figure 2 shows the closeness of fit. The average deviation for the scheme from the true vector-root value was -0.0008, where the standard deviation was 0.0052. For selective branching implementation, we set the branching factor and depth to 33 and 20 respectively. The branches were equally spaced, and reduction of branches occurred at depth $5, 9, 17$.

From Table 3 we observe the selective reducing scheme can produce output very close to accurate , and among all interpolation algorithm, 'gridpoint weight' strategy works better.

Figure 3 represents the histogram for deviation from actual calculation of vector root. The bias parameter 'b was set to 6 to generate the random vectors used for the histogram.

On an average, the execution time for any scheme was around 10 times faster than the real-time vector-root evaluation.

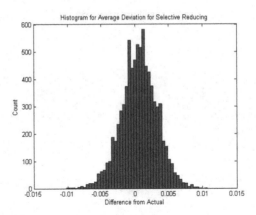

Fig. 3. Performance analysis for the selective branched grid

6   Conclusions and Future Work

In this paper we have developed a special purpose data structure for storing sample points of a function f so that the values of f at other points can be interpolated efficiently. This data structure can be used to store the result of computation intensive function values for faster remote computing. Our main concern is not space efficiency [6,7] rather faster retrieval of function evaluation. As the function is not evaluated in real time, this suits the requirement of embedded system, where both the processing power and conservation of energy is vital [8]. Along with these benefits, this data structure is around 10 times faster and provides good approximation in comparison to the real-time evaluation of various functions.

References

1. Campbell-Kelly, M., Croarken, M., Flood, R., Robson, E.: The History of Mathematical Tables. Oxford University Press, Oxford (2003)
2. Muthukrishnan, S.: Data Streams: Algorithms and Applications. Now Publishers (2005)
3. Michie, D.: Memo functions and machine learning. Nature 218, 19–22
4. Liang, F.: Stochastic approximation Monte Carlo for MLP learning. In: Encyclopedia of Artificial Intelligence, pp. 1482–1489. Wiley (2009)
5. Regan, K., Haworth, G.McC.: Intrinsic chess ratings. In: Proceedings of AAAI 2011, San Francisco (2011)
6. Khoshnevisan, H., Afshar, M.: Space-efficient memo-functions. Journal of Systems and Software 35 (1996)
7. Khoshnevisan, H.: Efficient memo-table management strategies. Acta Informatica 28, 43–81
8. Alidina, M., Monteiro, J.C., Devadas, S., Ghosh, A., Papaefthymiou, M.C.: Precomputation-based sequential logic optimization for low power. IEEE Trans. VLSI Syst. 2(4), 426–436

Partially Dynamic Single-Source Shortest Paths on Digraphs with Positive Weights

Wei Ding[1] and Guohui Lin[2,*]

[1] Zhejiang University of Water Resources and Electric Power
Hangzhou, Zhejiang 310018, China
dingweicumt@163.com

[2] Department of Computing Science, University of Alberta
Edmonton, Alberta T6G 2E8, Canada
guohui@ualberta.ca

Abstract. We examine several structural properties of single-source shortest paths and present a local search algorithm for the partially dynamic single-source shortest paths problem. Our algorithm works on both deterministic digraphs and undirected graphs. For a deterministic digraph with positive arc weights, our algorithm handles a single arc weight increase in $O(n + \frac{n^2 \log n}{m})$ expected time, where n is the number of nodes and m is the number of edges in the digraph. Specifically, our algorithm is an $O(n)$ expected time algorithm when $m = \Omega(n \log n)$. This solves partially an open problem proposed by Demetrescu and Italiano (Journal of the ACM. 51(2004), 968–992).

Keywords: partially dynamic, single-source shortest paths, local search, expected time.

1 Introduction

An *all-pairs shortest paths* (APSP) algorithm computes the shortest paths between every pair of nodes in a given digraph, and a *single-source shortest paths* (SSSP) algorithm computes the shortest paths from a given source node to all the other nodes. When dynamic changes occur to the digraph, a *dynamic* APSP (SSSP, respectively) algorithm updates the shortest paths. One can recompute the shortest paths using the *static* algorithms, but a truly dynamic algorithm seeks for updating operations using fundamental properties of the shortest paths, and is expected to run faster than recomputation by the static algorithms. We present a *partially* dynamic SSSP algorithm for digraphs with arbitrary positive arc weights. We use arc and edge interchangeably in this paper.

Dynamic changes of a digraph (a.k.a. *edge update*) include *topology update* and *edge weight update*. Topology update includes *edge insertions* and *deletions*, and edge weight update includes weight *increase* and *decrease*. Note that topology update can be realized by edge weight update, and vice versa. When dealing with

---

* Correspondence author.

Q. Gu, P. Hell, and B. Yang (Eds.): AAIM 2014, LNCS 8546, pp. 197–207, 2014.

only edge weight updates, an algorithm is said to be *fully dynamic* if it can handle both edge weight increases and decreases, but *partially dynamic* if it only can handle edge weight decreases or increases but not both [1,4,5,10,13,16,17]. When dealing with only topology updates, an algorithm is *fully dynamic* if it can handle both edge insertions and deletions, is *incremental* if it can handle only edge insertions but not deletions, and is *decremental* if it can handle only deletions but not insertions [2,3,7,8,11,14,15]. Incremental and decremental algorithms are sometimes collectively called *partially dynamic*.

Demetrescu and Italiano [4,5] studied a generalization of dynamic changes, in which the weights of all the edges incident at a given node are changed in one update. Such an update is called a *node update*, see also [16,17]. Clearly, dynamic APSP and SSSP algorithms for node updates work for edge updates too. Throughout this paper, n and m denote the numbers of nodes and edges (arcs) in the input digraphs, respectively.

The dynamic APSP problem with edge weight updates has been studied extensively since it was proposed [12]. Ausiello *et al.* [1] devised a partially dynamic APSP algorithm with an $O(Cn \log n)$ amortized update time for digraphs of which all edge weights are less than or equal to a constant C. King [11] studied the fully dynamic APSP problem on the same class of digraphs and presented an algorithm with $O(n^{\frac{5}{2}}\sqrt{C \log n})$ worst-case update time. Demetrescu and Italiano [4,5] considered node updates on digraphs with non-negative real-valued edge weights, and designed a fully dynamic APSP algorithm with an $O(n^2 \log^3 n)$ amortized time. Thorup [16,17] presented an improved algorithm with a worst-case $O(n^{2.75}\text{polylog}(n))$ time. The fully dynamic APSP problem on *random* graphs was considered by Friedrich and Hebbinghaus [10], who gave an $O(n^{\frac{4}{3}+\epsilon})$ expected time per update algorithm, for any $\epsilon > 0$, for random graphs $G(n,p)$ with uniform random edge weights, and by Peres *et al.* [13], who presented an $O(\log^2 n)$ expected time per update algorithm for complete digraphs with edge weights selected independently at random from the uniform distribution on interval $[0,1]$. There are also a number of *approximate* incremental and decremental APSP algorithms [2,3,14] for undirected graphs with positive edge weights.

For the dynamic SSSP problem with topology updates, Even and Shiloach [8] and Dinitz [7] presented $O(n)$ amortized time decremental algorithms for unweighted and undirected graphs. King [11] presented a decremental algorithm to maintain shortest paths of distance up to d in $O(md)$ time, for digraphs with positive integer edge weights. Bernstein and Roditty [3] devised an $O(\frac{n^2}{m})$ amortized time decremental algorithm to maintain $(1+\epsilon)$-approximate SSSP on unweighted and undirected graphs. This is the first algorithm that breaks the long-standing $O(n)$ update time barrier on decremental SSSP problem, on not-too-sparse graphs. On the other hand, Roditty and Zwick [15] showed that the incremental and decremental SSSP problems on edge weighted digraphs are at least as hard as the static APSP problem; by similar reductions they showed that the incremental and decremental SSSP problems on edge unweighted digraphs are at least as hard as the Boolean matrix multiplication problem. These hardness results hint that it will be difficult to improve the known best algorithms of Even and Shiloach [8].

For the dynamic SSSP problem with edge weight updates, Fakcharoemphol and Rao [9] studied the fully dynamic variant on planar digraphs, and devised an $O(n^{\frac{4}{5}} \log^{\frac{13}{5}} n)$ amortized time algorithm. Demetrescu and Italiano [4] raised the open problem on whether or not we can solve efficiently (i.e., better than recomputation) fully dynamic SSSP problem on general graphs. We address partially this open problem by presenting a partially dynamic SSSP algorithm for handling arc weight increase on digraphs with positive weights. For deterministic digraphs, our algorithm can handle a single arc weight increase in $O(n + \frac{n^2 \log n}{m})$ expected time. When $m = \Omega(n \log n)$, our algorithm is an $O(n)$ expected time algorithm. Moreover, our algorithm can also work on undirected graphs with positive weights.

The rest of the paper is organized as follows. In Sect. 2, we define some notations frequently used in this paper and the partially dynamic SSSP problem formally. In Sect. 3, we show several fundamental properties. In Sect. 4, we present a local search algorithm based on these properties. In Sect. 5, we analyze the worst-case expected update time of our algorithm handling a single arc weight increase for deterministic digraphs. In Sect. 6, we conclude the paper with some future work.

2   Problem Statements and Notations

Let $D = (V, A, w, s)$ be a weighted digraph, where V is the node set, A is the arc set, s is a designated node called *source*, and $w(\cdot)$ is a weight function $w : A \to \mathbb{R}^+$. Suppose that the weight of an arc a increases from $w(a)$ up to $w'(a)$ and all the other arc weights stay unchanged. Let $\delta = w'(a) - w(a)$. Note that $w'(a)$ always remains positive. The resultant digraph is denoted as $D' = (V, A, w', s)$. We use $d_D^*(s, v)$ (resp. $d_{D'}^*(s, v)$) to denote the shortest path distance from s to node v in D (resp. D'), and use T_s (resp. T_s') to denote the *single-source shortest paths tree* in D (resp. D').

Problem 1. Given $D = (V, A, w, s)$ and T_s in D, we replace $w(a)$ with $w'(a)$ for one arc a to obtain a new digraph $D' = (V, A, w', s)$. The problem of updating T_s to T_s' in D' is called the *partially dynamic SSSP problem with a single arc weight increase*.

The essence of Problem 1 is to maintain SSSP, that is, to update the given T_s in D to T_s' in D'. Obviously, T_s and T_s' are both out-trees from s, and both can be taken as rooted trees at s. For any node of V, let T_u (resp. T_u') denote the subtree of T_s (resp. T_s') rooted at u. We use $\pi_T(s, v)$ to denote the s-to-v simple path along T_s, and use $d_T(s, v)$ to denote the length of $\pi_T(s, v)$ which is equal to the sum of the weights on all the arcs of $\pi_T(s, v)$. Clearly, $d_T(s, v) = d_D^*(s, v)$ and $d_{T'}(s, v) = d_{D'}^*(s, v)$ for every $v \in V$. In addition, we use $V(\cdot)$ and $A(\cdot)$ to denote the node set and arc set of one digraph or its subgraph respectively. Let $S(D, U)$ denote the subgraph of D induced by the subset $U \subseteq V$ of nodes.

For any subset $U \subset V$, we let $C[U, V \setminus U]$ denote the subset of arcs with tail in U and head in $V \setminus U$.

Let $a = (u, v)$ denote the arc of D from u to v and $w(u, v)$ (or sometimes $w(a)$) denote its weight. We call u the *tail* of a and denote u as t_a, and call v the *head* of a and denote v as h_a. Thus, $a = (t_a, h_a)$, which is called an *outgoing arc* from t_a and an *incoming arc* to h_a. The set of all the outgoing arcs from a node u is denoted as $\mathcal{O}(u)$ and the set of all the incoming arcs to v is denoted as $\mathcal{I}(v)$.

3   Fundamental Properties

In the section, we show several fundamental properties which will play an important role in the design of our *local search algorithm*, described as PSAI, for updating SSSP under a single arc weight increase.

Theorem 1. *For any $a \notin T_s$, no matter how much $w(a)$ increases by to $w'(a)$, it always holds that $d_{D'}^*(s, v) = d_T(s, v)$ for any $v \in V$.*

Proof. No matter how much $w(a)$ increases by to $w'(a)$, the length of every simple s-to-v_1 path in D' passing through a is larger than its length in D, and the length of every simple s-to-v_2 path in D' not passing through a is the same as its length in D. So the length of every simple path in D' is either larger than or equal to its length in D. When $a \notin T_s$, $\pi_T(s, v)$ obviously does not pass through a, and thus the length of $\pi_T(s, v)$ stays unchanged. Considering that $d_T(s, v) = d_D(s, v)$, we conclude that $\pi_T(s, v)$ is always the s-to-v shortest path in D' for any $v \in V$. So, $d_{D'}^*(s, v) = d_T(s, v), \forall v \in V$. The proof is complete.   □

Theorem 2. *For any $a \in T_s$, it always holds that $d_{D'}^*(s, u) = d_T(s, u)$ for any $u \in V \setminus V(T_{h_a})$ regardless of how much $w(a)$ increases by to $w'(a)$.*

Proof. When $a \in T_s$, we observe that $\pi_T(s, u)$ does not pass through a for any $u \in V \setminus V(T_{h_a})$. Thus, the length of $\pi_T(s, u)$ stays unchanged and thus $\pi_T(s, u)$ is always the s-to-u shortest path in D' regardless of how much $w(a)$ increases by to $w'(a)$. This implies that $d_{D'}^*(s, u) = d_T(s, u), \forall u \in V \setminus V(T_{h_a})$. The proof is complete.   □

When $a \in T_s$ and its weight increases to $w'(a)$, one observes that since $\pi_T(s, v)$ passes through a for any $v \in V(T_{h_a})$, the length of $\pi_T(s, v)$ in D' is equal to its length in D plus $w'(a) - w(a)$. So, the length of $\pi_T(s, v)$ in D' is larger than $d_T(s, v)$. Since the length of every s-to-v simple path not passing through a stays unchanged, it may occur that some s-to-v simple paths in D' not passing through a have a smaller length than $\pi_T(s, v)$. We conclude that every s-to-v simple path in D' having a smaller length than $\pi_T(s, v)$ is surely composed of one s-to-u simple path in D', followed by arc (u, v') and the path v'-to-v in T_{h_a}, where $u \in \mathcal{I}(v') \setminus V(T_{h_a})$. Such an s-to-v' path can minimize its length by selecting $\pi_T(s, u)$ as the s-to-u simple path. Let

$$b(v) = \arg \min_{u \in \mathcal{I}(v) \setminus V(T_{h_a})} \{d_T(s, u) + w(u, v)\}, \quad \forall v \in V(T_{h_a}), \qquad (1)$$

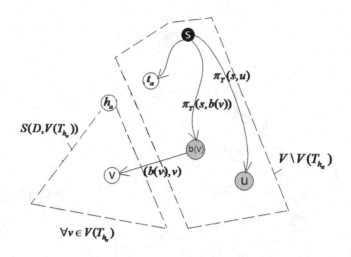

Fig. 1. Illustration of D_a

and $b(v)$ is called a *bridge node* of v. Let $b(v) = $ null if $b(v)$ does not exist in D' and thus $(b(v), v) = $ null. In addition, we use an arc $\epsilon(\pi_T(s, v))$ to represent $\pi_T(s, v)$.

Accordingly, we can construct an auxiliary graph $D_a = (V, A_a, w_a(\cdot))$ from D based on a in the following way, see Fig. 1. The new arc set A_a is composed of all the arcs in the subgraph of D induced by $V(T_{h_a})$ and all $\epsilon(\pi_T(s, u)), u \in V \setminus V(T_{h_a})$ and all $(b(v), v), v \in V(T_{h_a})$. For any arc $r \in A_a$, the weight of r, $w_a(r)$, is equal to the length of $\pi_T(s, u_0)$ if r represents $\pi_T(s, u_0)$, or otherwise $w_a(r) = w(r)$. That is,

$$A_a = A(S(D, V(T_{h_a}))) \cup \{(b(v), v) : v \in V(T_{h_a})\}$$
$$\cup \{\epsilon(\pi_T(s, u)) : u \in V \setminus V(T_{h_a})\}, \tag{2}$$

and

$$w_a(r) = \begin{cases} d_T(s, u_0) & \text{if} \quad r = \epsilon(\pi_T(s, u_0)), \\ w(r) & \text{otherwise,} \end{cases} \quad \forall r \in A_a. \tag{3}$$

Lemma 1. *For any $a \in T_s$ and any $v \in V(T_{h_a})$, it always holds that any s-to-v shortest path in D' contains exactly one arc in $C[V \setminus V(T_{h_a}), V(T_{h_a})]$.*

Proof. We conclude from $s \in V \setminus V(T_{h_a})$ and $v \in V(T_{h_a})$ that any s-to-v simple path in D' contains at least one arc in $C[V \setminus V(T_{h_a}), V(T_{h_a})]$. Suppose that $\pi^\triangle(s, v)$ is an s-to-v shortest path in D' containing two arcs in $C[V \setminus V(T_{h_a}), V(T_{h_a})]$. In details, $\pi^\triangle(s, v)$ consists of $\pi_T(s, u_1)$, two arcs (u_1, v_1) and (u_3, v_3) in $C[V \setminus V(T_{h_a}), V(T_{h_a})]$ and (v_2, u_2), two disjoint paths $\pi(v_1, v_2)$ and $\pi(v_3, v)$ in $S(D', V(T_{h_a}))$, and one path $\pi(u_2, u_3)$ in $S(D', V \setminus V(T_{h_a}))$, where $u_1, u_2, u_3 \in V \setminus V(T_{h_a})$ and $v_1, v_2 \in V(T_{h_a})$ (see Fig. 2). Note that it is

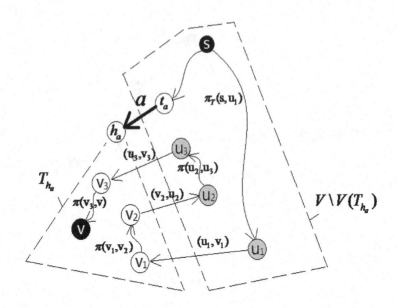

Fig. 2. Illustrate the proof of Lemma 1

possible that $v_1 = v_2, u_2 = u_3$ and $v_3 = v$. Clearly, the seven parts of $\pi^{\triangle}(s, v)$ are disjoint. The sub-path $\pi^{\triangle}(s, u_3)$ of $\pi^{\triangle}(s, v)$ is an s-to-u_3 simple path in D'. By Theorem 2, $\pi_T(s, u_3)$ is still an s-to-u_3 shortest path in D'. So, the length of $\pi_T(s, u_3)$ is less than the length of $\pi^{\triangle}(s, u_3)$. Therefore, a new path $\pi^*(s, v)$ composed of $\pi_T(s, u_3)$, (u_3, v_3) and $\pi(v_3, v)$ has a smaller length than $\pi^{\triangle}(s, v)$. This causes a contradiction.   □

Theorem 3. *For any $a \in T_s$, it always holds that $d_{D'}^*(s, v) = d_{D_a}^*(s, v)$ for any $v \in V$ regardless of how much $w(a)$ increases by to $w'(a)$.*

Proof. When $a \in T_s$, Theorem 2 shows that $\pi_T(s, u)$ is always an s-to-u shortest path in D' for any $u \in V \setminus V(T_{h_a})$ and also an s-to-u shortest path in D_a. So, $d_{D'}^*(s, u) = d_{D_a}^*(s, u)$. The work left is to prove $d_{D'}^*(s, v) = d_{D_a}^*(s, v)$ for any $v \in V(T_{h_a})$.

We observe that the weight of a increases from $w(a)$ up to $w'(a)$ and thus the length of $\pi_T(s, v)$ in D' increases by $w'(a) - w(a)$. So, it is certain that an s-to-v shortest path in D' is one of such simple paths as composed of three disjoint parts $\pi_T(s, b(v_0)), (b(v_0), v_0)$ and a v_0-to-v path $\pi(v_0, v)$ where $v_0 \in V(T_{h_a})$. We conclude from Lemma 1 and $(b(v_0), v_0) \in C[V \setminus V(T_{h_a}), V(T_{h_a})]$ that $\pi(v_0, v)$ contains no arc in $C[V \setminus V(T_{h_a}), V(T_{h_a})]$. So, $\pi(v_0, v)$ is a v_0-to-v shortest path in $S(D, V(T_{h_a}))$.

We need to visit all the incoming arcs to v in order to find $b(v)$ for all $v \in V(T_{h_a})$ and need to traverse $S(D, V(T_{h_a}))$ to compute a v_0-to-v shortest path in $S(D, V(T_{h_a}))$. Therefore, the problem of finding an s-to-v shortest path in D' is equivalent to the problem of finding an s-to-v shortest path in D_a. This implies that $d_{D'}^*(s, v) = d_{D_a}^*(s, v)$. The proof is complete. $\qquad\square$

4 Local Search Algorithm

From the properties shown in Sect. 3, we need to discuss the situation of the arc a with its weight increased, i.e., discuss whether $a \notin T_s$ or $a \in T_s$. When $a \notin T_s$, we conclude from Theorem 1 that $\pi_T(s, v)$ is also an s-to-v shortest path in D' for any $v \in V$. Therefore, T_s is also a single-source shortest paths tree in D' with s as the origin. So we need no work. When $a \in T_s$, we conclude from Theorem 2 that $\pi_T(s, u)$ is also an s-to-u shortest path in D' for any $u \in V \setminus V(T_{h_a})$, and from Theorem 3 that $\pi(s, v)$ is an s-to-v shortest path in D' iff it is an s-to-v shortest path in D_a for any $v \in V(T_{h_a})$. So, we only need to update the s-to-v shortest path for all $v \in V(T_{h_a})$. Above discussions can be described as a local search algorithm PSAI.

For every $v \in V(T_{h_a})$, we conclude from Eq. (1) that we need to visit all the nodes in $\mathcal{I}(v)$ and judge whether one node is in T_{h_a} or not. We define a 0-1 variable $p(v)$. In details, $p(v) = 0$ means that v is not in T_{h_a} and $p(v) = 1$ means that v is in T_{h_a}. Initially, we set $p(v) = 0$ for all $v \in V$. In order to make PSAI facilitate implementing its local search procedure, we use appropriate data structures to store the input digraph and single-source shortest paths trees.

Algorithm PSAI:
Input: $D = (V, A, w, s)$, T_s and $w'(a)$;
Output: T_s' in D'.
Step_0: If $a \notin T_s$, then return T_s; if $a \in T_s$, then goto Step_1;
Step_1: Use DFS to traverse T_{h_a} twice from h_a; in the first one, we let
$\qquad p(v) \leftarrow 1, \forall v \in V(T_{h_a})$; in the second one, for every $v \in V(T_{h_a})$,
$\qquad$ we visit all nodes in $\mathcal{I}(v)$ and find $b(v)$ using Eq. (1);
Step_2: Construct D_a based on Eqs. (2) and (3);
Step_3: Use Dijkstra's algorithm in D_a to compute the single-source
$\qquad$ shortest paths tree with s as the origin, and record it as T_s';
Step_4: Use DFS to traverse T_{h_a} and reset $p(v) \leftarrow 0, \forall v \in V(T_{h_a})$;

Theorem 4. *PSAI takes $O(m + n \log n)$ time in the worst case.*

Proof. Let $|V(T_{h_a})| = n'$, $|A(S(D, V(T_{h_a})))| = m'$ and $\sum_{v \in V(T_{h_a})} |\mathcal{I}(v)| = K$. Step_1 first takes $O(n')$ time to use DFS in T_{h_a} to do preliminaries, and then uses DFS in T_{h_a} the second time to find $b(v), \forall v \in V(T_{h_a})$ which takes $O(K)$ time. Obviously, $n' \leq K$. So, Step_1 runs $O(K)$ time. Since T_{h_a} has at most n' bridge nodes, D_a has n nodes and at most $m' + n$ arcs. Clearly, $m' \leq K$. So, Step_2 spends $O(K)$ time to construct D_a. Step_3 uses Dijkstra's algorithm [6] in D_a to compute T_s', whose running time is $O(|A(D_a)| + |V| \log |V|)$ and $O(K + n \log n)$. Step_4 takes $O(n')$ time to use DFS in T_{h_a} to reset all the values of $p(v), v \in V(T_{h_a})$. Since $K \leq m$, PSAI takes $O(m + n \log n)$ time in the worst case. $\qquad\square$

5   Average Case Analysis

Let $\{\cdot\}$ denote an *event*, e.g., $\{h_a = u\}$ represents the event of a having a head u, and $\{a \in T_s\}$ (*resp.* $\{a \notin T_s\}$) represents the event that T_s contains a (*resp.* T_s does not contain a). Let $\omega = \langle \omega_1, \omega_2 \rangle$ be an unchangeable couple, where $\omega_1 \in \Omega_1, \omega_2 \in \Omega_2$ and Ω_1, Ω_2 are two *event spaces* as follows

$$\Omega_1 = \{\{h_a = u\} : u \in V\}, \quad \Omega_2 = \{\{a \in T_s\}, \{a \notin T_s\}\}. \tag{4}$$

Suppose that $\{\omega_1 \in \Omega_1\}$ and $\{\omega_2 \in \Omega_2\}$ are *independent*. Let

$$\Omega = \Omega_1 \times \Omega_2 = \{\langle \omega_1, \omega_2 \rangle : \omega_1 \in \Omega_1, \omega_2 \in \Omega_2\}. \tag{5}$$

Given any $D = (V, A, w, s)$, we denote by G the topology of D and by T_s the SSSP tree of D with s as the origin. In fact, a given D means that both G and T_s of D are given. When D is given, every $\omega \in \Omega$ represents a situation of a single arc weight increase, and thus acts as an elementary *conditional event* of time costs of PSAI. So, Ω is just the *conditional event space* induced by D.

Let $\mathbb{E}[Z|X = x, Y = y]$ denote the *conditional expectation* of Z when $X = x$ and $Y = y$. Lemma 2 shows an important formula on conditional expectation.

Lemma 2. *Given two discrete random variables X, Y and another random variable Z, provided that $\{X = x\}$ and $\{Y = y\}$ are the condition events of Z, we have*

$$\mathbb{E}[Z] = \sum_{x,y} \mathbb{E}[Z|X = x, Y = y] \cdot \Pr[X = x, Y = y]. \tag{6}$$

Let $\mathbb{E}[time]$ denote the expected time of PSAI dealing with a single arc weight increase in D, $\mathbb{E}[time|h_a = u, a \in T_s]$ denote the expected time of PSAI dealing with the increase of a with $h_a = u$ and $a \in T_s$, and $\mathbb{E}[time|h_a = u, a \notin T_s]$ denote the expected time of PSAI dealing with the increase of a with $h_a = u$ and $a \notin T_s$. Suppose that a single arc weight increase of D occurs at random from the *uniform distribution* on A, i.e.,

$$\Pr[\text{an increase occurs to } a] = \frac{1}{m}, \quad \forall a \in A. \tag{7}$$

Theorem 5. *Given any $D = (V, A, w, s)$, the expected update time of PSAI dealing with a single arc weight increase is*

$$O\Big(\frac{1}{m} \sum_{u \in V} \sum_{v \in V(T_u)} |\mathcal{I}(v)| + \frac{1}{m} \sum_{u \in V} |V| \log |V| + \frac{1}{m} \sum_{u \in V} |\mathcal{I}(u)|\Big). \tag{8}$$

Proof. For any $u \in V$, D has a single incoming arc to u which is in T_s, and $|\mathcal{I}(u)| - 1$ incoming arcs to u which are not in T_s when $1 \le |\mathcal{I}(u)| \le n - 1$. According to Eq. (7), we get

$$\Pr[h_a = u, a \in T_s] = \frac{1}{m}, \tag{9}$$

and

$$\Pr[h_a = u, a \notin T_s] = \frac{|\mathcal{I}(u)| - 1}{m}. \tag{10}$$

Theorem 1 implies that PSAI needs no work when $h_a = u$ and $a \notin T_s$, and thus takes only $O(1)$ time in this case. So,

$$\mathbb{E}[time|h_a = u, a \notin T_s] = O(1). \tag{11}$$

From the description of PSAI and the proof of Theorem 4, we conclude the update time of PSAI for any $v \in V(T_u)$ when $h_a = u$ and $a \in T_s$ is

$$\mathbb{E}[time|h_a = u, a \in T_s] = O(\sum_{v \in V(T_u)} |\mathcal{I}(v)| + |V| \log |V|), \tag{12}$$

According to Eq. (5), we can further rewrite Ω to be

$$\Omega = \bigcup_{u \in V} \{h_a = u, a \in T_s\} \cup \{h_a = u, a \notin T_s\}. \tag{13}$$

and then derive from Lemma 2 that

$$\begin{aligned}
\mathbb{E}[time] &= \sum_{\omega \in \Omega} \mathbb{E}[time|\omega] \cdot \Pr[\omega] \\
&= \sum_{u \in V} \Big(\mathbb{E}[time|h_a = u, a \in T_s] \cdot \Pr[h_a = u, a \in T_s] \Big) + \\
&\quad \sum_{u \in V} \Big(\mathbb{E}[time|h_a = u, a \notin T_s] \cdot \Pr[h_a = u, a \notin T_s] \Big).
\end{aligned}$$

We take Eqs. (9), (10), (11) and (12) into above equality to obtain

$$\begin{aligned}
\mathbb{E}[time] &= \sum_{u \in V} \Big(O(\sum_{v \in V(T_u)} |\mathcal{I}(v)| + |V| \log |V|) \cdot \frac{1}{m} + O(1) \cdot \frac{|\mathcal{I}(u)| - 1}{m} \Big) \\
&= O\Big(\frac{1}{m} \sum_{u \in V} \sum_{v \in V(T_u)} |\mathcal{I}(v)| + \frac{1}{m} \sum_{u \in V} |V| \log |V| + \frac{1}{m} \sum_{u \in V} |\mathcal{I}(u)| \Big). \quad \square
\end{aligned}$$

Theorem 6. *Given any $D = (V, A, w, s)$, the worst-case expected update time of PSAI dealing with a single arc weight increase is $O(n + \frac{n^2 \log n}{m})$.*

Proof. Obviously, we have

$$\sum_{u \in V} |\mathcal{I}(u)| = m \quad \text{and} \quad \frac{1}{m} \sum_{u \in V} |V| \log |V| = \frac{n^2 \log n}{m}.$$

Combining with

$$\frac{1}{m} \sum_{u \in V} \sum_{v \in V(T_u)} |\mathcal{I}(v)| = \sum_{u \in V} \Big(\frac{1}{m} \sum_{v \in V(T_u)} |\mathcal{I}(v)| \Big) \le \sum_{u \in V} O(1) = O(n),$$

we conclude that algorithm PSAI runs in $O(n + \frac{n^2 \log n}{m})$ expected time in the worst case. $\square$

6 Concluding Remarks

We have presented a local search algorithm PSAI for handling a single arc weight increase to maintain SSSP on digraphs. The worst-case update time of PSAI is $O(m+n\log n)$, and the worst-case expected update time of PSAI is $O(n+\frac{n^2\log n}{m})$. When $m = \Omega(n\log n)$, PSAI has an $O(n)$ expected updated time. To the best of our knowledge, we are the first one to propose almost linear time algorithm for maintaining SSSP. Also, PSAI applies to undirected graphs with positive weights.

When a single arc weight reduces, whether a linear time algorithm exists for maintaining SSSP on digraphs remains open.

Acknowledgement. This research was supported in part by NSERC.

References

1. Ausiello, G., Italiano, G.F., Marchetti-Spaccamela, A., Nanni, U.: Incremental algorithms for minimal length paths. Journal of Algorithms 12, 615–638 (1991)
2. Bernstein, A.: Fully dynamic $(2 + \epsilon)$ approximate all-pairs shortest paths with fast query and close to linear update time. In: Proc. of 50th FOCS, pp. 693–702 (2009)
3. Bernstein, A., Roditty, L.: Improved dynamic algorithms for maintaining approximate shortest paths under deletions. In: Proc. of 22th SODA, pp. 1355–1365 (2011)
4. Demetrescu, C., Italiano, G.F.: A new approach to dynamic all pairs shortest paths. Journal of the ACM 51, 968–992 (2004)
5. Demetrescu, C., Italiano, G.F.: Experimental analysis of dynamic all pairs shortest path algorithms. ACM Transactions on Algorithms 2, 578–601 (2006)
6. Dijkstra, E.W.: A note on two problems in connection with graphs. Numerische Mathematik 1, 269–271 (1959)
7. Dinitz, Y.: Dinitz' algorithm: The original version and Even's version. In: Essays in Memory of Shimon Even, pp. 218–240 (2006)
8. Even, S., Shiloach, Y.: An on-line edge-deletion problem. Journal of the ACM 28, 1–4 (1981)
9. Fakcharoemphol, J., Rao, S.: Planar graphs, negative weight edges, shortest paths, and near linear time. In: Proc. of 42nd FOCS, pp. 232–241 (2001)
10. Friedrich, T., Hebbinghaus, N.: Average update times for fully-dynamic all-pairs shortest paths. In: Hong, S.-H., Nagamochi, H., Fukunaga, T. (eds.) ISAAC 2008. LNCS, vol. 5369, pp. 692–703. Springer, Heidelberg (2008)
11. King, V.: Fully dynamic algorithms for maintaining all-pairs shortest paths and transitive closure in digraphs. In: Proc. of 40th FOCS, pp. 81–99 (1999)
12. Murchland, J.: The effect of increasing or decreasing the length of a single arc on all shortest distances in a graph. Technical report, LBS-TNT-26. London Business School, Transport Network Theory Unit, London, UK (1967)
13. Peres, Y., Sotnikov, D., Sudakov, B., Zwick, U.: All-pairs shortest paths in $O(n^2)$ time with high probability. In: Proc. of 51th FOCS, pp. 663–672 (2010)
14. Roditty, L., Zwick, U.: Dynamic approximate all-pairs shortest paths in undirected graphs. In: Proc. of 45th FOCS, pp. 499–508 (2004)

15. Roditty, L., Zwick, U.: On dynamic shortest paths problems. In: Albers, S., Radzik, T. (eds.) ESA 2004. LNCS, vol. 3221, pp. 580–591. Springer, Heidelberg (2004)
16. Thorup, M.: Fully-dynamic all-pairs shortest paths: Faster and allowing negative cycles. In: Hagerup, T., Katajainen, J. (eds.) SWAT 2004. LNCS, vol. 3111, pp. 384–396. Springer, Heidelberg (2004)
17. Thorup, M.: Worst-case update times for fully-dynamic all-pairs shortest paths. In: Proc. of 37th STOC, pp. 112–119 (2005)

Obtaining Split Graphs by Edge Contraction

Chengwei Guo and Leizhen Cai*

Department of Computer Science and Engineering,
The Chinese University of Hong Kong, Hong Kong S.A.R., China
{cwguo,lcai}@cse.cuhk.edu.hk

Abstract. We study the parameterized complexity of the following SPLIT CONTRACTION problem: Given a graph G and an integer k as parameter, determine whether G can be modified into a split graph by contracting at most k edges. We show that SPLIT CONTRACTION can be solved in FPT time $2^{O(k^2)}n^5$, but admits no polynomial kernel unless $NP \subseteq coNP/poly$.

1 Introduction

Graph modification problems constitute a fundamental and well-studied family of problems in algorithmic graph theory, and many classical graph problems can be formulated as graph modification problems. A graph modification problem takes a graph G and an integer k as input, and asks whether G can be modified into a graph belonging to a specified graph class, using at most k operations of a given type, such as vertex deletion, edge deletion, or edge addition. The number k of operations measures how close a graph is to such a specified class of graphs.

Recently the study of modifying graphs by *edge contraction* has been initiated from the parameterized point of view, yielding several results for the following Π-CONTRACTION problem: Given a graph G and a positive integer k as parameter, determine whether G can be modified into a Π-graph (i.e. a graph belonging to class Π) by contracting at most k edges. The Π-CONTRACTION problem has been proved to be FPT for Π being bipartite graphs (Heggernes *et al.* [15], Guillemot and Marx [12]), trees and paths (Heggernes *et al.* [14]), planar graphs (Golovach *et al.* [11]), graphs with degree constraints (Golovach *et al.* [10], Belmonte *et al.* [1]), complete graphs (Cai and Guo [4], Lokshtanov *et al.* [17]), and cographs (Lokshtanov *et al.* [17]). On the other hand, very recently Cai and Guo [4], and Lokshtanov *et al.* [17] independently showed that Π-CONTRACTION is *W[2]-hard* for Π being chordal graphs. Furthermore, Cai and Guo [4] also proved W[2]-hardness of Π-CONTRACTION for Π being H-free for any fixed 3-connected graph H.

In this paper, we study the parameterized complexity of Π-CONTRACTION when Π is the class of split graphs, which forms a subclass of chordal graphs.

---

* Partially supported by GRF grant CUHK410409 of the Research Grants Council of Hong Kong.

Q. Gu, P. Hell, and B. Yang (Eds.): AAIM 2014, LNCS 8546, pp. 208–218, 2014.
© Springer International Publishing Switzerland 2014

SPLIT CONTRACTION
Instance: Graph $G = (V, E)$, positive integer k.
Question: Can we obtain a split graph from G by contracting at
             most k edges?
Parameter: k.

The edge deletion and vertex deletion variants of this problem, known as
SPLIT DELETION and SPLIT VERTEX DELETION, asks whether an input graph
can be modified into a split graph by deleting at most k edges or at most k vertices respectively. Both problems have been shown to be FPT (Cai [3]) and have
polynomial kernels (Guo [13]). Recently, faster FPT algorithms and improved
kernels have been constructed (Ghosh *et al.* [9], Cygan and Pilipczuk [7]). As
split graphs are characterized by forbidden induced subgraphs $\{2K_2, C_4, C_5\}$,
the solution set of SPLIT DELETION or SPLIT VERTEX DELETION must hit every induced copy of these forbidden subgraphs in the input graph, implying that
choices for branching can be bounded. This observation directly or indirectly
yields the above FPT algorithms and kernelization reduction rules for SPLIT
DELETION and SPLIT VERTEX DELETION. Unfortunately, such observation is
no longer true for SPLIT CONTRACTION as contractions can occur for edges not
involved in any induced copies of forbidden subgraphs, making SPLIT CONTRAC-
TION much harder than its edge deletion and vertex deletion variants.

Although most known techniques for SPLIT DELETION and SPLIT VERTEX
DELETION seems unavailable for SPLIT CONTRACTION due to the above reason,
there is a simple relationship between SPLIT VERTEX DELETION and SPLIT
CONTRACTION: Every yes-instance (G, k) of SPLIT CONTRACTION is a yes-
instance $(G, 2k)$ of SPLIT VERTEX DELETION. Therefore all known FPT algorithms for SPLIT VERTEX DELETION can be used to obtain $2k$ vertices whose
deletion results in a split graph) in a yes-instance of SPLIT CONTRACTION, which
will be used as a starting point for obtaining our FPT algorithm for SPLIT CON-
TRACTION.

Our Contributions. We show that SPLIT CONTRACTION can be solved in time
$2^{O(k^2)} n^5$. Our algorithm starts by finding a large split subgraph H in the input
graph and then considers two cases in terms of the clique size of the split subgraph H. If the clique of H is large, then we show that almost all vertices in
this clique are finally included in the clique of some target split graph. We use
a branch-and-search algorithm to enumerate all edge contractions and reduce
our instance to several instances of CLIQUE CONTRACTION that is known to
be FPT. If the clique of H is small, then there will be a large independent set
in the input graph. We develop reduction rules based on a variant of "modular
decomposition" of the input graph: Partition the set of vertices into groups such
that each group induces an independent set and all vertices in each group have
the same neighbors. We can bound the number of such groups, delete "irrelevant" vertices in each group, and reduce the input graph to an "equivalent"
graph with bounded number of vertices. We note that these reduction rules are

useful for obtaining kernelization algorithms of other contraction problems such as CLIQUE CONTRACTION and BICLIQUE CONTRACTION.

On the other hand, we prove, by a polynomial parameter transformation from the RED-BLUE DOMINATING SET problem, that SPLIT CONTRACTION admits no polynomial kernel unless $NP \subseteq coNP/poly$. This is in contrast to that SPLIT DELETION and SPLIT VERTEX DELETION have polynomial kernels.

2   Preliminaries

Graphs. We consider simple and undirected graphs $G = (V, E)$, where V is the vertex set and E is the edge set. Two vertices $u, v \in V$ are *adjacent* iff $uv \in E$. A vertex v is *incident* with an edge e iff v is an endpoint of e. The *neighbor set* $N_G(v)$ of a vertex $v \in V$ is the set of vertices that are adjacent to v in G. The *closed neighbor set* of v is denoted by $N_G[v] = N_G(v) \cup \{v\}$. For a set X of vertices or edges in G, we use $G - X$ to denote the graph obtained by deleting X from G. For a set of vertices $V' \subseteq V$, we write $G[V']$ to denote the subgraph of G induced by V' and write $E[V']$ to denote the set of edges in G whose both endpoints are in V'. A graph G is a *split graph* if its vertex set can be partitioned into a clique K and an independent set I, where $(K; I)$ is called a *split partition* of G. The class of split graphs is hereditary and is characterized by the set $\{2K_2, C_4, C_5\}$ of forbidden induced subgraphs.

Edge Contraction. The *contraction* of edge $e = uv$ in G removes u and v from G, and replaces them by a new vertex adjacent to precisely vertices that were adjacent to at least one of u or v. The resulting graph is denoted by G/e. For a set of edges $F \subseteq E(G)$, we write G/F to denote the graph obtained from G by sequentially contracting all edges from F.

For a graph H, if H can be obtained from G by a sequence of edge contractions, then G is *contractible* to H, or called H-*contractible*. Let $V(H) = \{h_1, \cdots, h_l\}$, then G is H-contractible iff it has a so-called H-*witness structure*: a partition of $V(G)$ into l sets $W(h_1), \cdots, W(h_l)$, called *witness sets*, such that each $W(h_i)$ induces a connected subgraph of G and for any two $h_i, h_j \in V(H)$, there is an edge between $W(h_i)$ and $W(h_j)$ in G iff $h_i h_j \in E(H)$. We obtain H from G by contracting vertices in each $W(h_i)$ into a single vertex.

Parameterized Complexity. A *paramerized problem* $\mathcal{Q}$ is a subset of $\Sigma^* \times \mathbb{N}$ for some finite alphabet Σ. The second component is called the *parameter*. The problem $\mathcal{Q}$ is *fixed-parameter tractable* (FPT) if it admits an algorithm deciding whether $(I, k) \in \mathcal{Q}$ in time $f(k)|I|^{O(1)}$, where $|I|$ is the size of I and f is a computable function depending only on k.

A *kernelization* of $\mathcal{Q}$ is a polynomial-time computable function that maps an instance (I, k) to an instance (I', k') such that (a) $(I, k) \in \mathcal{Q} \Leftrightarrow (I', k') \in \mathcal{Q}$, and (b) $|I'|, k' \leq g(k)$ for some computable function g. If g is a polynomial function then we say that $\mathcal{Q}$ admits a *polynomial kernel*. A problem $\mathcal{Q}$ is *incompressible* if it admits no polynomial kernel unless $NP \subseteq coNP/poly$.

A *polynomial parameter transformation* from a problem $\mathcal{Q}$ to a problem $\mathcal{Q}'$ is a polynomial-time computable function that maps (I, k) to (I', k') such that (a) $(I, k) \in \mathcal{Q} \Leftrightarrow (I', k') \in \mathcal{Q}'$, and (b) $k' \leq h(k)$ for some polynomial function h.

3   FPT Algorithm

In this section, we address the parameterized complexity of the SPLIT CONTRACTION problem. We first point out that this problem is NP-complete by reducing from another NP-complete edge contraction problem CLIQUE CONTRACTION [4]: Given a graph G and an integer k, can we modify G into a clique by contracting at most k edges? We construct a graph G' from G by adding an independent set of $k + 2$ new vertices and making new vertices adjacent to all vertices in G. Observe that at least two of these new vertices are not involved in any edge contractions, therefore at least one of them belongs to the independent set of the target split graph, which implies that all old vertices belong to the clique of the target graph. Thus we conclude that G' can be modified into a split graph using k edge contractions iff there exists a set of k edges in G whose contraction makes G into a clique.

Theorem 1. SPLIT CONTRACTION *is NP-complete.*

We now present an FPT algorithm for SPLIT CONTRACTION based on an $k^{O(k)} + O(m)$ time algorithm for CLIQUE CONTRACTION.

Theorem 2 (Cai & Guo [4]). CLIQUE CONTRACTION *can be solved in time* $O(2^{7k}k^{2k+5} + m)$.

Note that an n-vertex graph G must contain an induced split subgraph of $(n - 2k)$ vertices if (G, k) is a yes-instance of SPLIT CONTRACTION, because k edge contractions can affect at most $2k$ vertices. We start by finding an $(n-2k)$-vertex induced split subgraph H in $O(2^{2k}n^5)$ time using a known algorithm for SPLIT VERTEX DELETION (Ghosh *et al.* [9]). Let $V_k = V(G) - V(H)$, and let $(K_H; I_H)$ be a split partition of H where K_H is a maximal clique and I_H is an independent set. We first consider case $|K_H| > 2k$. In this case, at least one vertex in K_H is not involved in any edge contraction. The case $|K_H| \leq 2k$ will be discussed in the last part of the algorithm.

We branch out by contracting every possible set $E' \subseteq E[V_k]$ of at most k edges and obtain a resulting instance (G', k') where $G' = G/E'$ and $k' = k - |E'|$. For each resulting instance (G', k'), let $V_k' \subseteq V(G')$ be the set obtained from V_k after contractions. We have $|V_k'| \leq |V_k| = 2k$.

Proposition 3. (G, k) *has a solution S iff there exists a resulting instance* (G', k') *such that* (G', k') *has a solution $F = S - E'$ satisfying $F \cap E[V_k'] = \emptyset$.*

Suppose that (G', k') is a yes-instance and has a solution F. Then the graph G'/F is a split graph and has a split partition $(K_F; I_F)$. We further branch on at most $3^{|V_k'|}$ ways to find a partition $V_k' = R \cup K_p \cup I_p$ such that R consists of

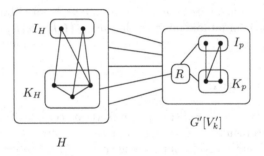

Fig. 1. An illustration of the structure of G'

exactly vertices in V'_k that are incident with some edges in F, $K_p \subseteq K_F$ induces a clique, and $I_p \subseteq I_F$ induces an independent set. See Fig. 1 for an illustration. It is easy to see that $|R| \leq k'$.

We consider the relationship between $\{K_H, I_H\}$ and $\{K_F, I_F\}$. It is clear that vertices in I_H that are adjacent to some vertices in I_p must be in the clique K_F of the target split graph, while other vertices in I_H can be in the independent set I_F after contractions. The following proposition states that there exists one solution F such that almost all vertices in K_H are finally in the clique K_F of the target graph.

Proposition 4. *If (G', k') is a yes-instance, then it has a solution F such that there is at most one vertex in K_H that is finally in the independent set I_F.*

Proof. For an arbitrary solution F of (G', k'), if there are at least two vertices in K_H that are finally in the independent set I_F, they must be contained in a same witness set W_0 because they are adjacent originally. Thus the number of such vertices is bounded by $k + 1$, implying that there is a vertex $u \in K_H$ that is in the clique K_F (i.e., not in I_F) since $|K_H| > 2k$.

By the definition of witness sets, we see that the induced subgraph $G'[W_0]$ has a spanning tree whose edges are all contained in F. Let x be an arbitrary leaf in this spanning tree. We can remove one edge from F to separate the vertex x from the witness set W_0, and add an edge ux into F to make x adjacent to all vertices in K_F after contracting F. It is easy to see that the resulting set obtained from F is also a solution of (G', k'), and x is no longer in the witness set W_0. We repeat this operation on F until there is exactly one vertex in the witness set W_0. Then we obtain a solution set F satisfying the requirement. □

Our algorithm further considers the following two cases. It outputs "YES" if either case outputs "YES".

Case 1. *There are no vertices in K_H that are finally in I_F.* Let T_1 be a subset of I_H containing exactly vertices that are adjacent to some vertices in I_p. It is clear that all vertices in T_1 are finally in K_F (i.e., not in I_F) after contractions. Therefore every vertex in T_1 must be involved in at least one edge contraction, implying that $|T_1| \leq k'$.

Remember that R consists of exactly vertices in V_k' that are incident with some edges in F, which implies that all vertices in R are merged into $K_H \cup I_H$ after contracting edges in F. In order to contract G' into a split graph, it is better to merge vertices R into $K_H \cup T_1$ than into $I_H - T_1$. Thus we may assume that all vertices in R are finally in K_F after contractions. Our goal becomes to check whether $T_1 \cup K_H \cup R \cup K_p$ induces a clique after contracting at most k' edges in G'.

Since $|K_H| > 2k$, there exists a vertex $u \in K_H$ that is not involved in any edge contraction. Obviously u is adjacent to all vertices in K_F. We can obtain an edge set F_1 from F by removing every edge of F whose endpoints are both outside $K_H \cup R$, and replacing every edge $ab \in F$ such that $a \in I_H - T_1$ and $b \in K_H \cup R$ by an edge ub in F. Since contracting such edge ab only affects vertex b and vertex b can also be merged into the clique K_F by contracting ub, it is clear that F_1 is also a solution set of G' which consists of edges between $K_H \cup R$ and $T_1 \cup K_H \cup R$.

Proposition 5. *(G', k') has a solution set that is entirely contained in $G'[T_1 \cup K_H \cup R \cup K_p]$ if it is a yes-instance for Case 1.*

By the above proposition, we first find the set T_1 in linear time and then apply an FPT algorithm for CLIQUE CONTRACTION (Theorem 2) to determine whether $G'[T_1 \cup K_H \cup R \cup K_p]$ can be made into a clique by at most k' edge contractions. If it outputs "YES", then (G', k') is a yes-instance. The running time is bounded by $k'^{O(k')} + O(m)$.

Case 2. *There is exactly one vertex in K_H that is finally in I_F.* For every $w \in K_H$ that is not adjacent to any vertex in I_p, we check whether (G', k') is a yes-instance by assuming that w is such a vertex, .

Let T_2 be a subset of I_H containing exactly vertices that are adjacent to some vertices in $I_p \cup \{w\}$. It is clear that every vertex in T_2 is finally in K_F (i.e., not in K_I) and thus is involved in some edge contraction. We have $|T_2| \le k'$. Our goal is to check whether $T_2 \cup (K_H - \{w\}) \cup R \cup K_p$ induces a clique after contracting at most k' edges in G'.

Since $|K_H| > 2k$, there exists a vertex $u \in K_H$ that is not involved in any edge contraction. Similar to Case 1, we can obtain an edge set F_2 from F by removing every edge of F whose both endpoints are outside $(K_H - \{w\}) \cup R$, and replacing every edge $ab \in F$ such that $a \in I_H \cup \{w\} - T_1$ and $b \in (K_H - \{w\}) \cup R$ by an edge ub in F. The resulting set F_2 is also a solution set of G'.

Proposition 6. *(G', k') has a solution set that is entirely contained in $G'[T_2 \cup (K_H - \{w\}) \cup R \cup K_p]$ if it is a yes-instance for Case 2.*

We also apply a $k'^{O(k')} + O(m)$ time algorithm for CLIQUE CONTRACTION (Theorem 2) to determine whether $G'[T_2 \cup (K_H - \{w\}) \cup R \cup K_p]$ can be made into a clique by contracting at most k' edges. If it outputs "YES", then (G', k') is a yes-instance.

Combining Case 1 and Case 2, we can decide whether a resulting instance (G', k') is a yes-instance in $k^{O(k)}n + O(mn)$ time when $|K_H| > 2k$.

Furthermore, we deal with the remaining case: $|K_H| \leq 2k$. We partition the vertex set $V(G)$ into disjoint sets $X_1, \cdots, X_d$ such that each X_i induces a maximal independent set and all vertices in each X_i have the same neighbors in G. This procedure is equivalent to partitioning the complement graph $\overline{G}$ into critical cliques, which can be done in linear time [6, 16]. A *critical clique* K in a graph is a clique such that all vertices in K have the same closed neighbor sets, and K is maximal under this property. It has been proved that all vertices in a graph can be uniquely partitioned into groups such that each group forms a critical clique [6, 16]. We now use the following reduction rules to obtain a smaller instance.

Rule 1. If $d > 2^{4k} + 4k$, then output "NO".

Rule 2. If there are more than $2k+5$ vertices in X_i for some i, then arbitrarily retain $2k + 5$ vertices among them and remove others in X_i from G.

It is clear that applying these reduction rules requires linear time. We now show the correctness of Rule 1 and Rule 2.

Lemma 7. *Rule 1 and Rule 2 are correct.*

Proof. Since $|K_H| \leq 2k$, we have $|V_k \cup K_H| \leq 4k$ and $|I_H| \geq n - 4k$. Note that vertices in I_H have at most 2^{4k} different connection configurations to vertices $V_k \cup K_H$. Thus I_H can be partitioned into at most 2^{4k} maximal independent sets such that vertices in each set have the same neighbors in G. Together with vertices $V_k \cup K_H$, the number d is bounded by $2^{4k} + 4k$ if (G, k) is a yes-instance. Thus Rule 1 is correct.

Moreover, we prove that the input graph G has a k-solution iff the graph G^* obtained after one application of Rule 2 has a k-solution. Let Y_i be the set of remaining vertices in X_i for $1 \leq i \leq d$.

Suppose that G has a solution $S \subseteq E(G)$. For every vertex a that is incident with some edge in S and is removed after applying Rule 2, there exists $1 \leq j \leq d$ such that $a \in X_j - Y_j$. Since $|Y_j| = 2k + 5 > 2k$, there exists a vertex $b \in Y_j$ that is not incident with any edge in S. Obviously a and b are not adjacent. We replace all edges $\{aw \in S : w \in V(G)\}$ by $\{bw : aw \in S\}$ in S for every such a, and then obtain a set $S' \subseteq E(G^*)$. Since $N_G(a) = N_G(b)$ for every such a, it is easy to see that G/S' is isomorphic to G/S that is a split graph. Note that G^*/S' is an induced subgraph of G/S' because S' is entirely included in $E(G^*)$. Therefore G^*/S' is a split graph.

Conversely, suppose that G^* has a solution S^*, i.e., G^*/S^* is a split graph. We claim that G/S^* is also a split graph. Assume that, to the contrary, G/S^* contains an induced subgraph D isomorphic to some graph in $\{2K_2, C_4, C_5\}$. For every vertex a that is contained in $V(D)$ and is removed after applying Rule 2, we know that a is in $X_j - Y_j$ for some j. Since $S^* \subseteq E(G^*)$ and $a \notin V(G^*)$, a is not incident with any edge in S^*. Note that $|Y_j| = 2k + 5 \geq 2k + |V(D)|$, there exists a vertex $b \in Y_j$ that is not contained in $V(D)$ and not incident with any edge in S^*. We replace a by b in D for every such a, and then obtain an induced subgraph D' whose vertices are clearly in $V(G/S^*)$. Note that all vertices in $V(D')$ remain after applying Rule 2. Therefore D' is an induced subgraph of G^*/S^*. Since $N_G(a) = N_G(b)$ for every such a, it is easy to see that D' is isomorphic to D, contradicting to the fact that G^*/S^* is a split graph that is D-free. Thus G/S^* is $\{2K_2, C_4, C_5\}$-free, implying that S^* is a solution of G. □

After applying Rule 1 and Rule 2, we reduce G to a graph of $O(2^{4k}k)$ vertices. Thus the problem can be solved in time $O\left(\binom{(2^{4k}k)^2}{k} + m\right) = 2^{O(k^2)} + O(m)$ by using brute-force search when $|K_1| \leq 2k$.

Our FPT algorithm for SPLIT CONTRACTION is summarized in Fig. 2:

Algorithm SPLIT CONTRACTION (G, k)

1   Find an induced split subgraph $H = (K_H; I_H)$ of size $(n - 2k)$ in G.
     if it does not exist **then**
        return "NO".
     else let $V_k = V(G) - V(H)$.

2   **if** $|K_H| > 2k$ **then**
2.1   Branch into instances (G', k') by contracting edges $E' \subseteq E[V_k]$.
2.2   Enumerate all partitions $V_k' = (R, K_p, I_p)$.
2.3   Let $T_1 = \{v \in I_H \mid \exists x \in I_p, vx \in E(G')\}$.
         if CLIQUE CONTRACTION$(G'[T_1 \cup K_H \cup R \cup K_p], k')$ = "YES" **then**
            return "YES".
2.4   **for each** $w \in K_H$ not adjacent to I_p **do**
         Let $T_2 = \{v \in I_H \mid \exists x \in I_p \cup \{w\}, vx \in E(G')\}$.
         if CLIQUE CONTRACTION$(G'[T_2 \cup (K_H - \{w\}) \cup R \cup K_p], k')$ = "YES" **then**
            return "YES".
2.5   Repeat 2.1 – 2.4, **return** "NO" if no (G', k') yields "YES".

3   **elseif** $|K_H| \leq 2k$ **then**
3.1   Partition $V(G)$ into disjoint sets $X_1, \cdots, X_d$: Each X_i induces a maximal independent set, and vertices in each X_i have the same neighbors.
3.2   Reduction Rule 1: If $d > 2^{4k} + 4k$, then output "NO".
3.3   Reduction Rule 2: If there are more than $2k + 5$ vertices in X_i for some i, then retain $2k + 5$ vertices among them and remove others in X_i.
3.4   Apply brute-force search to find a solution in the reduced graph G^*.

Fig. 2. Outline of algorithm for SPLIT CONTRACTION

Theorem 8. SPLIT CONTRACTION *can be solved in time* $2^{O(k^2)}n^5$.

Proof. In the above algorithm, we use an $O(2^{2k}n^5)$ time algorithm to find an $(n - 2k)$-vertex induced split subgraph H. If $|K_H| > 2k$, we branch into at most $|E[V_k]|^k = k^{O(k)}$ instances (G', k') and enumerate at most $3^{|V_k'|} = 3^{O(k)}$ partitions $V_k' = (R, K_p, I_p)$, and for each resulting instance with a specific partition it costs $k^{O(k)}n + O(mn)$ time to determine whether the instance is a yes-instance. Thus the running time of this case is bounded by $k^{O(k)}mn$. If $|K_H| \leq 2k$, the problem is solvable in time $2^{O(k^2)} + O(m)$ by applying reduction rules and brute-force search. Therefore the total running time of the algorithm is $2^{O(k^2)}n^5$. □

4   Incompressibility

To give a complete picture, we show that SPLIT CONTRACTION is very unlikely
to have polynomial kernels. To this end, we give a polynomial parameter transformation from the following RED-BLUE DOMINATING SET problem, which has
been proved to be incompressible using the Colors and IDs technique (Dom *et
al.* [8]).

RED-BLUE DOMINATING SET
Instance: Bipartite graph $G = (X, Y; E)$ and an integer t.
Question: Does Y have a subset of at most t vertices that dominates X?
Parameter: $|X|, t$.

Without loss of generality, we may assume that every vertex in X has at least
one neighbor in Y and thus $t \leq |X|$. Our method is inspired by the reduction
for TREE CONTRACTION (Heggernes *et al.* [14]).

Theorem 9. SPLIT CONTRACTION *admits no polynomial kernel unless* $NP \subseteq$
$coNP/poly$.

Proof. Given a bipartite graph $G = (X, Y; E)$ and a positive integers t, we
construct a graph G' from G by creating a clique C of size $|X| + t + 3$, making
a designated vertex $u \in C$ adjacent to all vertices of Y, and for every $v \in X$
appending $|X| + t + 1$ new leaves to v. See Fig. 3 for an illustration.

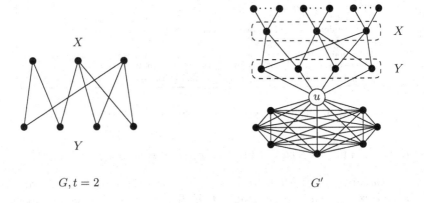

$G, t = 2$ G'

Fig. 3. A transformation from RED-BLUE DOMINATING SET to SPLIT CONTRACTION

We claim that Y has a subset of at most t vertices that dominates X iff G'
can be made into a split graph by contracting at most $|X| + t$ edges.

Suppose that Y' is a t-subset of Y that dominates X. Since vertices of $\{u\} \cup Y' \cup X$ induce a connected graph, we can merge these $|X| + t + 1$ vertices into a single vertex by using $|X| + t$ edge contractions. The subgraph of G' induced by $\{u\} \cup Y \cup X$ is contracted to a star, and thus G' is modified into a split graph after contractions.

Conversely, suppose that G' contains at most $|X| + t$ edges F whose contraction results in a split graph. Note that there exist two vertices a and b other than u in C such that a, b, and u are in the different witness sets because $|C| \geq |F| + 3$. If there exists some $x \in X$ such that x and u are in different witness sets, then x is not adjacent to neither a nor b in the target graph. Note that there exists a leaf x' appending to x that is not involved in any edge contraction by the construction. Therefore $\{x, x', a, b\}$ form an induced $2K_2$, contradicting to the fact that G'/F is a split graph. Thus all vertices in $X \cup \{u\}$ must be in the same witness set. Observe that each path starting from one vertex in X to u must go through some vertices in Y, implying that X is dominated by a subset I of Y where I consists of exactly vertices in Y that share the same witness set with u. Therefore we obtain a solution set I of G with $|I| \leq |F| - |X| \leq t$ vertices.

We have given a polynomial parameter transformation from RED-BLUE DOMINATING SET to SPLIT CONTRACTION. Based on a general result (Bodlaender et al. [2]) for kernelization transformation, SPLIT CONTRACTION admits no polynomial kernel unless $NP \subseteq coNP/poly$.   □

5   Concluding Remarks

In this paper we have shown that SPLIT CONTRACTION is fixed-parameter tractable, but admits no polynomial kernel unless $NP \subseteq coNP/poly$. We believe that the running time $2^{O(k^2)}n^5$ for SPLIT CONTRACTION can be improved. The bottleneck in the current algorithm lies in the case $|K_H| \leq 2k$, which costs $2^{O(k^2)} + O(m)$ time comparing with time $2^{O(k \log k)}mn$ for $|K_H| > 2k$. It seems possible to design faster algorithm for the case $|K_H| \leq 2k$, since in this case the input graph contains a large independent set of $n - 4k$ vertices.

Conjecture 10. SPLIT CONTRACTION *can be solved in time* $2^{O(k \log k)}n^{O(1)}$.

It will be also interesting to study Π-CONTRACTION for other subclasses Π of chordal graphs. In particular, INTERVAL CONTRACTION deserves special attention, as its vertex deletion variation, INTERVAL VERTEX DELETION, has been shown to be FPT by Cao and Marx [5] recently.

Problem 11. *Determine whether* INTERVAL CONTRACTION *is fixed-parameter tractable.*

References

1. Belmonte, R., Golovach, P.A., van 't Hof, P., Paulusma, D.: Parameterized Complexity of Two Edge Contraction Problems with Degree Constraints. In: Gutin, G., Szeider, S. (eds.) IPEC 2013. LNCS, vol. 8246, pp. 16–27. Springer, Heidelberg (2013)
2. Bodlaender, H.L., Downey, R.G., Fellows, M.R., Hermelin, D.: On problems without polynomial kernels. Journal of Computer and System Sciences 75(8), 423–434 (2009)
3. C.L.: Fixed-parameter tractability of graph modification problems for hereditary properties. Information Processing Letters 58, 171–176 (1996)
4. Cai, L., Guo, C.: Contracting few edges to remove forbidden induced subgraphs. In: Gutin, G., Szeider, S. (eds.) IPEC 2013. LNCS, vol. 8246, pp. 97–109. Springer, Heidelberg (2013)
5. Cao, Y., Marx, D.: Interval deletion is fixed-parameter tractable. In: Chekuri, C. (ed.) SODA 2014, pp. 122–141. SIAM (2014)
6. Chen, J., Meng, J.: A 2k kernel for the cluster editing problem. Journal of Computer and System Sciences 78(1), 211–220 (2012)
7. Cygan, M., Pilipczuk, M.: On fixed-parameter algorithms for split vertex deletion. arXiv:1208.1248 (2012)
8. Dom, M., Lokshtanov, D., Saurabh, S.: Incompressibility through colors and iDs. In: Albers, S., Marchetti-Spaccamela, A., Matias, Y., Nikoletseas, S., Thomas, W. (eds.) ICALP 2009, Part I. LNCS, vol. 5555, pp. 378–389. Springer, Heidelberg (2009)
9. Ghosh, E., Kolay, S., Kumar, M., Misra, P., Panolan, F., Rai, A., Ramanujan, M.S.: Faster parameterized algorithms for deletion to split graphs. In: Fomin, F.V., Kaski, P. (eds.) SWAT 2012. LNCS, vol. 7357, pp. 107–118. Springer, Heidelberg (2012)
10. Golovach, P.A., Kamiński, M., Paulusma, D., Thilikos, D.M.: Increasing the minimum degree of a graph by contractions. Theoretical Computer Science 481, 74–84 (2013)
11. Golovach, P.A., van't Hof, P., Paulusma, D.: Obtaining planarity by contracting few edges. Theoretical Computer Science 476, 38–46 (2013)
12. Guillemot, S., Marx, D.: A faster FPT algorithm for bipartite contraction. In: Gutin, G., Szeider, S. (eds.) IPEC 2013. LNCS, vol. 8246, pp. 177–188. Springer, Heidelberg (2013)
13. Guo, J.: Problem kernels for NP-complete edge deletion problems: Split and related graphs. In: Tokuyama, T. (ed.) ISAAC 2007. LNCS, vol. 4835, pp. 915–926. Springer, Heidelberg (2007)
14. Heggernes, P., van 't Hof, P., Lévêque, B., Lokshtanov, D., Paul, C.: Contracting graphs to paths and trees. In: Marx, D., Rossmanith, P. (eds.) IPEC 2011. LNCS, vol. 7112, pp. 55–66. Springer, Heidelberg (2012)
15. Heggernes, P., van't Hof, P., Lokshtanov, D., Paul, C.: Obtaining a bipartite graph by contracting few edges. In: Supratik, C., Amit, K. (eds.) FSTTCS 2011. LIPIcs, vol. 13, pp. 217–228. Schloss Dagstuhl, Leibniz-Zentrum für Informatik (2011)
16. Lin, G.-H., Kearney, P.E., Jiang, T.: Phylogenetic k-root and steiner k-root. In: Lee, D.T., Teng, S.-H. (eds.) ISAAC 2000. LNCS, vol. 1969, pp. 539–551. Springer, Heidelberg (2000)
17. Lokshtanov, D., Misra, N., Saurabh, S.: On the hardness of eliminating small induced subgraphs by contracting edges. In: Gutin, G., Szeider, S. (eds.) IPEC 2013. LNCS, vol. 8246, pp. 243–254. Springer, Heidelberg (2013)

Parameterized Complexity
of Connected Induced Subgraph Problems

Leizhen Cai[*] and Junjie Ye

Department of Computer Science and Engineering,
The Chinese University of Hong Kong, Shatin, New Territories, Hong Kong
{lcai,jjye}@cse.cuhk.edu.hk

Abstract. For a graph property Π, i.e., a collection Π of graphs, the
CONNECTED INDUCED Π-SUBGRAPH problem asks whether a graph G
contains k vertices V' such that the induced subgraph $G[V']$ is connected
and belongs to Π.

In this paper, we regard k as a parameter and study the parameterized complexity of CONNECTED INDUCED Π-SUBGRAPH for hereditary
properties Π. We give an almost complete characterization in terms of
whether Π includes all complete graphs, all stars, or all paths: FPT if
Π includes all complete graphs and stars, or excludes some complete
graphs, stars and paths; and W[1]-hard otherwise (except the case that
Π includes all complete graphs and paths but exclude some stars). For
the remaining case, we show that it is W[1]-hard if Π includes all complete graphs K_t, excludes a star $K_{1,s}$ but includes all trees of maximum
degree less than s. Our results imply a complete characterization for Π
being H-free graphs for a fixed graph H: W[1]-hard if H is K_t with $t \geq 3$
or $K_{1,s}$ with $s \geq 2$, and FPT otherwise.

1 Introduction

Subgraph problems are central to graph algorithms and have been studied extensively under frameworks of both traditional complexity and parameterized
complexity [4] [3]. For a graph property Π, i.e., a collection Π of graphs, any
graph in Π is a Π-*graph* and the INDUCED Π-SUBGRAPH problem asks whether
the input graph contains an induced Π-subgraph with k vertices. Property Π is
hereditary if all induced subgraphs of a Π-graph are Π-graphs.

A classical result of Lewis and Yannakakis [6] states that INDUCED Π-
SUBGRAPH is NP-hard for any "interesting" hereditary property Π (i.e., Π holds
for infinitely many graphs but not for all graphs), and the problem remains NP-
hard if we require the induced Π-subgraph to be connected. Khot and Raman [5]
give a complete characterization of the parameterized complexity of INDUCED
Π-SUBGRAPH, with k being the parameter, depending on whether Π includes
all complete graphs or trivial graphs (i.e., graphs without edges): W[1]-complete

---

[*] Partially supported by GRF grant CUHK410409 of the Research Grants Council of
Hong Kong.

Q. Gu, P. Hell, and B. Yang (Eds.): AAIM 2014, LNCS 8546, pp. 219–230, 2014.

if Π includes all trivial graphs but not all complete graphs or vice versa, and FPT otherwise. In connection with this, Cai [1] showed earlier that the parametric dual of INDUCED Π-SUBGRAPH (i.e., determining whether an n-vertex G contains an induced Π-graph on $n - k$ vertices, instead of k vertices) is FPT whenever Π can be characterized by a finite set of forbidden induced subgraphs.

In this paper, we investigate the parameterized complexity of the following induced Π-subgraph problems with an additional requirement that the k-vertex induced Π-graph is connected. We will mainly focus on hereditary properties Π.

CONNECTED INDUCED Π-SUBGRAPH
Instance: Graph G, positive integer k as parameter.
Question: Does G contain a connected induced Π-subgraph on k vertices?

We note that the work of Khot and Raman [5] does not address the issue of connectedness in induced Π-graphs. On the other hand, FPT algorithms of Cai [1] for the parametric dual of INDUCED Π-SUBGRAPH also work when the required Π-subgraphs on $n - k$ vertices need to be connected, and therefore the parametric dual of CONNECTED INDUCED Π-SUBGRAPH is FPT whenever Π is characterized by a finite set of forbidden induced subgraphs.

It turns out that the situation for CONNECTED INDUCED Π-SUBGRAPH is more complicated than that for INDUCED Π-SUBGRAPH, and the parameterized complexity of CONNECTED INDUCED Π-SUBGRAPH depends on whether Π includes all complete graphs, stars, and paths (instead of complete graphs and trivial graphs for INDUCED Π-SUBGRAPH). Table 1 summarizes our results for hereditary properties Π into six cases.

Table 1. The parameterized complexity of CONNECTED INDUCED Π-SUBGRAPH for hereditary properties Π

Property Π	Include all complete graphs	Exclude some complete graphs
Include all stars	FPT	W[1]-hard
Exclude some stars include all paths	unknown, but W[1]-hard if include all degree-bounded trees	W[1]-hard
Exclude some stars exclude some paths	W[1]-hard	FPT

For the remaining unknown case (Π includes all complete graphs and paths but excludes some stars), we are able to establish its W[1]-hardness when paths are replaced by trees of maximum degree less than s in the condition, where $K_{1,s}$ is the smallest star excluded by Π.

Our results settle the parameterized complexity of CONNECTED INDUCED Π-SUBGRAPH for many well-known hereditary properties Π. See Table 2 for some examples. Furthermore, our results also imply a complete characterization when Π is H-free graphs for a fixed graph H: W[1]-hard if H is K_t with $t \geq 3$ or $K_{1,s}$ with $s \geq 2$, and FPT otherwise.

All graphs in the paper are simple undirected graphs. For a graph G, we use $V(G)$ to denote its vertex set and $E(G)$ its edge set. We use n and m, resp., to denote the numbers of vertices and edges of G. For a subset $V' \subseteq V$, $N_G(V')$ denotes the neighbours of V' in $V(G) - V'$, and $G[V']$ represents the subgraph induced by V'. A universal vertex v of G is a vertex adjacent to all other vertices in G. We use $R(t, s)$ to denote the Ramsey number, i.e., any graph with $R(t, s)$ vertices contains either a t-clique or an independent s-set. We use $M_{\Delta,D}$ to denote the Moore's bound [7] which is the maximum number of vertices in a connected graph G with maximum degree Δ and diameter D:

$$M_{\Delta,D} = 1 + \Delta \sum_{i=0}^{D-1} (\Delta - 1)^i < \Delta^{D+1} \text{ for } \Delta \geq 2.$$ A property Π is *hereditary* iff it has a forbidden induced subgraph characterization, i.e., there is a smallest forbidden set $Forb(\Pi)$ of graphs such that G is a Π-graph iff G contains no graph in $Forb(\Pi)$ as an induced subgraph. For any Π-graphs, co-Π graphs denote complement graphs of Π-graphs.

Table 2. Some well-known hereditary properties Π for CONNECTED INDUCED Π-SUBGRAPH problems settled by our results

Property Π	Include all complete graphs	Exclude some complete graphs
Include all stars	perfect graphs chordal graphs interval graphs	3-colorable graphs bipartite graphs planar graphs
Exclude some stars include all paths	claw-free graphs line graphs line graphs of bipartite graphs	degree-bounded graphs
Exclude some stars exclude some paths	co-planar graphs co-bipartite graphs co-forest graphs	degree-bounded graphs with small vertex cover

In the rest of the paper, we present FPT algorithms in Section 2, give W[1]-hardness proofs in Section 3, and consider the remaining case in Section 4. We discuss some open problems in Section 5.

2   FPT Algorithms

We start with the two fixed-parameter tractable cases in Table 1. To obtain these FPT algorithms, we use a combination of Ramsey's theorem, Moore's bound, and the random separation method of Cai, Chan and Chan [2].

Theorem 1. *Let Π be a decidable property. Then* CONNECTED INDUCED Π-SUBGRAPH *is FPT whenever*

1. Π *includes all complete graphs and stars, or*
2. Π *is hereditary and excludes some complete graphs, some stars, and some paths.*

Proof. By the assumption that Π is decidable, we may assume that it takes $T(k)$ time to determine whether a k-vertex graph is a Π-graph.

Case 1. Π includes all complete graphs and stars.

We show that G always has a solution if it has a vertex of degree at least $R(k-1, k-1)$, and otherwise we can use the random separation method to determine if it has a solution.

If G contains a vertex v of degree at least $R(k-1, k-1)$, then by Ramsey's theorem, $N(v)$ contains $k-1$ vertices V' that induce either a complete graph or an independent set. Therefore $G[V' \cup \{v\}]$ is either a complete graph K_k or a star $K_{1,k-1}$, which is a connected Π-graph. Otherwise, the maximum degree of G is at most d for $d = R(k-1, k-1) - 1$.

Since d is bounded above by a function of k only, we use the random separation method to design an FPT algorithm[1]. First, we randomly color each vertex of G independently by red or green with probability $\frac{1}{2}$. A set V' of k vertices is a *well-colored solution* if

1. $G[V']$ is a connected Π-graph, and
2. all vertices in V' are green and all vertices in $N(V')$ are red.

Let V_g be the set of green vertices of G. Then a well-colored solution is a connected component of $G[V_g]$ that is a k-vertex Π-graph. Therefore, given a red-green coloring of G, we can easily determine whether there is a well-colored solution in $O(m + n + \frac{n}{k}T(k)) = O(dn + nT(k))$ time.

It follows that, when G has a solution V', we can find it with probability at least 2^{-dk} in $O(dn + nT(k))$ time, since $|V' \cup N(V')| \leq dk$ and a random red-green coloring has probability at least 2^{-dk} to make V' a well-colored solution.

To derandomize the algorithm, we use a family of (n, dk)-universal sets of size $\leq 2^{O(dk)} \log n$ [9] and obtain a deterministic algorithm that runs in time $O(2^{O(dk)}(dn + nT(k)) \log n)$, which is an FPT algorithm as $d = R(k-1, k-1) - 1$.

Case 2. Π excludes some complete graphs, some stars, and some paths.

Let K_t, $K_{1,s}$, and P_l, respectively, be the smallest complete graph, smallest star and shortest path excluded by Π. If a Π-graph H has a vertex v of degree at least $R(t-1, s)$, then $N_H(v)$ contains either a $(t-1)$-clique or an independent set of size s, and thus $H[N_H(v) \cup \{v\}]$ contains either complete graph K_t or star $K_{1,s}$, contradicting to H being a Π-graph as Π is hereditary. Therefore the maximum degree of any Π-graph is less than $R(t-1, s)$.

Since P_l is not a Π-graph, the diameter of any connected Π-graph is at most $l - 1$. By Moore's bound, any graph of maximum degree Δ and diameter D

---

[1] Bounded search tree also works in FPT time $f(k)n^{O(1)}$ but with much worse $f(k) = O(2^{d^k} T(k))$.

has at most Δ^{D+1} vertices. Therefore any connected Π-graph contains less than $c = R(t-1, s)^l$ vertices. As t, s and l are constants, c is a constant independent of k and n.

If $k \geq c$, then the answer to the problem is always "NO". Otherwise we can use exhaustive search to determine whether G contains a connected induced Π-subgraph on k vertices. The running time for the algorithm is $O(T(k)n^k) = O(T(k)n^c)$, which is FPT time. ∎

We can use the above theorem to deduce that CONNECTED INDUCED Π-SUBGRAPH is FPT for many well-known properties. The following corollary lists a few of them, where the last two are not hereditary.

Corollary 1. CONNECTED INDUCED Π-SUBGRAPH *is FPT for Π being perfect graphs, chordal graphs, interval graphs, cographs, split graphs, permutation graphs, degree bounded graphs with bounded-size vertex cover, graphs of bounded diameter d with $d \geq 2$ and graphs of bounded-size dominating set.*

3   W[1]-Hardness

In this section, we will establish W[1]-hardness of three cases in Table 1. We need the following theorem of Khot and Raman [5] in our proofs.

Theorem 2. (Khot and Raman) *For any hereditary property Π, INDUCED Π-SUBGRAPH is W[1]-complete if Π includes all trivial graphs but not all complete graphs or vice versa, and FPT otherwise.*

We start with the case that Π includes all complete graphs but excludes some stars and paths. For a graph F, let F^* be the graph obtained from F by deleting all universal vertices of F. For a hereditary property Π, let Π^* be the property with

$$Forb(\Pi^*) = \{F^* : F \in Forb(\Pi)\}$$

(e.g. $Forb(\Pi^*) = \{2K_1, C_4\}$ for Π defined by $Forb(\Pi) = \{P_3, C_4, K_4 - e\}$). Note that stars in $Forb(\Pi)$ become independent sets in $Forb(\Pi^*)$, and independent sets in $Forb(\Pi)$ remain independent sets in $Forb(\Pi^*)$.

Theorem 3. *If a hereditary property Π includes all complete graphs but excludes some stars and paths, CONNECTED INDUCED Π-SUBGRAPH is W[1]-hard.*

Proof. Since Π excludes some stars, $Forb(\Pi)$ contains a star or an independent set, and thus $Forb(\Pi^*)$ contains an independent set. Also, since Π includes all complete graphs, $Forb(\Pi)$ contains no complete graph and thus $Forb(\Pi^*)$ contains no complete graph. Then Π^* includes all complete graphs and excludes some independent sets, and therefore it follows from Theorem 2 that INDUCED Π^*-SUBGRAPH is W[1]-hard.

We now prove the theorem by an FPT reduction from INDUCED Π^*-SUBGRAPH. Let (G, k) be an arbitrary instance of INDUCED Π^*-SUBGRAPH. Let P_l and $K_{1,s}$, respectively, be the smallest path and smallest star excluded by Π. And let

$$d = \max\{|V(F)| - |V(F^*)| : F \in Forb(\Pi) \text{ and } |V(F^*)| \leq k\}$$

and $r = R(k, s)^l + d$. For a graph F^*, it is easy to see that we have only one corresponding $F \in Forb(\Pi)$. Therefore d is a finite number related to k and property Π but independent of n. We construct an instance (G', k') of CONNECTED INDUCED Π-SUBGRAPH from (G, k) as follows:

1. Add to G a complete graph on r new vertices U,
2. add all possible edges between G and U to form graph G', and set $k' = k + r$.

We claim that G contains a k-vertex induced Π^*-subgraph iff G' contains a k'-vertex connected induced Π-subgraph. Suppose that G contains a k-vertex induced Π^*-graph H. Then $G'[V(H) \cup U]$ is connected with $k + r = k'$ vertices. If it contains an induced subgraph $F \in Forb(\Pi)$, then F^* resides entirely inside H as all vertices of $V(F) \cap U$ are universal vertices of F. But $F^* \in Forb(\Pi^*)$ contradicting to H being a Π^*-graph. Therefore $G'[V(H) \cup U]$ is a k'-vertex connected induced Π-graph.

Conversely, suppose that G' contains a k'-vertex connected induced Π-graph H'. We show that G contains a k-vertex induced Π^*-graph by considering two cases. Let $H = H' \cap G$, which is also a Π-graph as Π is hereditary.

Case 1. H contains at least $R(k, s)^l$ vertices.

If H' contains a vertex v of U, then H has no independent set of size s as v is adjacent to every vertex of H and H' is $K_{1,s}$-free. Since H has $R(k, s)^l > R(k, s)$ vertices, it follows from Ramsey's theorem that H contains a complete graph K_k, which is a k-vertex Π^*-subgraph in G.

Otherwise $H = H'$ and thus H is connected. Since Π excludes path P_l, the diameter of H is at most $l - 1$ and hence, by Moore's bound, H has at most Δ^l vertices, where Δ is the maximum degree of H. It follows that H contains a vertex v of degree at least $R(k, s)$ as $V(H) \geq R(k, s)^l$. Since H is a Π-graph, it is $K_{1,s}$-free. Therefore $G[N_H(v)]$ contains no independent s-set. According to Ramsey's theorem, $G[N_H(v)]$ contains a complete graph K_k.

Case 2. H contains less than $R(k, s)^l$ vertices.

Since $k' = k + r$, H contains at least k vertices and H' contains at least d vertices of U. We arbitrarily select k vertices S from H and claim that $G[S]$ is a Π^*-subgraph. Suppose to the contrary that $G[S]$ contains a graph $F^* \in Forb(\Pi^*)$ for some $F \in Forb(\Pi)$, then $|V(F^*)| \leq |S| = k$. We can arbitrarily choose $|V(F) - V(F^*)|$ vertices U' from $V(H') \cap U$ as $|V(H') \cap U| \geq d$ and $|V(F) - V(F^*)| \leq d$. Since in G' all vertices of U' are adjacent to all vertices of G and thus $G'[V(F^*) \cup U']$ is isomorphic to F, contradiction to H' being a Π-graph. Therefore $G[S]$ contains no graphs in $Forb(\Pi^*)$, and is a k-vertex Π^*-subgraph. ∎

Corollary 2. CONNECTED INDUCED Π-SUBGRAPH *is W[1]-hard for Π being the following properties: co-planar graphs, co-bipartite graphs, and co-forests.*

For a graph F, let F^- denote the graph obtained from F by removing a universal vertex of F and also all isolated vertices in the resulting graph. For a hereditary property Π, let Π^- be the property with $Forb(\Pi^-) = \{F^- : F \in Forb(\Pi)\}$. By a proof similar with that of Theorem 3, we can use Π^- to obtain the following theorem (the proof will appear in the full paper):

Theorem 4. *If Π includes all stars but excludes some complete graphs, then* CONNECTED INDUCED Π-SUBGRAPH *is W[1]-hard.*

For the case that Π includes all paths, but excludes some complete graphs and some stars, we use an FPT reduction from INDEPENDENT SET problem to prove the W[1]-hardness.

Theorem 5. *If a hereditary property Π includes all paths, but excludes some complete graphs and stars,* CONNECTED INDUCED Π-SUBGRAPH *is W[1]-hard.*

Proof. We prove the theorem by an FPT reduction from the INDEPENDENT SET problem. Let K_t and $K_{1,s}$ be the smallest clique and star excluded by Π. Then both s and t are at least 3 as paths are Π-graphs. Set $r = 4sR(t,k)$, which is a function of k only, as s and t are constants.

For an arbitrary instance (G, k) of INDEPENDENT SET, we construct an instance (G', k') of CONNECTED INDUCED Π-SUBGRAPH by setting $k' = k(r + 1) - 1$ and constructing graph G' as follows:

1. For each vertex v of graph G, replace it by a path $P(v) = v(1), \ldots, v(r)$ on r vertices. We refer to $P(v)$ as a *vertex-path*.
2. For each edge uv of G, add all possible edges between $P(u)$ and $P(v)$.
3. For each nonedge uv of G, connect $u(r)$ and $v(1)$ by a path $P(uv)$ of length two, and also connect $u(1)$ and $v(r)$ by a path $P(vu)$ of length two. We refer to the middle vertices of $P(uv)$ and $P(vu)$ as *nonedge-vertices*.

The construction of (G', k') takes $O(R(t,k)(m+n)) = O(f(k)(m+n))$ time which is FPT. We claim that G has an independent k-set iff G' contains a connected induced Π-subgraph with $k' = k(r+1) - 1$ vertices. If $I = \{v_1, \ldots, v_k\}$ is an independent k-set in G, then

$$P(v_1), P(v_1v_2), P(v_2), P(v_2v_3), \ldots, P(v_k)$$

is an induced path in G' with $k(r+1) - 1 = k'$ vertices, which is a connected induced Π-graph in G'.

Conversely, suppose that G' contains a connected induced Π-subgraph $G'[S']$ on k' vertices S'. Let S be vertices in G whose vertex-paths in G' contain vertices of S'. We show that $G[S]$ has an independent k-set.

By the construction of G' and the assumption that $G'[S']$ is K_t-free, we see that $G[S]$ is also K_t-free. Therefore if $|S| \geq R(t,k)$, we deduce from Ramsey's theorem that $G[S]$ has an independent k-set.

Otherwise $|S| < R(t,k)$. A vertex $v \in S$ is a *large-vertex* if $P(v)$ contains at least $2s$ vertices from S', and *small-vertex* otherwise. We show that large-vertices form an independent set in G. Consider two arbitrary large-vertices u and v. If uv is an edge of G, then in G' we have all edges between $P(u)$ and $P(v)$. Since

$P(u)$ is a path and contains at least $2s$ vertices of S', it has an independent s-set I in G'. But I together with any vertex of $P(v)$ in S' induce star $K_{1,s}$ in G', contradicting to $G'[S']$ being $K_{1,s}$-free. Therefore u and v are not adjacent in G, and large-vertices form an independent set in G.

It remains to show that there are at least k large-vertices. First we bound the number of nonedge-vertices in S'. Since $G'[S']$ is connected, we see that, by the construction of G', each nonedge-vertex in S' is adjacent to an endpoint of some vertex-path $P(v)$. Since $G'[S']$ is $K_{1,s}$-free, the endpoint of any vertex-path $P(v)$ is adjacent to at most $s - 1$ nonedge-vertex, implying that S' contains at most $2(s-1)|S| < 2(s-1)R(t,k)$ nonedge-vertices.

Now let p be the number of large-vertices. Since every vertex-path has r vertices, S' contains at most pr vertices from vertex-paths of large-vertices. Furthermore, S' contains $2s|S| < 2sR(t,k)$ vertices from vertex-paths of small-vertices. Since S' consist of vertices from vertex-paths of large-vertices, vertex-paths of small-vertices, and nonedge-vertices, we have $pr > k' - 2sR(t,k) - 2(s-1)R(t,k) > (k(r+1)-1) - 4sR(t,k) = (k-1)(r+1)$, implying $p \geq k$. ∎

Corollary 3. CONNECTED INDUCED Π-SUBGRAPH *is W[1]-hard for Π being degree-bounded graphs.*

4   The Remaining Case

In this section, we present a partial result for the remaining case: Π *includes all complete graphs and paths, but excludes some stars.* We show that the case is W[1]-hard when Π includes all complete graphs, excludes a star $K_{1,s}$ but includes all trees of maximum degree less than s. For this purpose, we need the W[1]-hardness of INDUCED EVEN PATH problem on bipartite graphs, which asks whether a bipartite graph contains an induced path on $2k$ vertices. We also need the following construction of *composition* (a.k.a. *lexicographic product*) $G[H]$ of two graphs G and H: replace each vertex v of G by a distinct copy H_v of H and each edge uv of G by a complete bipartite graph joining all vertices of H_u with all vertices of H_v.

Theorem 6. INDUCED EVEN PATH *is W[1]-hard on bipartite graphs.*

Proof. We give an FPT reduction from INDUCED MATCHING on bipartite graphs which asks whether a bipartite graph contains an induced matching with k edges and is W[1]-hard [8]. For an arbitrary instance (G, k) of INDUCED MATCHING with $G = (X, Y; E)$ being a bipartite graph, we construct an instance (G', k') of INDUCED EVEN PATH as follows:

1. Take the composition graph $G[\overline{K_k}]$ as *base graph* B, where $\overline{K_k}$ is the trivial graph on k vertices. For a vertex v in G, let $\{v_1, \ldots, v_k\}$ be its corresponding vertices in B. Let $X_i = \{x_i : x \in X\}$ and $Y_i = \{y_i : y \in Y\}$ for $1 \leq i \leq k$. See Figure 1 for an example.
2. For each X_i (resp., Y_i), add a vertex x_i^* (resp., y_i^*) and make it adjacent to all vertices of X_i (resp., Y_i) by adding edges. Call x_i^* and y_i^* *external-vertices*.

3. Set $l = 4k + 1$. Attach to x_1^* (resp., y_k^*) a path P_0 (resp., P_k) of length l, and for each $1 \le i \le k - 1$ connect y_i^* with x_{i+1}^* by a path $P_i = y_i^*, \ldots, x_{i+1}^*$ of length l. All internal vertices of P_i's are new vertices. Call each P_i an *external-path*.
4. Set $k' = k(l + 3) + l + 1 = 4k(k + 2) + 2$, which is even.

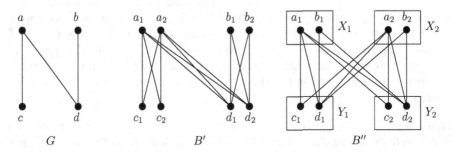

Fig. 1. B' and B'' are two different ways viewing base graph for k being 2. Each subgraph induced by $X_i \cup Y_i$ is a copy of G.

Each induced subgraph $B[X_i \cup Y_i]$ in the base graph is a copy of G. It is easy to see that G has an induced k-matching iff B has an induced k-matching.

It is clear that the above construction takes polynomial time, and that G' is a bipartite graph as every external-path has odd length. We claim that G has an induced k-matching iff G' has an induced k'-path.

Suppose that $I = \{x^1 y^1, \ldots, x^k y^k\}$ is an induced k-matching of $G = (X, Y; E)$ where each $x^i \in X$ and $y^i \in Y$. Then obviously

$$P = P_0, x_1^1, y_1^1, P_1, x_2^2, y_2^2, P_2, \ldots, x_k^k, y_k^k, P_k$$

is a path on $k(l + 3) + l + 1 = k'$ vertices. To see that P is an induced path of G', we note that, by the construction of G', the only possible chords of P are edges inside the base graph

$$B[x_1^1, y_1^1, \ldots x_k^k, y_k^k].$$

Since $\{x^i y^i : 1 \le i \le k\}$ is an induced k-matching of G, there is no edge in G connecting x^i with any y^j for $i \ne j$, implying that there is no edge between x_i^i and y_j^j in B, hence in G'. Therefore P is an induced k'-path in G'.

Conversely, suppose that P is an induced k'-path of G'. It suffices to show that B has an induced k-matching as we mentioned earlier that G has an induced k-matching iff B has an induced k-matching. A path connecting two external-vertices is an *internal-path* if all its internal vertices are in the base graph B. If P contains $k - 1$ external-paths $\{P_i : 1 \le i \le k - 1\}$ as well as vertices x_1^* and y_k^*, then these external-paths are linked together with x_1^* and y_k^* by k internal-paths to form P. Consider the $[x_1^*, y_k^*]$-section of P: it alternates between internal-paths and external-paths, and each internal-path contains at least one distinct edge inside the base graph B. The k internal-paths give us k such edges in B, one for each internal-path. It is easy to see that these k edges form an induced k-matching of B.

Otherwise, either P misses one of x_1^* and y_k^*, or P contains both x_1^* and y_k^* but misses all internal vertices of an external-path. In both cases, P contains at least $l = 4k + 1$ edges of the base graph B, implying an induced matching of size at least $(4k + 1)/3 > k$ in B as a path of length l has an induced matching of size at least $l/3$. Therefore B has an induced k-matching and so does G. ∎

Note that we can use a similar construction to show the W[1]-hardness of INDUCED CYCLE on bipartite graphs, which asks whether a bipartite graph contains an induced cycle on k vertices. Now we are ready to prove the following theorem.

Theorem 7. CONNECTED INDUCED Π-SUBGRAPH *is W[1]-hard if Π includes all complete graphs, excludes a star $K_{1,s}$ but includes all trees of maximum degree less than s.*

Proof. We give an FPT reduction from INDUCED EVEN PATH on bipartite graphs. Let $(G, 2k)$ be an instance of INDUCED EVEN PATH with $G = (X, Y; E)$ being a bipartite graph. To construct an instance (G', k') of our CONNECTED INDUCED Π-SUBGRAPH, we use the composition $G[\overline{K_k}]$ of G as the skeleton in our reduction, and then attach some *selection-trees* to ensure that G admits an induced $2k$-path iff G' contains a connected induced Π-subgraph on k' vertices. A *selection-tree* T is the star-shaped tree consisting of a vertex as the *root* of T and $s - 2$ root-to-leaf paths each of length $l = 2(s - 1)^{2k+2}k - 2k$. Note that T contains $(s - 2)l + 1$ vertices. We now give the construction of (G', k') as follows:

1. Take the composition graph $G[\overline{K_k}]$ as our base graph B. For a vertex v in G, let $\{v_1, \ldots, v_k\}$ be its corresponding vertices in B. For $1 \le i \le k$, let $X_i = \{x_i : x \in X\}$ and $Y_i = \{y_i : y \in Y\}$.
2. For X_1 (resp., Y_1) in B, make $s - 2$ vertex-disjoint selection-trees, and add all possible edges between their roots and all vertices of X_1 (resp., Y_1).
3. For each X_i (resp., Y_i) in B ($2 \le i \le k$), make $s - 3$ vertex-disjoint selection-trees, and add all possible edges between their roots and all vertices of X_i (resp., Y_i).
4. Set $k' = [2k(s - 3) + 2][(s - 2)l + 1] + 2k$.

Clearly, G' is bipartite and the construction takes FPT time as $k' = O(s^2 lk)$. Note that, for $k \ge 4$, G has an induced k-path iff B has an induced k-path. Also note that $k' - 2k$ equals the total number of vertices in all selection-trees as G' contains $2k(s - 3) + 2$ selection-trees. We claim that G has an induced $2k$-path iff G' has a connected induced Π-graph on k' vertices.

If G has an induced path $P = \{x^1, y^2, x^3, y^4, \ldots, x^{2k-1}, y^{2k}\}$, then obviously $P' = \{x_1^1, y_2^2, x_3^3, y_4^4, \ldots, x_k^{2k-1}, y_1^{2k}\}$ is an induced path of G'. Let S' be the set of vertices in all selection-trees plus all vertices in P'. Then $G'[S']$ is a tree with all selection-trees attached to P' and its maximum degree is $s - 1$. Therefore $G'[S']$ is a connected induced Π-graph with k' vertices.

Conversely, suppose that $G'[S']$ is a connected induced Π-graph with $|S'| = k'$. Let $V_b = S' \cap V(B)$, i.e., V_b contains vertices of base graph in S'. It suffices to show that $G'[V_b]$ contains an induced $2k$-path. Since G' is bipartite and $G'[S']$ is a Π-graph (recall that a Π-graph is $K_{1,s}$-free), the maximum degree of $G'[S']$ is at most $s - 1$. For a selection-tree, we refer to the $s - 2$ neighbors of its root as *selection-vertices*.

If S' misses some selection-vertices, then V_b contains at least $l + 2k = 2(s - 1)^{2k+2}k$ vertices. Let V_r be the set of root vertices in S', then $|V_r| \leq 2(s - 3)k + 2 \leq 2(s - 2)k$ as there are $2(s - 3)k + 2$ selection-trees. Since $G'[S']$ is connected and $S' - \{V_b \cup V_r\}$ only contains non-root vertices of selection-trees, $G'[V_b \cup V_r]$ is connected. Since the maximum degree of $G'[S']$ is at most $s - 1$, the maximum degree of $G'[V_b \cup V_r]$ is also bounded by $s - 1$. It follows that $G'[V_b]$, which is obtained by deleting $|V_r|$ vertices from $G'[V_b \cup V_r]$, has at most $(s - 1)|V_r| \leq 2(s - 1)(s - 2)k$ components. Let G'_1 be the largest component of $G'[V_b]$, then G'_1 contains at least $\frac{|V_b|}{2(s-1)(s-2)k} > \frac{l+2k}{2(s-1)^2k} \geq (s - 1)^{2k}$ vertices. As G'_1's maximum degree is $\leq s - 1$ and G'_1 is connected, according to Moore's bound, there is an induced $2k$-path.

Otherwise S' contains all selection-vertices and roots of selection-trees, we show that $G'[V_b]$ is just an induced $2k$-path. If S' contains more than one vertex of X_i, then the degree of a root attached to X_i will exceed $s - 1$, contradicting to $G'[S']$'s maximum degree being $\leq s - 1$. Since the connectedness of $G'[S']$ forces S' to contain at least one vertex of each X_i, S' contains exactly one vertex of X_i, i.e., $|V_b \cap X_i| = 1$. Similarly, we have $|V_b \cap Y_i| = 1$ and then $|V_b| = 2k$. Now each root is adjacent to only one vertex in V_b and $G'[S']$ is connected, which implies that $G'[V_b]$ is connected. Let x_1^1 and y_1^{2k} be the vertices in $V_b \cap X_1$ and $V_b \cap Y_1$ respectively. There are already $s - 2$ roots attached to x_1^1, which means $|N_{G'[V_b]}(x_1^1)| \leq 1$, i.e., x_1^1 is adjacent to at most one vertex in V_b. Similarly, $|N_{G'[V_b]}(y_1^{2k})| \leq 1$. Furthermore we have $|N_{G'[V_b]}(v)| \leq 2$ for every v in $\{v : v \in V_b, v \neq x_1^1 \text{ and } v \neq y_1^{2k}\}$ as v is in X_i or Y_i for $2 \leq i \leq k$ and v has $s - 3$ roots attached to it in $G'[S']$. Since $G'[V_b]$ is connected, $G'[V_b]$ must be an induced $2k$-path with x_1^1 and y_1^{2k} being endpoints. ∎

Corollary 4. CONNECTED INDUCED Π-SUBGRAPH *is $W[1]$-hard for Π being claw-free graphs, line graphs, line graphs of bipartite graphs, or line graphs of multigraphs.*

5 Concluding Remarks

We have obtained an almost complete characterization of the parameterized complexity of CONNECTED INDUCED Π-SUBGRAPH in terms of whether Π contains all complete graphs, all stars, or all paths.

Theorem 8. *Let Π be a hereditary property. Then CONNECTED INDUCED Π-SUBGRAPH is*

1. *FPT if Π includes all complete graphs and stars, or excludes some complete graphs, stars and paths; and*
2. *W[1]-hard if Π*
    (a) includes all complete graphs, but excludes some stars and paths,
    (b) includes all stars, but excludes some complete graphs, or
    (c) includes all paths, but excludes some complete graphs and stars.

Corollary 5. CONNECTED INDUCED H-FREE SUBGRAPH *is W[1]-hard if H is a complete graph K_t for some fixed $t \geq 3$ or star $K_{1,s}$ for some fixed $s \geq 2$, and FPT otherwise.*

For the remaining case, we believe that it is also W[1]-hard.

Conjecture 1. CONNECTED INDUCED Π-SUBGRAPH is W[1]-hard for any hereditary Π that includes all complete graphs and paths, but excludes some stars.

We note that Raman and Sikdar [10] have studied the parameterized complexity of INDUCED Π-SUBGRAPH for digraphs. In light of the work in this paper, it seems quite interesting to study INDUCED Π-SUBGRAPH on digraphs for strong connectivity.

Problem 1. Determine the parameterized complexity of INDUCED Π-SUBGRAPH on digraphs with the requirement: a Π-subgraph need to be strongly connected.

References

1. Cai, L.: Fixed-parameter tractability of graph modification problems for hereditary properties. Information Processing Letters 58(4), 171–176 (1996)
2. Cai, L., Chan, S.M., Chan, S.O.: Random separation: a new method for solving fixed-cardinality optimization problems. In: Bodlaender, H.L., Langston, M.A. (eds.) IWPEC 2006. LNCS, vol. 4169, pp. 239–250. Springer, Heidelberg (2006)
3. Downey, R.G., Fellows, M.R.: Parameterized Complexity. Springer, New York (1999)
4. Garey, M.R., Johnson, D.S.: Computers and Intractability: A Guide to the Theory of NP-Completeness. WH Freeman, New York (1979)
5. Khot, S., Raman, V.: Parameterized complexity of finding subgraphs with hereditary properties. Theoretical Computer Science 289(2), 997–1008 (2002)
6. Lewis, J.M., Yannakakis, M.: The node-deletion problem for hereditary properties is NP-complete. Journal of Computer and System Sciences 20(2), 219–230 (1980)
7. Miller, M., Širáň, J.: Moore graphs and beyond: a survey of the degree/diameter problem. Electronic Journal of Combinatorics 61, 1–63 (2005)
8. Moser, H., Sikdar, S.: The parameterized complexity of the induced matching problem. Discrete Applied Mathematics 157(4), 715–727 (2009)
9. Naor, M., Schulman, L.J., Srinivasan, A.: Splitters and near-optimal derandomization. In: Proceedings of the 36th Annual Symposium of Foundations of Computer Science, pp. 182–191 (1995)
10. Raman, V., Sikdar, S.: Parameterized complexity of the induced subgraph problem in directed graphs. Information Processing Letters 104(3), 79–85 (2007)

Semi-online Hierarchical Load Balancing Problem with Bounded Processing Times

Taibo Luo[1] and Yinfeng Xu[1,2]

[1] Business School, Sichuan University, Chengdu 610065, China
[2] State Key Lab. for Manufacturing Systems Engineering, Xi'an 710049, China
luotaibo@126.com, yfxu@scu.edu.cn

Abstract. In this paper, we consider the online hierarchical scheduling problem on two parallel machines, with the objective of maximizing the minimum machine load. Since no competitive algorithm exists for this problem, we consider the semi-online version with bounded processing times, in which the processing times are bounded by an interval $[1, \alpha]$ where $\alpha \geq 1$. We prove that no algorithm can have a competitive ratio less than $1 + \alpha$ and give an optimal algorithm with the competitive ratio of $1 + \alpha$. Moreover, if we further know the sum of jobs' processing time in advance, we prove that no algorithm can have a competitive ratio less than α where $1 \leq \alpha < 2$, and we also propose an algorithm which is shown to be optimal for the case $1 \leq \alpha < 2$.

Keywords: Scheduling, Semi-online, Load balancing, Competitive ratio, Hierarchy.

1 Introduction

Hierarchical scheduling on m parallel machines problem has been widely studied. The typical goal is to minimize the maximum load of any machine (i.e., minimize makespan). Hwang et al. [3] studied the offline version and proposed an approximation algorithm LG-LPT with the makespan no more than $\frac{5}{4}$ times the optimum for $m = 2$, and $2 - \frac{1}{m-1}$ times the optimum for $m \geq 3$. For the online version, Park et al. [7] and Jiang et al. [4] independently presented an optimal algorithm with a competitive ratio of $\frac{5}{3}$ for the case of two machines. Besides, there are also many papers focus on the semi-online hierarchical scheduling on two machines [5],[6],[7],[9].

Meanwhile, another goal that to maximize the minimum load of any machine (i.e., load balancing problem) was also studied well. Chassid and Epstein [1] considered the hierarchical load balancing model on two machines of possibly different speeds. They proved that no competitive algorithm exists for this problem, and they overcame this barrier by two ways. The first one is a fractional assignment model where each job can be arbitrarily split between the machines. The second one is a semi-online model where the sum of jobs' processing time is known in advance. They designed algorithms of best possible competitive ratios for both cases. Wu et al. [10] proved that no competitive algorithm exists for the semi-online version where the largest processing time of all jobs is known in

Q. Gu, P. Hell, and B. Yang (Eds.): AAIM 2014, LNCS 8546, pp. 231–240, 2014.

advance. However, when the hierarchy of the largest job is further known, Wu et al. [8] designed an optimal algorithm with a competitive ratio of $1 + \frac{\sqrt{2}}{2}$ for the case where the largest job belongs to the higher hierarchy and designed an optimal algorithm with competitive ratio of β for the case where the largest job belongs to lower hierarchy, β is the largest root of equation $x^3 - 2x^2 - 2x + 2 = 0$. For the version where the optimal offline value is known in advance, Wu et al. [10] proposed an optimal algorithm with competitive ratio of 2. Then Hou and Kang [2] investigated this problem on m parallel uniform machines with two hierarchies. They also proved that there exists no competitive algorithm and overcame this barrier by the way of fractional assignment, and they designed a best possible algorithm with competitive ratio of $\frac{2ks+m-k}{ks+m-k}$ for any speed s.

In this paper, we consider the online hierarchical load balancing problem on two parallel machines with bounded processing times. Jobs arrive one by one over a list and each job has a positive processing time and hierarchy 1 or 2. The machines have different capability, i.e., the first machine M_1 can process all the jobs while the second machine M_2 can process only jobs with hierarchy 2. As we know, Chassid and Epstein [1] proved that there exists no competitive algorithm for this problem, this paper overcomes this barrier by a semi-online version where the processing times are bounded by an interval $[1, \alpha]$ where $\alpha \geq 1$. We prove that no algorithm can have a competitive ratio less than $1 + \alpha$ and give an optimal algorithm with the competitive ratio of $1 + \alpha$. Moreover, if we further know the sum of jobs' processing time in advance, we prove that no algorithms can have a competitive ratio less than α and propose an algorithm which is shown to be optimal for the case $1 \leq \alpha < 2$. Note that for the case where $\alpha \geq 2$, the lower bound of 2 has been proved in [8], and they have also presented an optimal algorithm. So, we just focus on the case where $1 \leq \alpha < 2$. For convenience, we call the first problem as hierarchical scheduling problem with bounded processing times, and call the second problem as hierarchical scheduling problem with total processing time.

The rest of this paper is organized as follows: Section 2 gives some basic definitions. In Section 3, we investigate the hierarchical scheduling problem with bounded processing times. In Section 4, we investigate the hierarchical scheduling problem with total processing time. Section 5 concludes the paper.

2   Problem Definition

We are given two machines and a series of jobs arriving online which are to be scheduled irrevocably at the time of their arrivals. The first machine can process all the jobs while the second one can process only part of the jobs. The arrival of a new job occurs only after the current job is scheduled. Let $\sigma = \{J_1, ..., J_n\}$ be the set of all jobs arranged in the order of arrival. We denote each job by $J_i = (p_i, g_i)$, where p_i is the processing time (also called job size) of job J_i and $g_i \in \{1, 2\}$ is the hierarchical level of job J_i. $g_i = 1$ if the job J_i must be processed by the first machine, and $g_i = 2$ if it can be processed by either of the two machines. p_i and g_i are not known until the arrival of job J_i.

The schedule can be seen as the partition of σ into two subsets, denoted by $\langle S_1, S_2 \rangle$, where S_1 and S_2 contain job indices assigned to the first and the second machine, respectively. Let $t(S_1) = \Sigma_{J_i \in S_1} p_i$ and $t(S_2) = \Sigma_{J_i \in S_2} p_i$ denote the loads of the first machine and the second machine, respectively. The following notations will be used.

T_i: total size of the first i jobs.

D_i: total size of jobs with hierarchy 1 in the first i jobs.

p_i^m: the largest job size among the first i jobs.

$t(S_1^i)$: total size of jobs scheduled on M_1 after job J_i is scheduled.

$t(S_2^i)$: total size of jobs scheduled on M_2 after job J_i is scheduled.

The minimum value of $t(S_1)$ and $t(S_2)$, i.e. $\min\{t(S_1), t(S_2)\}$, is defined as the minimum machine load of the schedule $\langle S_1, S_2 \rangle$. The objective is to find a schedule $\langle S_1, S_2 \rangle$ that maximizes the minimum machine load.

We define $L_i = \min\{T_i - D_i, \frac{T_i}{2}, T_i - p_i^m\}$ as the standard upper bound of the optimal minimum machine load of the sequence containing the first i jobs.

Lemma 1. *The optimal minimum machine load is at most L_i at any moment i.*

For a job sequence σ and an algorithm A, let c_A denote the objective function value produced by A and let c_{opt} denote the optimal objective function value in an offline version. Then the competitive ratio of A is defined as the smallest number r such that for any σ, $c_{opt} \leq r \cdot c_A$.

3   Hierarchical Scheduling with Bounded Processing Times

In this section, we study the hierarchical load balancing problem on two machines with bounded processing times. The processing times are bounded by an interval $[1, \alpha]$ where $\alpha \geq 1$.

3.1   Lower Bounds of Competitive Ratio

When the processing times are bounded by an interval $[1, \alpha]$, we prove that no online algorithm can have a competitive ratio less than $1 + \alpha$.

Theorem 1. *There exists no algorithm with a competitive ratio less than $1 + \alpha$.*

Proof. Consider an algorithm A and the following sequence of jobs. The first job is with $g_1 = 2$ and $p_1 = 1$. If algorithm A schedules J_1 on M_1, we further generate the last job with $g_2 = 1$ and $p_2 = 1$. Therefore, we have $c_{opt} = 1$ and $c_A = 0$, which lead to $\frac{c_{opt}}{c_A} = \infty$. Otherwise, if algorithm A schedules J_1 on M_2, we generate job $J_2 = (\alpha, 2)$. Algorithm A must schedule it on M_1, otherwise, $c_{opt} = 1$ and $c_A = 0$, which again makes $\frac{c_{opt}}{c_A} = \infty$. Then we further generate jobs $J_3 = (1, 1)$ and $J_4 = (\alpha, 1)$, these two jobs must be scheduled on M_1. Since the optimal algorithm will schedule jobs J_1 and J_2 on M_2 and schedule jobs J_3 and J_4 on M_1, we have $c_{opt} = 1 + \alpha$ and $c_A = 1$. Hence, we cannot have an algorithm with a competitive ratio less than $1 + \alpha$. $\qquad \square$

3.2   An Optimal Algorithm

In this subsection, we present an optimal online algorithm with a competitive ratio of $1+\alpha$ to match the lower bound. First, we describe our algorithm as follows.

Algorithm M:
Step 1. Let $S_1^0 = \emptyset$, $S_2^0 = \emptyset$, $i = 1$;
Step 2. Receive job $J_i = (p_i, g_i)$, update T_i, D_i, p_i^m and L_i;
Step 3. If $g_i = 1$, schedule it to M_1. Go to **Step 5**;
Step 4. If $g_i = 2$. Schedule J_i on M_1 if $t(S_1^{i-1}) < \frac{1}{1+\alpha}L_i$, else schedule it on
$\qquad$ M_2. Go to **Step 5**;
Step 5: If there is a new job, let $i = i + 1$ and go to **Step 2**. Else, output
$\qquad$ S_1 and S_2.

Let c_M be the cost of algorithm M and c_{opt} be the cost of the optimal offline algorithm.

Theorem 2. *If* $c_M = \min\{t(S_1^n), t(S_2^n)\} = t(S_1^n)$, *then* $\frac{c_{opt}}{c_M} \le 1 + \alpha$.

Proof. If no jobs are scheduled on M_2 by algorithm M, obviously, $t(S_1^n) = T_n \ge c_{opt}$ holds. Otherwise, assume J_i is the last job scheduled on M_2. According to the rules of algorithm M, we have $t(S_1^{i-1}) \ge \frac{1}{1+\alpha}L_i$. Since $L_n - L_i \le T_n - T_i$ and all the jobs arrived after J_i will be scheduled on M_1, we have

$$t(S_1^n) = t(S_1^{i-1}) + (T_n - T_i) \ge \frac{1}{1+\alpha}L_i + (L_n - L_i) \ge \frac{1}{1+\alpha}L_n \ge \frac{1}{1+\alpha}c_{opt}.$$

$\square$

Lemma 2. *If job* J_i *with* $g_i = 2$ *is scheduled on* M_1 *by algorithm* M, *then the case of* $L_i = T_i - D_i$ *cannot happen.*

Proof. If job J_i with $g_i = 2$ is scheduled on M_1 by algorithm M, according to the rules of algorithm M, we have $t(S_1^{i-1}) < \frac{1}{1+\alpha}L_i$. If $L_i = \min\{T_i - D_i, \frac{T_i}{2}, T_i - p_i^m\} = T_i - D_i$, then we have $T_i - D_i \le \frac{T_i}{2}$, which means $D_i \ge \frac{T_i}{2}$. Since $t(S_1^{i-1}) \ge D_{i-1} = D_i$, then we have $t(S_1^{i-1}) \ge \frac{T_i}{2} > \frac{1}{1+\alpha}L_i$, that is contradicted with $t(S_1^{i-1}) < \frac{1}{1+\alpha}L_i$. Hence, the case of $L_i = T_i - D_i$ cannot happen. $\square$

Lemma 3. *If job* J_i *with* $g_i = 2$ *is scheduled on* M_1 *by algorithm* M *and* $T_i < 4$, *then* $t(S_2^i) \ge \frac{1}{1+\alpha}(T_i - D_i)$.

Proof. In this case, $\frac{1}{1+\alpha}L_i \le \frac{1}{1+\alpha} \cdot \frac{T_i}{2} \le \frac{T_i}{4} < 1$ must hold. This implies that no jobs are scheduled on M_1 before job J_i, i.e., $t(S_1^{i-1}) = 0$. Otherwise, $t(S_1^{i-1}) \ge 1 > \frac{1}{1+\alpha}L_i$. Moreover, there is at least one job that scheduled on M_2, otherwise, $L_i = T_i - p_i^m = 0 = t(S_1^{i-1})$ must hold. This is contradicted with job J_i is scheduled on M_1. Therefore, we have $t(S_2^i) \ge 1$. Since $p_i \le \alpha$, we have

$$\frac{T_i - D_i}{t(S_2^i)} \le \frac{T_i}{t(S_2^i)} \le \frac{t(S_2^i) + p_i}{t(S_2^i)} \le 1 + \frac{p_i}{t(S_2^i)} \le 1 + \alpha.$$

$\square$

Lemma 4. *If job J_i with $g_i = 2$ is scheduled on M_1 by algorithm M and $L_i = \frac{T_i}{2}$, then $t(S_2^i) \geq \frac{1}{1+\alpha}(T_i - D_i)$.*

Proof. We will prove the lemma by discussing the following two cases according to the values of α.

Case 1. $\alpha \geq 2$.

Since job J_i with $g_i = 2$ is scheduled on M_1, we have $t(S_1^{i-1}) < \frac{1}{1+\alpha}L_i = \frac{1}{2(1+\alpha)}T_i$. Since $L_i = \min\{T_i - D_i, \frac{T_i}{2}, T_i - p_i^m\} = \frac{T_i}{2}$, $T_i - p_i^m \geq \frac{T_i}{2}$ must hold, which means $p_i^m \leq \frac{T_i}{2}$. Of course, $p_i \leq p_i^m \leq \frac{T_i}{2}$ must hold. Then we have

$$t(S_2^i) = T_i - t(S_1^{i-1}) - p_i > T_i - \frac{1}{2(1+\alpha)}T_i - \frac{T_i}{2} = \frac{\alpha}{2(1+\alpha)}T_i.$$

Combined with $\alpha \geq 2$ and $T_i \geq T_i - D_i$, we have

$$t(S_2^i) > \frac{\alpha}{2(1+\alpha)}T_i \geq \frac{1}{1+\alpha}(T_i - D_i).$$

Case 2. $1 \leq \alpha < 2$.

In this case, we have $\frac{2\alpha(1+\alpha)}{2\alpha-1} \leq 4$. We discuss the following two subcases according to the values of T_i.

Subcase 2.1. $T_i \geq 4$.

In this subcase, we have $T_i \geq \frac{2\alpha(1+\alpha)}{2\alpha-1}$ since $\frac{2\alpha(1+\alpha)}{2\alpha-1} \leq 4$. Then we have

$$2\alpha T_i - 2\alpha(1 + \alpha) \geq T_i.$$

Since $p_i \leq \alpha$ and $t(S_1^{i-1}) < \frac{1}{2(1+\alpha)}T_i$, we have

$$t(S_2^i) = T_i - t(S_1^{i-1}) - p_i > T_i - \frac{1}{2(1+\alpha)}T_i - \alpha = \frac{2\alpha+1}{2(1+\alpha)}T_i - \alpha.$$

Then we have

$$\frac{t(S_2^i)}{t(S_1^{i-1})} > \frac{\frac{2\alpha+1}{2(1+\alpha)}T_i - \alpha}{\frac{1}{2(1+\alpha)}T_i} = \frac{2\alpha T_i - 2\alpha(1+\alpha) + T_i}{T_i} \geq \frac{T_i + T_i}{T_i} = 2$$

which further implies that

$$\frac{T_i - D_i}{t(S_2^i)} \leq \frac{T_i}{t(S_2^i)} = \frac{t(S_1^{i-1}) + t(S_2^i) + p_i}{t(S_2^i)} = 1 + \frac{t(S_1^{i-1}) + p_i}{t(S_2^i)}.$$

Now, we need to prove that $t(S_2^i) \geq 2$ will hold in this subcase. If $t(S_1^{i-1}) = 0$, then we have $t(S_2^i) = T_i - t(S_1^{i-1}) - p_i > 2$ since $T_i \geq 4$ and $p_i \leq \alpha < 2$. Otherwise, we have $t(S_1^{i-1}) \geq 1$. Combined with $t(S_2^i) > 2t(S_1^{i-1})$, $t(S_2^i) > 2t(S_1^{i-1}) \geq 2$ must hold. Hence, we have $t(S_2^i) \geq 2$ in this subcase.

We continue to prove the lemma. If $t(S_1^{i-1}) \geq p_i$, combined with $t(S_2^i) > 2t(S_1^{i-1})$ and $\alpha \geq 1$, we have

$$\frac{T_i - D_i}{t(S_2^i)} \leq 1 + \frac{t(S_1^{i-1}) + p_i}{t(S_2^i)} \leq 1 + \frac{2t(S_1^{i-1})}{t(S_2^i)} < 2 \leq 1 + \alpha.$$

Otherwise, $t(S_1^{i-1}) < p_i$, combined with $t(S_2^i) \geq 2$ and $p_i \leq \alpha$, we have

$$\frac{T_i - D_i}{t(S_2^i)} \leq 1 + \frac{t(S_1^{i-1}) + p_i}{t(S_2^i)} < 1 + \frac{2p_i}{2} \leq 1 + \frac{2\alpha}{2} = 1 + \alpha.$$

So, in this subcase, we have $t(S_2^i) > \frac{1}{1+\alpha}(T_i - D_i)$.

Subcase 2.2. $T_i < 4$.
  Based on Lemma 3, we have $t(S_2^i) > \frac{1}{1+\alpha}(T_i - D_i)$.   □

Lemma 5. *If job J_i with $g_i = 2$ is scheduled on M_1 by algorithm M and $L_i = T_i - p_i^m$, then $t(S_2^i) \geq \frac{1}{1+\alpha}(T_i - D_i)$.*

Proof. As $L_i = T_i - p_i^m$, we have $T_i - p_i^m \leq \frac{T_i}{2}$, which implies $p_i^m \geq \frac{T_i}{2}$. If p_i^m belongs to S_1^{i-1}, we have $t(S_1^{i-1}) \geq \frac{T_i}{2} > \frac{1}{1+\alpha}L_i$. This is contradicted with job J_i is scheduled on M_1 by algorithm M. If p_i^m belongs to S_2^i, we have $t(S_2^i) \geq \frac{T_i}{2} \geq \frac{1}{1+\alpha}T_i \geq \frac{1}{1+\alpha}(T_i - D_i)$.
  If $p_i^m = p_i$, then we have $T_i - p_i^m = t(S_1^{i-1}) + t(S_2^i)$. Since job J_i is scheduled on M_1 by algorithm M, we have $t(S_1^{i-1}) < \frac{1}{1+\alpha}L_i = \frac{1}{1+\alpha}(T_i - p_i^m)$. Hence,

$$t(S_2^i) = T_i - p_i^m - t(S_1^{i-1}) > T_i - p_i^m - \frac{1}{1+\alpha}(T_i - p_i^m) = \frac{\alpha}{1+\alpha}(T_i - p_i^m).$$

Then we discuss the following two cases according to the values of α.

Case 1. $\alpha \geq 2$.
  In this case, since $t(S_1^{i-1}) + t(S_2^i) = T_i - p_i = T_i - p_i^m$ and $t(S_1^{i-1}) < \frac{1}{1+\alpha}(T_i - p_i^m)$, we have

$$t(S_2^i) > \frac{\alpha}{1+\alpha}(T_i - p_i^m) > 2t(S_1^{i-1}).$$

If no jobs are scheduled on M_1 before job J_i, then there is at least one job that scheduled on M_2, otherwise, $L_i = T_i - p_i^m = 0 = t(S_1^{i-1})$ must hold. This is contradicted with job J_i is scheduled on M_1. Therefore, $t(S_2^i) \geq 1$. Since $p_i \leq \alpha$, we have

$$\frac{T_i - D_i}{t(S_2^i)} \leq \frac{T_i}{t(S_2^i)} \leq \frac{t(S_2^i) + p_i}{t(S_2^i)} \leq 1 + \frac{p_i}{t(S_2^i)} \leq 1 + \alpha.$$

If there is at least one job that scheduled on M_1 before job J_i, then we have $t(S_2^i) > 2t(S_1^{i-1}) \geq 2$. Since $p_i \geq \frac{T_i}{2} > t(S_1^{i-1})$ and $p_i \leq \alpha$, we have

$$\frac{T_i - D_i}{t(S_2^i)} \leq \frac{T_i}{t(S_2^i)} = 1 + \frac{t(S_1^{i-1}) + p_i}{t(S_2^i)} < 1 + \alpha.$$

Case 2. $1 \leq \alpha < 2$.
  Since $\frac{T_i}{2} \leq p_i^m < 2$, we have $T_i < 4$. Based on Lemma 3, we have $t(S_2^i) > \frac{1}{1+\alpha}(T_i - D_i)$.

                    □

Theorem 3. *Algorithm M is $(1 + \alpha)$-competitive which is shown to be optimal.*

Proof. Based on Theorem 2, if $c_M = \min\{t(S_1^n), t(S_2^n)\} = t(S_1^n)$, then $\frac{c_{opt}}{c_M} \leq 1 + \alpha$. Therefore, we just need to prove $\frac{c_{opt}}{c_M} \leq 1 + \alpha$ will hold when $c_M = \min\{t(S_1^n), t(S_2^n)\} = t(S_2^n)$. We discuss the following two cases.

Case 1. No jobs with hierarchy 2 are scheduled on M_1.

In this case, we have $t(S_2^n) = T_n - D_n$. Since $T_n - D_n \geq L_n \geq c_{opt}$, $t(S_2^n) \geq c_{opt} > \frac{1}{1+\alpha} c_{opt}$ holds.

Case 2. At least one job with hierarchy 2 is scheduled on M_1.

Let J_f denote the last job with $g_f = 2$ scheduled on M_1. Based on Lemma 2-5, we have $t(S_2^f) \geq \frac{1}{1+\alpha}(T_f - D_f)$. Since all the jobs with hierarchy 2 are scheduled on M_2 after job J_f, then we get

$$t(S_2^n) = t(S_2^f) + ((T_n - D_n) - (T_f - D_f))$$

which implies that

$$\begin{aligned}
t(S_2^n) &= t(S_2^f) + ((T_n - D_n) - (T_h - D_f)) \\
&\geq \tfrac{1}{1+\alpha}(T_f - D_f) + ((T_n - D_n) - (T_f - D_f)) \\
&\geq \tfrac{1}{1+\alpha}(T_n - D_n).
\end{aligned}$$

Since $T_n - D_n \geq L_n \geq c_{opt}$, $t(S_2^n) \geq \frac{1}{1+\alpha} c_{opt}$ hold.

Since the lower bound of this problem is $1 + \alpha$, so algorithm M is $(1 + \alpha)$-competitive which is shown to be optimal.   □

4   Hierarchical Scheduling with Total Processing Time

In this section, we further know jobs' total processing time in advance. Note again, for the case where $\alpha \geq 2$, the lower bound of 2 has been proved in Wu et al. [8], and they have also presented two optimal algorithms. So, we just focus on the case where $1 \leq \alpha < 2$. Let Σ denote the jobs' total processing time.

4.1   Lower Bounds of Competitive Ratio

In this subsection, we prove that no online algorithm can have a competitive ratio less than α where $1 \leq \alpha < 2$.

Theorem 4. *There exists no online algorithm with a competitive ratio less than α for $1 \leq \alpha < 2$.*

Proof. Consider an algorithm A and the following sequence of jobs. We know $\Sigma = 2 + \alpha$ in advance. The first job is $J_1 = (1, 1)$, then it will be scheduled on M_1. We further generate job $J_2 = (1, 2)$. If algorithm A schedule it on M_1, then the last job with $g_3 = 1$ and $p_3 = \alpha$ will arrive. Therefore, we have $c_{opt} = \alpha$ and $c_A = 0$, which lead to $\frac{c_{opt}}{c_A} = \infty$. So, algorithm A will schedule J_2 on M_2, then the last job with $g_2 = 2$ and $p_2 = \alpha$ arrives. No matter which machine that J_3 is scheduled on, we have $c_A = 1$. Since the optimal algorithm will schedule jobs J_1 and J_2 on M_1 and schedule job J_3 on M_2, we have $\frac{c_{opt}}{c_A} = \alpha$. So, we cannot have an algorithm with a competitive ratio less than α.   □

4.2 An Optimal Algorithm

In this subsection, we present an optimal online algorithm with a competitive ratio of α to match the lower bound. Let $L = \frac{\Sigma}{2}$. First, we describe our algorithm as follows.

Algorithm H:
Step1. Let $S_1^0 = \emptyset$, $S_2^0 = \emptyset$, $i = 1$;
Step2. Receive job $J_i = (p_i, g_i)$;
Step3. If $g_i = 1$, let $S_1^i = S_1^{i-1} \bigcup \{J_i\}$. Go to **Step 5**;
Step4. If $g_i = 2$,
4.1. If $t(S_2^{i-1}) + p_i < \frac{1}{\alpha}L$, let $S_2^i = S_2^{i-1} \bigcup \{J_i\}$. Go to **Step 5**;
4.2 (Stopping criterion 1). If $t(S_2^{i-1}) + p_i \geq \frac{1}{\alpha}L$ and $\Sigma - t(S_2^{i-1}) - p_i \geq \frac{1}{\alpha}L$, schedule job J_i on M_2 and schedule all the remaining jobs to M_1. Stop and output M_1 and M_2.
4.3 (Stopping criterion 2). If $t(S_2^{i-1}) < \frac{1}{\alpha}L$, $\Sigma - t(S_2^{i-1}) - p_i < \frac{1}{\alpha}L$ and $\Sigma - t(S_2^{i-1}) - p_i < t(S_2^{i-1})$, schedule job J_i and all the remaining jobs to M_1. Stop and output M_1 and M_2.
4.4 (Stopping criterion 3). If $t(S_2^{i-1}) < \frac{1}{\alpha}L$, $\Sigma - t(S_2^{i-1}) - p_i < \frac{1}{\alpha}L$ and $\Sigma - t(S_2^{i-1}) - p_i \geq t(S_2^{i-1})$, schedule job J_i to S_2 and schedule all the remaining jobs to M_1. Stop and output S_1 and S_2.
Step5: If no more jobs arrive, stop and output S_1 and S_2; Else, let $i = i + 1$ and go to **Step 2**.

Before we prove algorithm H is optimal, we give a lemma first.

Lemma 6. *Suppose $1 \leq \alpha < 2$, for an arbitrary job sequence σ, if the number of jobs in σ, denoted by n, is an even number, then the total processing time of arbitrary $\frac{n}{2}$ jobs in σ is at least $\frac{1}{\alpha}L$.*

Proof. Let S_h be a job set of arbitrary $\frac{n}{2}$ jobs in σ, and S_l be the set of the other $\frac{n}{2}$ jobs where $\Sigma = t(S_h) + t(S_l)$. Then we have

$$t(S_h) - \tfrac{1+\alpha}{2}L = t(S_h) - \tfrac{1+\alpha}{2} \cdot \tfrac{t(S_h)+t(S_l)}{2} = \tfrac{3-\alpha}{4}t(S_h) - \tfrac{1+\alpha}{2} \cdot \tfrac{t(S_l)}{2}.$$

As $\alpha < 2$, we have $\frac{3-\alpha}{4} > 0$. Combined with $t(S_h) \leq \frac{n}{2}\alpha$ and $t(S_l) \geq \frac{n}{2}$, we have

$$t(S_h) - \tfrac{1+\alpha}{2}L \leq \tfrac{3-\alpha}{4} \cdot \tfrac{n}{2} \cdot \alpha - \tfrac{1+\alpha}{2} \cdot \tfrac{n}{4} = -\tfrac{n}{8} \cdot (\alpha - 1)^2 \leq 0$$

which means that the total processing times of arbitrary $\frac{n}{2}$ jobs in σ is at most $\frac{1+\alpha}{2}L$. Further, we get

$$t(S_l) \geq \Sigma - \tfrac{1+\alpha}{2}L = \tfrac{3-\alpha}{2}L.$$

Since $\frac{3-\alpha}{2}L - \frac{1}{\alpha}L \geq 0$ will hold for $1 \leq \alpha < 2$, then we have $t(S_l) \geq \frac{1}{\alpha}L$. By the same way, we can get $t(S_h) \geq \frac{1}{\alpha}L$. So, the total processing time of arbitrary $\frac{n}{2}$ jobs in σ is at least $\frac{1}{\alpha}L$. □
Straightforwardly, we have the following corollary.

Corollary 1. *Suppose* $1 \leq \alpha < 2$, *for an arbitrary job sequence* σ *which contains* n *jobs, the total processing time of arbitrary* n' *jobs in* σ *is at least* $\frac{1}{\alpha}L$ *where* $n' \geq \lceil \frac{n}{2} \rceil$.

Theorem 5. *Algorithm* H *is* α-*competitive for* $1 \leq \alpha < 2$.

Proof. Suppose the theorem is false, then there must exist at least one instance I which makes $\frac{c_{opt}}{c_H} > \alpha$. Let n be the number of jobs in I. We distinguish the following two cases according to the value of n.

Case 1: n is an even number.

In this case, we have two subcases. The first subcase is that no jobs with hierarchy 2 are scheduled to S_1. If no jobs with hierarchy 2 are scheduled on M_1, according to the rules of algorithm H, we have $t(S_2) = \Sigma - D_n \geq c_{opt}$. Also, we have $t(S_1) \geq \frac{1}{\alpha}L \geq \frac{1}{\alpha}c_{opt}$, otherwise, based on Lemma 6, at least one job with hierarchy 2 will be scheduled on M_1. The other subcase is that there is at least one job with hierarchy 2 is scheduled on M_1. Based on Lemma 6, algorithm H schedules at most $\frac{n}{2}$ jobs of I on M_2, which implies that algorithm H will stop at Step 4.2, which further implies that $t(S_1) \geq \frac{1}{\alpha}L$ and $t(S_2) \geq \frac{1}{\alpha}L$. Again, we have $c_H = \min\{t(S_1), t(S_2)\} \geq \frac{1}{\alpha}c_{opt}$.

Case 2: n is an odd number.

Divide I into two job sets I_1 and I_2 where I_1 contains $\frac{n+1}{2}$ jobs and I_2 contains $\frac{n-1}{2}$ jobs. The processing time of any job in I_1 is not greater than the processing time of any job in I_2. Let $t(I_1)$ and $t(I_2)$ denote the total processing time of jobs in I_1 and I_2, respectively, where $t(I_1) + t(I_2) = \Sigma$. As the processing times are bounded in the interval $[1, \alpha]$, we have $t(I_1) \geq \frac{n+1}{2}$ and $t(I_2) \leq \frac{n-1}{2}\alpha$. Since the optimal algorithm must schedule at most $\frac{n-1}{2}$ jobs in I to one of the two machines, and according to the definition of I_2, we know that I_2 contains the $\frac{n-1}{2}$ jobs which have the most largest processing time, so $c_{opt} \leq t(I_2)$ holds.

If algorithm H stops at Step 4.2, $c_H = \min\{t(S_1), t(S_2)\} \geq \frac{1}{\alpha}c_{opt}$ can be got directly. If algorithm H stops at Step 5, then no jobs with hierarchy 2 are scheduled to S_1. Also we have $c_M = \min\{t(S_1), t(S_2)\} \geq \frac{1}{\alpha}c_{opt}$. Therefore, it must stop at Step 4.3 or Step 4.4.

Suppose algorithm H stops at Step 4.3, and job J_i is the job makes $t(S_2^{i-1}) < \frac{1}{\alpha}L$, $\Sigma - t(S_2^{i-1}) - p_i < \frac{1}{\alpha}L$ and $\Sigma - t(S_2^{i-1}) - p_i < t(S_2^{i-1})$ hold. In this case, we have $t(S_2) = t(S_2^{i-1})$ and $t(S_1) = \Sigma - t(S_2^{i-1})$. Since $t(S_2) = t(S_2^{i-1}) < \frac{1}{\alpha}L$, based on Corollary 1, there are at most $\frac{n-1}{2}$ jobs are scheduled on M_2 and at least $\frac{n+1}{2}$ jobs are scheduled on M_1. This means that $t(S_1) \geq \frac{1}{\alpha}c_{opt}$.

So, $c_H = \min\{t(S_1), t(S_2)\} = t(S_2) < \frac{1}{\alpha}c_{opt}$. As $\Sigma - t(S_2) - p_i < t(S_2) < \frac{1}{\alpha}c_{opt}$, we get $t(S_1) - p_i < t(S_2) < \frac{1}{\alpha}c_{opt}$. Then we have $t(S_1) + t(S_2) - p_i < \frac{2}{\alpha}c_{opt}$ or $\alpha(t(S_1) - p_i + t(S_2)) < 2c_{opt}$. As $t(S_1) + t(S_2) = t(I_1) + t(I_2)$, we have

$$\alpha(t(S_1) - p_i + t(S_2)) = \alpha(t(I_1) + t(I_2) - p_i) \geq (n-1)\alpha.$$

Since $t(I_2) \leq \frac{n-1}{2}\alpha$ and $c_{opt} \leq t(I_2)$, we have

$$2c_{opt} \leq 2t(I_2) \leq (n-1)\alpha.$$

These two inequalities imply that

$$(n-1)\alpha \le \alpha(t(S_1) + t(S_2)) < 2c_{opt} + p_i \le (n-1)\alpha$$

which is a contradiction.

If algorithm H stops at Step 4.4, we can the get same contradiction like Step 4.3. Hence, we know that such an example which makes $\frac{c_{opt}}{c_H} > \alpha$ do not exist. □

5   Conclusion

In this paper, we consider the semi-online version of hierarchical scheduling problem on two parallel machines with the objective of maximizing the minimum machine load. If the processing times are bounded in an interval $[1, \alpha]$, we show the lower bound of competitive ratio is $1 + \alpha$ and present an algorithm which is shown to be optimal. If we further know the sum of jobs' processing times (i.e., the total processing time), we show a lower bound of α for the case where $1 \le \alpha < 2$ and present an optimal algorithm for the case where $1 \le \alpha < 2$.

Acknowledgements. This work is supported by National Natural Science Foundation of China under Grants 71071123, 60921003 and 71371129 and Program for Changjiang Scholars and Innovative Research Team in University under Grant IRT1173.

References

1. Chassid, O., Epstein, L.: The hierarchical model for load balancing on two machines. Journal of Combinatorial Optimization 15, 305–314 (2008)
2. Hou, L., Kang, L.: Online and semi-online hierarchical scheduling for load balancing on uniform machines. Theoretical Computer Science 412, 1092–1098 (2011)
3. Hwang, H., Chang, S., Lee, K.: Parallel machine scheduling under a grade of service provision. Computers and Operations Research 31, 2055–2061 (2004)
4. Jiang, Y., He, Y., Tang, C.: Optimal online algorithms for scheduling on two identical machines under a grade of service. Journal of Zhejiang University. Science A 7, 309–314 (2006)
5. Liu, M., Chu, C., Xu, Y., Zheng, F.: Semi-online scheduling on 2 machines under a grade of service provision with bounded processing times. Journal of Combinatorial Optimization 21, 138–149 (2011)
6. Luo, T., Xu, Y., Luo, L., He, C.: Semi-online scheduling with two GoS levels and unit processing time. Theoretical Computer Science 521, 62–72 (2014)
7. Park, J., Chang, S.K., Lee, K.: Online and semi-online scheduling of two machines under a grade of service provision. Operations Research Letters 34, 692–696 (2006)
8. Wu, Y., Cheng, T.C.E., Ji, M.: Optimal algorithm for semi-online machine covering on two hierarchical machines. Theoretical Computer Science 531, 37–46 (2014)
9. Wu, Y., Ji, M., Yang, Q.: Optimal semi-online scheduling algorithms on two parallel identical machines under a grade of service provision. International Journal of Production Economics 135, 367–371 (2012)
10. Wu, Y., Ji, M., Yang, Q.: Semi-Online Machine Covering under a Grade of Service Provision. Applied Mechanics and Materials 484, 101–102 (2011)
11. Zhang, A., Jiang, Y., Fan, L., Hu, J.: Optimal online algorithms on two hierarchical machines with tighly-grouped processing times. Journal of Combinatorial Optimization (2013), doi:10.1007/s10878-013-9627-7

Restricted Bipartite Graphs:
Comparison and Hardness Results[*]

Tian Liu

Key Laboratory of High Confidence Software Technologies, Ministry of Education,
Institute of Software, School of Electronic Engineering and Computer Science,
Peking University, Beijing 100871, China
lt@pku.edu.cn

Abstract. Convex bipartite graphs are a subclass of circular convex bipartite graphs and chordal bipartite graphs. Chordal bipartite graphs are a subclass of perfect elimination bipartite graphs and tree convex bipartite graphs. No other inclusion among them is known. In this paper, we make a thorough comparison on them by showing the nonemptyness of each region in their Venn diagram. Thus no further inclusion among them is possible, and the known complexity results on them are incomparable. We also show the $\mathcal{NP}$-completeness of treewidth and feedback vertex set for perfect elimination bipartite graphs.

Keywords: Perfect elimination bipartite graphs, tree convex bipartite graphs, circular convex bipartite graphs, chordal bipartite graphs, convex bipartite graphs, $\mathcal{NP}$-completeness, treewidth, feedback vertex set.

1 Introduction

Some $\mathcal{NP}$-complete graph problems, such as treewidth and feedback vertex set, are still $\mathcal{NP}$-complete for bipartite graphs, but tractable for restricted bipartite graphs, such as convex bipartite graphs, chordal bipartite graphs, circular convex bipartite graphs, and so on. Exploring the properties of these restricted bipartite graphs and the boundary between $\mathcal{NP}$-completeness and tractability are well established research directions, see e.g. [2]. In this paper, we show some separation results for restricted bipartite graphs, including perfect elimination bipartite graphs, chordal bipartite graphs, convex bipartite graphs, tree convex bipartite graphs, and circular convex bipartite graphs. We also show the $\mathcal{NP}$-completeness of treewidth and feedback vertex set for perfect elimination bipartite graphs.

Perfect elimination bipartite graphs, chordal bipartite graphs, and convex bipartite graphs are well studied bipartite graph classes [2]. In a *convex bipartite* graph $G = (V_1, V_2, E)$, there is a linear ordering L defined on V_1, such that for each vertex in V_2, its neighborhood induces an interval under L [4]. Given a cycle,

---

[*] Partially supported by National 973 Program of China (Grant No. 2010CB328103) and Natural Science Foundation of China (Grant Nos. 61370052 and 61370156).

Q. Gu, P. Hell, and B. Yang (Eds.): AAIM 2014, LNCS 8546, pp. 241–252, 2014.

an edge with two endpoints nonconsecutive in the cycle is called a chord. In a *chordal bipartite* graph, each cycle of length at least six must have a chord [3]. An edge in a bipartite graph is *bisimplicial*, if its endpoint neighborhoods induce a complete bipartite subgraph. A perfect elimination ordering of a bipartite graph is a linear ordering on a subset of nonadjacent edges, such that each edge in this subset is bisimplicial in the remaining bipartite subgraphs when all endpoints of preceding edges are removed, and finally no edge is left in the graph. In a *perfect elimination bipartite* graph, there is a perfect elimination ordering [3].

Circular convex bipartite graphs and tree convex bipartite graphs are two natural generalizations to convex bipartite graphs [13,5,7,19,22,17,15,18,21,14]. In a *circular convex bipartite* graph $G = (V_1, V_2, E)$, there is a circular ordering R defined on V_1, such that for each vertex in V_2, its neighborhood induces a circular arc under R [13]. In a *tree convex bipartite* graph $G = (V_1, V_2, E)$, there is a tree T defined on V_1, such that for each vertex in V_2, its neighborhood induces a subtree on T [5]. When T is a path, G is just a convex bipartite graph. When T is a star, G is called a *star convex bipartite* graph [5]. When T is a triad, which is three paths with a common endpoint, G is called a *triad convex bipartite* graph [7].

It has been known that chordal bipartite graphs is sandwiched between convex bipartite graphs and perfect elimination bipartite graphs [3], and also between convex bipartite graphs and tree convex bipartite graphs [6]. Convex bipartite graphs are a subclass of circular convex bipartite graphs [13]. No other inclusion of them is known. So our first question is

– *Is there any other inclusion among perfect elimination bipartite graphs, tree convex bipartite graphs, circular convex bipartite graphs, chordal bipartite graphs, and convex bipartite graphs?*

In this paper, we give a negative answer by showing the nonemptyness of each region in their Venn diagram. Thus no further inclusion among them is possible, and the known complexity results on them in literatures are incomparable.

Treewidth and feedback vertex set are two well studied $\mathcal{NP}$-complete problems. They are also $\mathcal{NP}$-complete for bipartite graphs [9,23] and for tree convex bipartite graphs [21,5,6], but tractable for chordal bipartite graphs [10,11]. Feedback vertex set is also tractable for circular convex bipartite graphs [18] and for triad convex bipartite graphs [7,6]. Our second question is

– *Where is the boundary between $\mathcal{NP}$-completeness and tractability to treewidth and feedback vertex set for these restricted bipartite graphs?*

In this paper, we give a partial answer by showing the $\mathcal{NP}$-completeness of treewidth and feedback vertex set for perfect elimination bipartite graphs. Therefore, the known tractability of them for chordal bipartite graphs [10,11] can not be extended to perfect elimination bipartite graphs, unless $\mathcal{NP} = \mathcal{P}$.

This paper is structured as follows. After introducing necessary definitions and facts in Section 2, separation results for restricted bipartite graph classes are shown in Section 3, $\mathcal{NP}$-completeness results for perfect elimination bipartite graphs are shown in Section 4, and finally are concluding remarks in Section 5.

2   Preliminaries

For a graph $G = (V, E)$, we denote the *neighborhood* of a vertex u by $N_G(u) = \{v|(u,v) \in E\}$. When G is clear from the context, we just write $N(u)$. A *complete bipartite* graph $G = (V_1, V_2, E)$ has $E = \{(u,v)|u \in V_1, v \in V_2\}$. For a bipartite graph $G = (V_1, V_2, E)$, a subset of pairwise nonadjacent edges $\{(u_1, v_1), (u_2, v_2), \cdots, (u_k, v_k)\}$ is a *perfect elimination ordering*, if each (u_i, v_i) is bisimplicial in G after removing $\{u_1, v_1, u_2, v_2, \cdots, u_{i-1}, v_{i-1}\}$, and there is no edge in G after removing $\{u_1, v_1, u_2, v_2, \cdots, u_k, v_k\}$. A *perfect elimination bipartite* graph has a perfect elimination ordering [3,2]. A hypergraph $H = (V, \mathcal{E})$ has the *Helly property*, if for every subset $\mathcal{E}' \subseteq \mathcal{E}$, if each pair of e_1, e_2 in $\mathcal{E}'$ has a nonempty intersection, then all the e's in $\mathcal{E}'$ have a nonempty intersection.

For a graph $G = (V, E)$, its *tree decomposition* is a tree $T = (B, F)$, with each vertex in B labeled by a subset of V, called *bag*, such that (1) each edge in E is contained in at least one bag; (2) for each vertex u in V, all bags containing u induce a subtree of T. The maximum size of bags minus one is the *width* of the tree decomposition. The minimum width over all tree decompositions of a graph is the *treewidth* of the graph [9,12]. The following Lemma is easy to prove by definition of treewidth [9].

Lemma 1. *Adding a new pendent vertex to a graph will not change its treewidth.*

A *feedback vertex set* is a subset of vertices whose removal renders the graph cycle-free. The minimum feedback vertex set problem is to decide whether a given graph has a feedback vertex set of size no more than a given integer [8]. The minimum size of feedback vertex sets is also called a *decycling number*.

3   Comparison Results

In this section, we make a thorough comparison on perfect elimination bipartite graphs, tree convex bipartite graphs, circular convex bipartite graphs, chordal bipartite graphs, and convex bipartite graphs, by showing the nonemptyness of each region in their Venn diagram, see Figure 1.

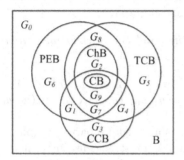

PEB: Perfect Elimination Bipartite
CCB: Circular Convex Bipartite
TCB: Tree Convex Bipartite
ChB: Chordal Bipartite
CB: Convex Bipartite

B: Bipartite

Fig. 1. Venn diagram of five restricted bipartite graph classes and graphs in each region

We use the following trick to deal with perfect elimination bipartite graphs. For a non-perfect elimination bipartite graph, we usually can add one pendent vertex or many pendent vertices to make it a perfect elimination bipartite graph, while to keep its other properties invariant.

Theorem 1. *There is a perfect elimination and circular convex bipartite graph* G_1 *which is not a tree convex bipartite graph.*

Proof. The graph $G_1 = (V_1, V_2, E)$, where $V_1 = \{x, y, z, u_a, u_b, u_c\}$, $V_2 = \{a, b, c, d_x, d_y, d_z\}$ and $E = \{(x, a), (a, y), (y, b), (b, z), (z, c), (c, x), (x, d_x), (y, d_y), (z, d_z), (a, u_a), (b, u_b), (c, u_c)\}$, is shown in Figure 2 (left).

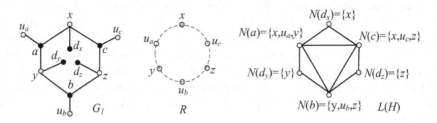

Fig. 2. A perfect elimination and circular convex bipartite graph G_1 which is not a tree convex bipartite graph

G_1 is a perfect elimination bipartite graph, since a perfect elimination ordering of G_1 is given by $\{(x, d_x), (y, d_y), (z, d_z), (a, u_a), (b, u_b), (c, u_c)\}$.

G_1 is a circular convex bipartite graph, since a circular ordering R on V_1 is given by $x \prec u_a \prec y \prec u_b \prec z \prec u_c \prec x$, as shown in Figure 2 (middle), such that the neighborhood of each vertex in V_2 induces a circular arc under R.

If G_1 is a tree convex bipartite graph with a tree associated on V_1, the hypergraph $H = (V_1, \mathcal{E})$ is a hypertree, where $\mathcal{E} = \{N(d) | d \in V_2\}$. Then $H = (V_1, \mathcal{E})$ has the Helly property and the line graph $L(H) = (\mathcal{E}, \mathcal{F})$ is chordal, where $\mathcal{F} = \{(N(d_1), N(d_2)) | N(d_1) \cap N(d_2) \neq \emptyset\}$ (Theorem 1.3.1, page 9, [2]). However, $H = (V_1, \mathcal{E})$ is not Helly, since $N(a), N(b), N(c)$ are pairwise intersect, but $N(a) \cap N(b) \cap N(c) = \emptyset$. This can be seen with the help of the line graph $L(H)$ shown in Figure 2 (right). The same holds for V_2, due to the symmetry of G_1. Therefore, G_1 is not a tree convex bipartite graph. $\square$

Theorem 2. *There is a chordal bipartite graph* G_2 *which is not a circular convex bipartite graph.*

Proof. The graph $G_2 = (V_1, V_2, E)$, where $V_1 = \{x, y, z, u_1, u_2, u_3\}$, $V_2 = \{a_0, a_1, a_2, a_3\}$ and $E = \{(x, a_1), (a_1, u_1), (u_1, a_0), (y, a_2), (a_2, u_2), (u_2, a_0), (z, a_3), (a_3, u_3), (u_3, a_0)\}$, is shown in Figure 3 (left).

There is no cycle in G_2 at all, so G_2 is a chordal bipartite graph.

Since $N(a_0) = \{u_1, u_2, u_3\}$, u_1, u_2 and u_3 must be consecutive in any circular ordering on V_1 for G_2 to be circular convex bipartite. The same reasoning applies

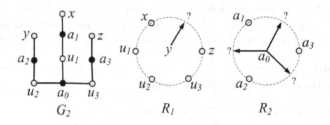

Fig. 3. A chordal bipartite graph G_2 which is not a circular convex bipartite graph

to $N(a_i)$ for $i = 1, 2, 3$. So x and u_1, y and u_2, z and u_3 respectively must be consecutive in any circular ordering, say R_1, on V_1, as shown in Figure 3 (middle). Due to the symmetry in G_2, without loss of generality, we can assume that $x \prec u_1 \prec u_2 \prec u_3 \prec z$ in R_1. Then the only possible place for y is at between y and z, but in this case, $N(a_2) = \{y, u_2\}$ is not a circular arc, since none of x, u_1, u_2, z is in $N(a_2)$. Thus, y can not be inserted into R_1 and G_2 is not circular convex bipartite with a circular ordering on V_1. A similar reasoning also applies to V_2, as shown in Figure 3 (right). Thus, G_2 is not a circular convex bipartite graph.   □

Theorem 3. *There is a circular convex bipartite graph G_3 which is neither a perfect elimination bipartite graph nor a tree convex bipartite graph.*

Proof. The graph $G_3 = (V_1, V_2, E)$, where $V_1 = \{x, y, z\}$, $V_2 = \{a, b, c\}$, and $E = \{(x, a), (a, y), (y, b), (b, z), (z, c), (c, x)\}$, is shown in Figure 4 (left).

G_3 is a circular convex bipartite graph, since a circular ordering R on V_1 can be defined by $x \prec y \prec z \prec x$, as shown in Figure 4 (right).

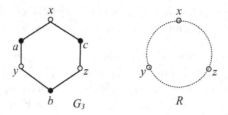

Fig. 4. A circular convex bipartite graph G_3 which is neither a perfect elimination bipartite graph nor a tree convex bipartite graph

G_3 is not a perfect elimination bipartite graph, since in any perfect elimination ordering of G_3, the first edge must be bisimplicial in G_3, but each edge of G_3 is not bisimplicial in G_3. For example, consider an edge (x, a). We have $N(x) = \{a, c\}$ and $N(a) = \{x, y\}$. Since there is no edge (c, y) in E, $N(x) \cup N(a)$ does not induce a biclique in G_3. Thus the edge (x, a) is not bisimplicial. The same holds for other five edges in E due to the symmetry of G_3.

G_3 is not a tree convex bipartite graph, since V_1 has only three vertices, any tree on V_1 is a path, say $x - y - z$. But then, the neighborhood of c, which is $N_{G_3}(c) = \{x, z\}$, does not induce a subtree. Same for V_2 by symmetry. □

Theorem 4. *There is a circular convex and tree convex bipartite graph G_4 which is not a perfect elimination bipartite graph.*

Proof. The graph $G_4 = (V_1, V_2, E)$, where $V_1 = \{x, y, z, u\}$, $V_2 = \{a, b, c\}$, and $E = \{(x, a), (a, y), (y, b), (b, z), (z, c), (c, x), (u, a), (u, b), (u, c)\}$, is shown in Figure 5 (left).

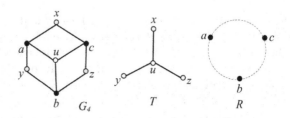

Fig. 5. A circular convex and tree convex bipartite graph G_4 which is not a perfect elimination bipartite graph

G_4 is a tree convex bipartite graph, since a tree $T = (V_1, F)$ on V_1 can be defined by $F = \{(x, u), (y, u), (z, u)\}$, as shown in Figure 5 (right), such that for each vertex in V_2, its neighborhood induces a subtree in T.

G_4 is a circular convex bipartite graph, since a circular ordering R on V_2 can be defined by $a \prec b \prec c \prec a$, as shown in Figure 5 (right), such that for each vertex in V_2, its neighborhood induces a circular arc in R.

G_4 is not a perfect elimination bipartite graph, similarly as G_3. □

Theorem 5. *There is a tree convex bipartite graph G_5 which is neither a perfect elimination bipartite graph nor a circular convex bipartite graph.*

Proof. The graph $G_5 = (V_1, V_2, E)$, where $V_1 = \{x_1, y_1, z_1, u_1, x_2, y_2, z_2, u_2\}$, $V_2 = \{a_1, b_1, c_1, d, a_2, b_2, c_2\}$, and $E = \{(x_1, a_1), (a_1, y_1), (y_1, b_1), (b_1, z_1), (z_1, c_1), (c_1, x_1), (u_1, a_1), (u_1, b_1), (u_1, c_1), (u_1, d), (d, u_2), (x_1, a_1), (a_1, y_1), (y_1, b_1), (b_1, z_1), (z_1, c_1), (c_1, x_1), (u_1, a_1), (u_1, b_1), (u_1, c_1)\}$, is shown in Figure 6 (left).

G_5 is not a circular convex bipartite graph, since G_5 is essentially two copies of G_4 linked by a vertex d. Though each copies of G_4 has a circular ordering, they can not be combined into a larger one for G_5, as readers can check it.

G_5 is not a perfect elimination bipartite graph, by the same reasoning as G_4, as well as the fact that the edges (u_1, d) and (d, u_1) are not bisimplicial.

G_5 is a tree convex bipartite graph, since a tree T on V_1 can be defined as shown in Figure 6 (right), such that for each vertex in V_2, its neighborhood induces a subtree in T. □

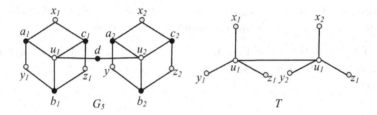

Fig. 6. A tree convex bipartite graph G_5 which is neither a perfect elimination bipartite graph nor a circular convex bipartite graph

Theorem 6. *(1) There is a bipartite graph G_0 which is neither a tree convex bipartite graph, a circular convex bipartite graph, nor a perfect elimination bipartite graph. (2) There is a perfect elimination bipartite graph G_6 which is neither a tree convex bipartite graph nor a circular convex bipartite graph.*

Proof. (1) The graph $G_0 = (V_1, V_2, E)$ is shown in Figure 7 (left).

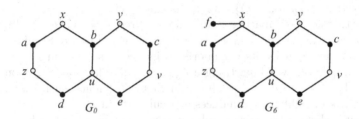

Fig. 7. A bipartite graph G_0 which is neither a tree convex bipartite graph, a circular convex bipartite graph, nor a perfect elimination bipartite graph, and a perfect elimination bipartite graph G_6 which is neither a tree convex bipartite graph nor a circular convex bipartite graph

G_0 is not a circular convex bipartite graph, since G_0 is essentially two copies of G_1 with a common edge (b, u). Though each copies of G_1 has a circular ordering, they can not be combined into a larger one for G_0.

G_0 is not a perfect elimination bipartite graph, by the same reasoning as G_1, as well as the fact that the edges (x, b) and (b, u) are not bisimplicial.

G_0 is not a tree convex bipartite graph, since any tree on V_1 must be a path $x - z - u - v - y$, due to the degree two vertices a, d, e, c. But then $N(b) = \{x, y, u\}$ does not induce a subtree. The same holds for V_2 due to symmetry.

(2) The graph $G_6 = (V_3, V_4, F)$ is shown in Figure 7 (right).

G_6 is neither a circular convex bipartite graph, nor a tree convex bipartite graph, by exactly the same reasoning as for G_0.

G_6 is a perfect elimination bipartite graph, since a perfect elimination ordering is given by $\{(f, x), (a, z), (d, u), (e, v), (c, y)\}$, as readers can check it.   □

Theorem 7. *There is a circular convex and tree convex and perfect elimination bipartite graph G_7 which is not a chordal bipartite graph.*

Proof. The graph $G_7 = (V_1, V_2, E)$, where $V_1 = \{x, y, z, u, w\}$, $V_2 = \{a, b, c\}$, and $E = \{(x, a), (a, y), (y, b), (b, z), (z, c), (c, x), (u, a), (u, b), (u, c), (w, a)\}$, is shown in Figure 8 (left).

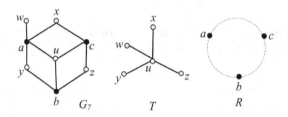

Fig. 8. A circular convex and tree convex and perfect elimination bipartite graph G_7 which is not a chordal bipartite graph

G_7 is a perfect elimination bipartite graph, since a perfect elimination ordering is given by $\{(w, a), (y, b), (z, c)\}$, as readers can check it.

G_7 is a tree convex bipartite graph, since a tree $T = (V_1, F)$ on V_1 can be defined by $F = \{(x, u), (y, u), (z, u), (w, u)\}$, as shown in Figure 8 (middle), such that for each vertex in V_2, its neighborhood induces a subtree in T.

G_7 is a circular convex bipartite graph, since a circular ordering R on V_2 can be defined by $a \prec b \prec c \prec a$, as shown in Figure 8 (right), such that for each vertex in V_2, its neighborhood induces a circular arc in R.

G_7 is not a chordal bipartite graph, since the cycle $x - a - y - b - z - c - x$ of length six has no chord. $\qquad\square$

Theorem 8. *There is a tree convex and perfect elimination bipartite graph G_8 which is neither a chordal bipartite graph nor a circular convex bipartite graph.*

Proof. The graph $G_8 = (V_1, V_2, E)$ is shown in Figure 9 (left).

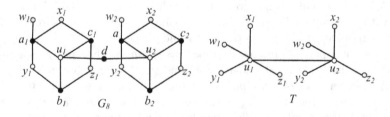

Fig. 9. A tree convex and perfect elimination bipartite graph G_8 which is neither a chordal bipartite graph nor a circular convex bipartite graph

G_8 is a perfect elimination bipartite graph, since a perfect elimination ordering is given by $\{(w_1, a_1), (y_1, b_1), (z_1, c_1), (w_2, a_2), (y_2, b_2), (z_2, c_2), (u_1, d_1)\}$, as readers can check it.

G_8 is a tree convex bipartite graph, similarly as G_5, see Figure 9 (right).

G_8 is not a circular convex bipartite graph, similarly as G_5.

G_8 is not a chordal bipartite graph, since the cycle $x_1 - a_1 - y_1 - b_1 - z_1 - c_1 - x_1$ of length six has no chord.   □

Theorem 9. *There is a circular convex and chordal bipartite graph G_9 which is not a convex bipartite graph.*

Proof. The graph $G_9 = (V_1, V_2, E)$, where $V_1 = \{x, y, z, u, w\}$, $V_2 = \{a, b, c\}$, and $E = \{(x, a), (a, y), (y, b), (b, z), (z, c), (c, x), (u, a), (u, b), (u, c), (w, a)\}$, is shown in Figure 10 (left).

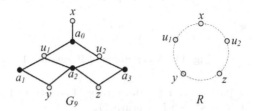

Fig. 10. A circular convex and chordal bipartite graph G_9 which is not a convex bipartite graph

G_9 is a chordal bipartite graph, since each cycle of length at least six has a chord, as readers can check it.

G_9 is a circular convex bipartite graph, since a circular ordering R on V_2 can be defined by $x \prec u_1 \prec y \prec z \prec u_2 \prec x$, as shown in Figure 10 (right), such that for each vertex in V_2, its neighborhood induces a circular arc in R.

G_9 is not a convex bipartite graph, since G_9 is a forbidden subgraph in Tucker's characterization of convex bipartite graphs [20].   □

4 Hardness Results

In this section, we show the $\mathcal{NP}$-completeness of treewidth and feedback vertex set for perfect elimination bipartite graphs. These two problems are known to be $\mathcal{NP}$-complete for bipartite graphs. We use a simple reduction from bipartite graphs to perfect elimination bipartite graphs, which keeps treewidth and decycling number invariant. The reduction just adds a different pendent vertex for each vertex in one side of the bipartite graph.

Theorem 10. *Treewidth is $\mathcal{NP}$-complete for perfect elimination bipartite graphs.*

Proof. Treewidth is well known in $\mathcal{NP}$ [1,9]. We reduce from Treewidth which is $\mathcal{NP}$-complete for bipartite graphs [9].

Reduction 1.

Input: A bipartite graph $G = (V_1, V_2, E)$ and a positive integer k, where $V_1 = \{x_1, x_2, \ldots, x_n\}$.

Output: A bipartite graph $G' = (V_1, V_2', E')$ and a positive integer k, where $V_2' = V_2 \cup \{a_1, a_2, \cdots, a_n\}$ and $E' = E \cup \{(x_k, a_k)|k = 1, 2, \cdots, n\}$.

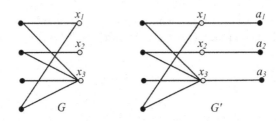

Fig. 11. An example of Reduction 1

Clearly, G' is bipartite and is computable from G in polynomial time. An example of G and G' is shown in Figure 11.

The graph G' is a perfect elimination bipartite graph, since a perfect elimination ordering of G' is given by $\{(x_1, a_1), (x_2, a_2), \cdots, (x_n, a_n)\}$. Indeed, these edges are pairwise nonadjacent. Each edge in them has a degree one endpoint b_i, thus are bisimplicial. These edges contain all the vertices in V_1, no edge in G' will be left after removing these edges and their endpoints.

By repeatedly applying Lemma 1 in the construction of G' from G, G has treewidth k if and only if G' has treewidth k.   □

Theorem 11. *Feedback vertex set is $\mathcal{NP}$-complete for perfect elimination bipartite graphs.*

Proof. Feedback vertex set problem is well known in $\mathcal{NP}$ [8]. We reduce from feedback vertex set which is $\mathcal{NP}$-complete for bipartite graphs [23]. The reduction is exact the same as Reduction 1 in proof of Theorem 10. The correctness of this reduction is shown as follows.

First, for any feedback vertex set D' of G', there is a feedback vertex set D'' of G', such that D'' only contains vertices in $V_1 \cup V_2$ and D'' is not larger than D'. Indeed, if there is a vertex a_i in D', then we can replace a_i by x_i, since a_i is a pendent vertex not on any cycle.

Second, for any $D \subseteq V_1 \cup V_2$, D is a feedback vertex set in G if and only if it is a feedback vertex set in G'. Therefore, G has a feedback vertex set of size at most k if and only if G' has a feedback vertex set of size at most k.   □

5   Conclusions

We have made a thorough comparison for perfect elimination bipartite graphs, chordal bipartite graphs, convex bipartite graphs, tree convex bipartite graphs, and circular convex bipartite graphs, showing the nonemptyness of each region in their Venn diagram (Figure 1), thus ruling out any further inclusion among them. We also show the $\mathcal{NP}$-completeness of treewidth and feedback vertex set for perfect elimination bipartite graphs.

A trick we used to obtain these results is that, for a bipartite graph, we usually can add one pendent vertex or many pendent vertices to make it a perfect elimination bipartite graph, while to keep its other properties invariant. This trick may be useful to obtain further results for perfect elimination bipartite graphs.

The complexity of feedback vertex set for restricted bipartite graphs is shown in Figure 12. The complexity of treewidth for triad convex bipartite graphs or circular convex bipartite graphs is unknown. We conjecture that treewidth is also tractable for these two classes of bipartite graphs, and thus the same picture as Figure 12 also holds for treewidth.

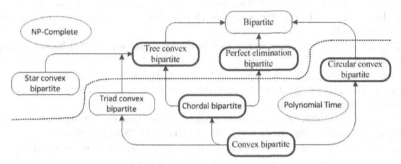

Fig. 12. The known inclusion among some restricted bipartite graphs and complexity classification of feedback vertex set for these bipartite graphs

A set system $(U, \mathcal{S})$ contains a universe set U and a family $\mathcal{S}$ of subsets of U. A set system $(U, \mathcal{S})$ can be represented by a bipartite graph $(U, \mathcal{S}, E)$, where $E = \{(x, Y) | x \in U, Y \in \mathcal{S}\}$. When the bipartite graphs are restricted, we also get the corresponding restricted set systems. Our separation results for the restricted bipartite graphs are also applicable to the restricted set systems. Recently, some complexity results on set cover, set packing and hitting set for tree convex, circular convex, tree-like and circular-like set systems are obtained in [16]. We can also define perfect elimination set systems and chordal set systems, and the complexity results for them is largely unknown.

Acknowledgments. The help of anonymous reviewers has improved our presentation greatly.

References

1. Arnborg, S., Corneil, D.G., Proskurowski, A.: Complexity of finding embeddings in a k-tree. SIAM J. Algebraic Discrete Methods 8, 277–284 (1987)
2. Brandstad, A., Le, V.B., Spinrad, J.P.: Graph Classes - A Survey. Society for Industrial and Applied Mathematics, Philadelphia (1999)

3. Golumbic, M.C., Goss, C.F.: Perfect elimination and chordal bipartite graphs. J. Graph Theory 2, 155–163 (1978)
4. Grover, F.: Maximum matching in a convex bipartite graph. Nav. Res. Logist. Q. 14, 313–316 (1967)
5. Jiang, W., Liu, T., Ren, T., Xu, K.: Two hardness results on feedback vertex sets. In: Atallah, M., Li, X.-Y., Zhu, B. (eds.) FAW-AAIM 2011. LNCS, vol. 6681, pp. 233–243. Springer, Heidelberg (2011)
6. Jiang, W., Liu, T., Wang, C., Xu, K.: Feedback vertex sets on restricted bipartite graphs. Theor. Comput. Sci. 507, 41–51 (2013)
7. Jiang, W., Liu, T., Xu, K.: Tractable feedback vertex sets in restricted bipartite graphs. In: Wang, W., Zhu, X., Du, D.-Z. (eds.) COCOA 2011. LNCS, vol. 6831, pp. 424–434. Springer, Heidelberg (2011)
8. Karp, R.: Reducibility among combinatorial problems. In: Complexity of Computer Computations, pp. 85–103. Plenum Press, New York (1972)
9. Kloks, T.: Treewidth: Computations and Approximations. Springer (1994)
10. Kloks, T., Kratsch, D.: Treewidth of chordal bipartite graphs. J. Algorithms 19, 266–281 (1995)
11. Kloks, T., Liu, C.H., Pon, S.H.: Feedback vertex set on chordal bipartite graphs. arXiv:1104.3915 (2011)
12. Kloks, T., Wang, Y.L.: Advances in Graph Algorithms (2013) (manuscript)
13. Liang, Y.D., Blum, N.: Circular convex bipartite graphs: Maximum matching and Hamiltonian circuits. Inf. Process. Lett. 56, 215–219 (1995)
14. Liu, T., Lu, Z., Xu, K.: Tractable connected domination for restricted bipartite graphs. J. Comb. Optim. (2014), , doi 10.1007/s10878-014-9729-x
15. Lu, M., Liu, T., Xu, K.: Independent domination: Reductions from circular- and triad-convex bipartite graphs to convex bipartite graphs. In: Fellows, M., Tan, X., Zhu, B. (eds.) FAW-AAIM 2013. LNCS, vol. 7924, pp. 142–152. Springer, Heidelberg (2013)
16. Lu, M., Liu, T., Tong, W., Lin, G., Xu, K.: Set cover, set packing and hitting set for tree convex and tree-like set systems. In: Gopal, T.V., Agrawal, M., Li, A., Cooper, S.B. (eds.) TAMC 2014. LNCS, vol. 8402, pp. 248–258. Springer, Heidelberg (2014)
17. Lu, Z., Liu, T., Xu, K.: Tractable connected domination for restricted bipartite graphs (extended abstract). In: Du, D.-Z., Zhang, G. (eds.) COCOON 2013. LNCS, vol. 7936, pp. 721–728. Springer, Heidelberg (2013)
18. Lu, Z., Lu, M., Liu, T., Xu, K.: Circular convex bipartite graphs: Feedback vertex set. In: Widmayer, P., Xu, Y., Zhu, B. (eds.) COCOA 2013. LNCS, vol. 8287, pp. 272–283. Springer, Heidelberg (2013)
19. Song, Y., Liu, T., Xu, K.: Independent domination on tree convex bipartite graphs. In: Snoeyink, J., Lu, P., Su, K., Wang, L. (eds.) AAIM 2012 and FAW 2012. LNCS, vol. 7285, pp. 129–138. Springer, Heidelberg (2012)
20. Tucker, A.: A structure theorem for the consecutive 1's property. Journal of Combinatorial Theory 12(B), 153–162 (1972)
21. Wang, C., Chen, H., Lei, Z., Tang, Z., Liu, T., Xu, K.: Tree convex bipartite graphs: $\mathcal{NP}$-complete domination, hamiltonicity and treewidth. In: Proc. of FAW (2014)
22. Wang, C., Liu, T., Jiang, W., Xu, K.: Feedback vertex sets on tree convex bipartite graphs. In: Lin, G. (ed.) COCOA 2012. LNCS, vol. 7402, pp. 95–102. Springer, Heidelberg (2012)
23. Yannakakis, M.: Node-deletion problem on bipartite graphs. SIAM J. Comput. 10, 310–327 (1981)

The Research on Controlling the Iteration of Quantum-Inspired Evolutionary Algorithms for Artificial Neural Networks

Fengmao Lv, Guowu Yang*, Shuangbao Wang, and Fuyou Fan

School of Computer Science and Engineering, University of Electronic Science
and Technology of China, Chengdu, Sichuan 611731, P.R. China
{yipiwa,reacerland}@gmail.com,
guowuy@126.com,
fanfuyou@163.com

Abstract. From recent research on optimizing artificial neural networks
(ANNs), quantum-inspired evolutionary algorithm (QEA) was proved to
be an effective method to design an ANN with few connections and high
classifications. Quantum-inspired evolutionary neural network (QENN)
is a kind of evolving neural networks. Similar to other evolutionary algorithms, it is important to control the iteration of QENN, otherwise it
will waste a lot of time when QENN has been convergent. This paper
proposes an appropriate termination criterion to control the iteration
of QENN. The proposed termination criterion is based on the probability of the best solution. Experiments about pattern classification on iris
have been done to demonstrate the effectiveness and applicability of the
termination criterion. The results show that the termination criterion
proposed in this paper could control the iteration of QENN effectively
and save a mass of computing time by decreasing the number of generations of QENN.

Keywords: Q-bit representation, quantum-inspired evolutionary algorithms (QEA), quantum-inspired evolutionary neural network (QENN),
termination criterion, convergence.

1 Introduction

Artificial neural networks (ANNs) has been proved to be a useful mathematical
tool in machine learning [4]. A neural network is very good at learning using
some learning algorithms such as genetic algorithm (GA) and back propagation
(BP). ANNs trained using learning algorithms are limited to search for a suitable
set of weights in a prior fixed network topology. So it is very important to design
a suitable network structure for an ANN when it is used for a given task [20].
A fixed structure of overall connectivity between neurons may not provide the
optimal performance within a given training period [12]. Moreover, a small ANN

---

* Corresponding author.

Q. Gu, P. Hell, and B. Yang (Eds.): AAIM 2014, LNCS 8546, pp. 253–262, 2014.

may not provide good performance due to its limited information processing capability. On the other hand, a large ANN tend to overfit the training data and cost a mass of time to accomplish a learning task.

Constructive algorithms and destructive algorithms have been used to obtain the network structure. Constructive algorithms start with the smallest possible network and gradually add neurons or connections [13]. Destructive algorithms start with the largest possible network and gradually delete unnecessary neurons or connections [15].

The design of a network structure can also be formulated into a stochastic search problem [19]. Evolutionary algorithms (EAs) were employed to obtain the solution due to the non-differentiability of the search for the optimal network structure [8]. In addition, as indicated in [9], it is a multimodal problem to design the structure of an ANN so that constructive algorithms or destructive algorithms are susceptible to being trapped at structural local optima [1]. However, EA is a global search problem so that it is widely used to design the structure of ANNs in recent years [17]. Leung et al. [10] used an improved genetic algorithm to tune the structure and parameters simultaneously. Tsai et al. [18] presented a hybrid Taguchi-genetic algorithm to solve the problem of tuning the structure and the connection weights for ANNs. Yao et al. [21] proposed an evolutionary system named EPNet to evolve neural networks.

As indicated in [11], Lu et al. proposed quantum-inspired evolutionary algorithm (QEA) to search the optimal structure of ANNs due to the permutation problems existing in traditional evolving neural networks [3], [16], as well as the rapid convergence and global search capability of QEA. Unlike most previous evolutionary algorithms, quantum bit representation is used in QEA to codify the network. As a result, the connectivity bits do not indicate the actual links but the probability of the existence of the connections.

However, regarding the research of Lu et al. there is no termination criterion in his evolutionary algorithms. The maximum number of generations of quantum-inspired evolutionary neural network (QENN) is simply set to 2000. If the evolutionary algorithm has been convergent before the number of generations reaches 2000, the subsequent computing would not be effective. Therefore it would cost a mass of time to run a quantum-inspired evolutionary algorithm if we just simply set the maximum number of generations to a large value. As indicated above, it is important to design an appropriate termination criterion to control the iteration of QENN automatically.

This paper proposed a termination criterion for QENN based on the probability of the best solution. QEA starts with a global search and changes automatically into a local search because of its inherent probabilistic mechanism. Along with the evolution of QENN, the probability of the solutions with better fitness increases gradually and the probability of the best solution converges to 1 eventually. Considering all above, the probability of the best solution could be used as an effective termination criterion for QENN. Experiments about pattern classification on iris have been done to demonstrate the effectiveness and applicability of the termination criterion. The results show that the termination criterion proposed in this paper could control the iteration of QENN effectively and save a mass of computing time by decreasing the number of generations of QENN.

This paper is organized as follows. Section 2 reviews the previous work on QEA and QENN. Section 3 proposes the termination criterion for QENN. Section 4 shows the experimental results of the proposed termination criterion. Finally, Section 5 summarizes this paper.

2 Preliminaries

In this section, QEA proposed in [5], [6] and QENN proposed in [11] will be described. The rapid convergence and global search capability make QEA an effective algorithm for optimal problems. So far, QEA has been effectively used in optimal problems [2], [7], [22]. QEA is a branch of study on evolutionary algorithms. Though QEA is characterized by certain principles of quantum mechanics, it is an evolutionary algorithm for a classical computer and it just adopts the idea of probabilistic representation of quantum bit.

2.1 Representation

In evolutionary algorithms, a number of representations can be used to encode the solutions to individuals. QEA uses a novel Q-bit representation which uses probability to present binary information. A Q-bit is defined as the smallest unit of information in QEA, which is defined with a pair of numbers (α, β) as

$$\begin{bmatrix} \alpha \\ \beta \end{bmatrix} \tag{1}$$

where $0 \le |\alpha| \le 1$, $0 \le |\beta| \le 1$, $|\alpha|^2 + |\beta|^2 = 1$, $|\alpha|^2$ is the probability that the Q-bit will be observed to be 1 and $|\beta|^2$ is the probability that the Q-bit will be observed to be 0. A Q-bit is in a linear superposition of the two states.

A Q-bit individual containing a string of q Q-bits can be defined as

$$\begin{bmatrix} \alpha_1 & \alpha_2 & \cdots & \alpha_q \\ \beta_1 & \beta_2 & \cdots & \beta_q \end{bmatrix} \tag{2}$$

where $0 \le |\alpha_i| \le 1$, $0 \le |\beta_i| \le 1$, $|\alpha_i|^2 + |\beta_i|^2 = 1$, $i = 1, 2, ..., q$. Since $|\alpha_i|^2 + |\beta_i|^2 = 1$, a Q-bit individual can be simplified as

$$\begin{bmatrix} \alpha_1 & \alpha_2 & \cdots & \alpha_q \end{bmatrix} \tag{3}$$

Q-bit representation has the advantage that it is able to represent any linear superposition of states probabilistically. For instance, a three-Q-bit individual such as

$$\begin{bmatrix} \frac{1}{\sqrt{2}} & \frac{1}{2} & \frac{\sqrt{3}}{2} \\ \frac{1}{\sqrt{2}} & \frac{\sqrt{3}}{2} & \frac{1}{2} \end{bmatrix} \tag{4}$$

can be represented as

$$\frac{\sqrt{3}}{4\sqrt{2}}\left|000\right\rangle + \frac{3}{4\sqrt{2}}\left|001\right\rangle + \frac{1}{4\sqrt{2}}\left|010\right\rangle + \frac{\sqrt{3}}{4\sqrt{2}}\left|011\right\rangle +$$

$$\frac{\sqrt{3}}{4\sqrt{2}}\left|100\right\rangle + \frac{3}{4\sqrt{2}}\left|101\right\rangle + \frac{1}{4\sqrt{2}}\left|110\right\rangle + \frac{\sqrt{3}}{4\sqrt{2}}\left|111\right\rangle. \tag{5}$$

The above result means that the probabilities that the three-Q-bit individual will be observed to be the state of 000, 001, 010, 011, 100, 101, 110 and 111 are 3/32, 9/32, 1/32, 3/32, 3/32, 9/32, 1/32, and 3/32 respectively.

2.2 QENN

Quantum-inspired evolutionary neural network (QENN) is based on the theory of QEA and ANNs. QENN utilizes QEA to design the structure and connection weights of ANNs. As indicated in [11], the basic network considered in this paper is a generalized multilayer perceptron (GMLP) network. QEA utilizes Q-bit to represent the probabilities of various network connectivity and connection weights. In QENN, network connectivity is represented by a quantum bit individual as

$$Q_c = \left[\alpha_1 \,|\, \alpha_2 \,|\, ... \,|\, \alpha_{c\,\text{max}}\right] \tag{6}$$

where α_i, $i = 1, 2, ..., c_{\text{max}}$, is a quantum bit and c_{max} is the maximum number of connections. Among the c_{max} maximum number of connections, some connections may be present and some may be absent. In the structure design, Q_c is observed to generate a binary string b_c, where 1 indicates the presence of a connection and 0 indicates the absence of a connection. If a connection is observed to be present, then we need to determine the weight of this connection.

In QENN, connection weight is determined by a quantum bit individual. A quantum bit individual

$$Q_{wi} = \left[\alpha_{i,1} \,|\, \alpha_{i,2} \,|\, ... \,|\, \alpha_{i,k}\right] \tag{7}$$

is used to determine the weight of the i-th connection, where $i = 1, 2, ..., c_{\text{max}}$, k is a parameter chosen by designer to divide the weight space into 2^k subspaces. Q_{wi} is used to represent the probability of the subspaces that render good weight values. For instance, if we set k to be 3 and Q_{wi} is observed to be 011, then the third subspace is chosen to generate the weight value of the i-th connection. The specific realization associated with the i-th connection of the weighting is governed by a normal random number generator with mean $\mu_{i,j}$ and variance $(\sigma_{i,j})^2$, $N(\mu_{i,j}, \sigma_{i,j})$, where $i = 1, 2, ..., c_{\text{max}}$, $j = 1, 2, ..., 2^k$.

Similar to other evolutionary algorithms, QENN is characterized by population dynamics. The population is divided into G structure subpopulations and each subpopulation contains L identical structure individuals. Each subpopulation searches for L optimal connection weight individuals under the same structure, that means L individuals in a subpopulation share the same Q_c.

In this paper, QENN is used to take the task of pattern recognition. So the percentage of incorrectly classified patterns is used as the fitness function of QEA. In every generation, rotation operator is adopted to update Q_c and Q_{wi} by comparing the fitness value in the current generation to the stored best fitness value as indicated in [11]. If the fitness value in the current generation is better than the stored fitness value, then it will be stored instead of the best fitness value stored in the past. The detail for QENN could be found in [11].

3 Termination Criterion

In this section, the termination criterion for QENN based on probability of the best solution will be proposed. Evolutionary algorithms always need many iterations to render a good fitness value. In most instances, a large fixed value is chosen to be the maximum number of generations of EAs. In QENN, this may cost a mass of computing time when QENN has been convergent. So it is important to control the iteration of QENN automatically with an effective termination criterion.

As analyzed in [6], QEA starts with a global search and changes automatically into a local search because of its inherent probabilistic mechanism, which leads to a good balance between exploration and exploitation. Along with the evolution of QENN, the probability of the solutions with better fitness increases gradually. Eventually, the probability of the best solution converges to 1.

To explain the reason for the convergence of the probability of the best solution, the best solution could be considered as an attractor [14]. The exploration of the search space is driven by attractors corresponding to the best solution found so far. If a solution is searched to be the best solution, then this solution starts to attract the entire population. As long as no better solution is found, all the Q-individuals converge towards this best solution, that means rotation operator makes each Q-bit (α, β) converge to the corresponding binary bit in the best solution. If the value of the i-th binary bit in the best solution takes 1, the rotation operator makes $|\alpha_i|^2$ convergent to 1. On the other hand, the rotation operator makes $|\beta_i|^2$ convergent to 1. The probability of the best solution $prob(b)$ is calculated by the multiplication of the probabilities of all the Q-bits in the Q-bit individual. For example, consider $b = 0101$, then the probability of b can be obtained by $prob(b) = |\beta_1|^2|\alpha_2|^2|\beta_3|^2|\alpha_4|^2$. When rotation operator makes all the Q-bits in the Q-bit individual convergent to 1, $prob(b)$ is also convergent to 1.

As indicated above, the probability of the best solution converges to 1 eventually. When the probability of the best solution has been convergent to 1, the quantum-individuals become unable to produce solutions different from the attractor and the subsequent computing would not be effective.

Due to the inherent probabilistic mechanism of QEA, the probability of the best solution can be employed as a termination criterion. In [6], the probability of the best solution b as

$$prob(b) = \frac{1}{n} \sum_{i=1}^{n} (\prod_{j=1}^{q} P_{ij}) > \gamma_0 \tag{8}$$

can be designed as the termination criterion for QEA, where q is the number of Q-bits of a Q-bit individual, n is the number of individuals in a population and P_{ij} is the probability for the j-th bit of the i-th individual to be observed to be the j-th bit of the best solution b. The latter part, $\prod_{j=1}^{q} P_{ij}$ gives the probability for the i-th Q-bit individual to be observed to be the best solution and then the average value of $\prod_{j=1}^{q} P_{ij}$ in a population is calculated to give the value of $prob(b)$. The value of P_{ij} is calculated as follows:

$$P_{ij} = \begin{cases} |\alpha_{ij}|^2, & \text{if } b_j = 1 \\ |\beta_{ij}|^2, & \text{if } b_j = 0 \end{cases} \tag{9}$$

where b_j is the j-th bit of the stored best solution b, $|\alpha_{ij}|^2$ is the probability that the j-th Q-bit of the i-th Q-bit individual will be found in the 1 state and $|\beta_{ij}|^2$ is the probability that the j-th Q-bit of the i-th Q-bit individual will be found in the 0 state.

In QENN, since the solution $s = \{Q_c, Q_{w1}, Q_{w2}, ..., Q_{wc_{\max}}\}$, that means the solution s represented by Q-bit is composed of Q_c and Q_{wj}, $j = 1, 2, ..., c_{\max}$, the probability of the best solution for QENN should be adjusted to

$$prob(s) = \frac{1}{n} \sum_{i=1}^{n} [\prod_{j=1}^{c_{\max}} P_{ij} \bullet \prod_{j \in E} (\prod_{l=1}^{k} P'_{ijl})] > \gamma_0 \tag{10}$$

where $c_{\max}$ is the maximum number of connections, k is a parameter chosen by designer to divide the weight space into 2^k subspaces, E is the set of the connections which are observed to be present, p_{ij} is the probability for the j-th bit of Q-bit individual Q_c in the i-th solution to be observed to be the j-th bit of the best solution b_c and P'_{ijl} is the probability for the l-th bit of the j-th connection Q_{wj} in the i-th solution to be observed to be the l-th bit of the best solution b_{wj}.

As the probability of the best solution $prob(s)$ converges to 1 eventually, it is appropriate to set the value of γ_0 to 0.99, that means the Q-bit individuals become nearly unable to produce solutions different from the best solution found so far.

4   Experimental Results

In this section, experiments about pattern classification on iris have been done to demonstrate the effectiveness of the termination criterion based on the probability of the best solution $prob(s)$. The experiment data is from the University of California Irvine Machine Learning Repository.

In this experiment, the population is divided into 3 structure populations and each subpopulation includes 30 identical structure individuals. Each subpopulation searches for 30 optimal connection weights under the same structure.

One hundred samples of iris are chosen randomly to train QENN and modify the structure and weights of QENN. There are 3 types of iris to be classified among the experimental samples. The number of attributes is 4, all of which are real values. So the number of input nodes of GMLP is 4 and the number of output nodes of GMLP is 3. With the number of hidden nodes of GMLP $n_h = 2, 4$, Fig. 1 shows the variation of the mean error rate and the best error rate in a population about classification of iris as generation T advances.

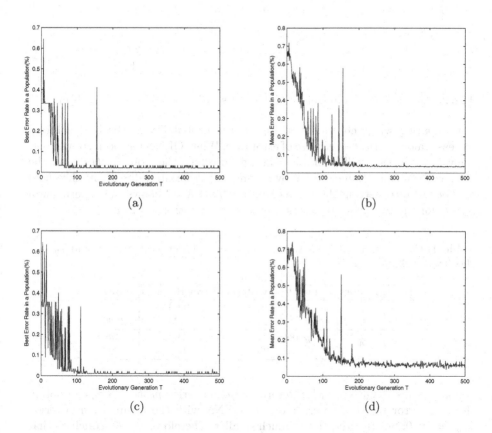

(a)

(b)

(c)

(d)

Fig. 1. Variation of best error rate and mean error rate in a population. (a) best error rate($n_h = 2$) (b) mean error rate($n_h = 2$) (c) best error rate($n_h = 4$) (d) mean error rate($n_h = 4$).

From Fig. 1, as generation T advances, both the mean error rate and the best error rate in a population decrease. When the generation T reaches a certain value, the error rate will not decrease and be stable near a fixed value. This means QENN has been convergent and the subsequent computing will not be effective. Therefore, it is necessary to control the iteration of QENN.

With the number of hidden nodes of GMLP $n_h = 2, 4$, Fig. 2 shows the variation of the probability of the best solution $prob(s)$ as generation T advances.

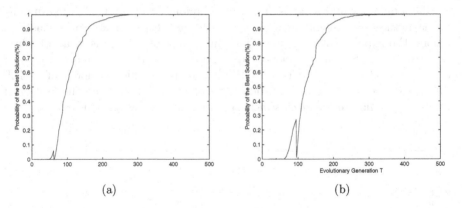

Fig. 2. Variation of the probability of the best solution $prob(s)$ (a)$n_h = 2$ (b)$n_h = 2$

Comparing with the results of Fig. 1, the probability of the best solution increases along with the decrease of error rate. When QENN has been convergent, the value of $prob(s)$ will be stable near a fixed value. Therefore, $prob(s)$ could be used as the termination criterion by setting a appropriate value to the threshold γ_0. For instance, setting the value of γ_0 to 0.99, Table 1 shows the experimental results for the termination criterion $prob(s) > \gamma_0$ with $n_h = 2, 4$.

Table 1. Comparison of QENN with termination criterion $prob(s) > \gamma_0$ and $T_{max} = 2000$ ($\gamma_0 = 0.99$, $n_h = 2, 4$)

Hidden Nodes	Termination Criterion	Error Rate(%)	Iteration
$n_h = 2$	$prob(s) > \gamma_0$	0.03	132
	$T_{max} = 2000$	0.03	2000
$n_h = 4$	$prob(s) > \gamma_0$	0.02	198
	$T_{max} = 2000$	0.02	2000

As we can see, QENN with the termination criterion $prob(s) > \gamma_0$ can obtain the same error rate to the error rate of QENN with the termination criterion $T_{max} = 2000$, but the iteration is much smaller. Therefore, we can conclude that the termination criterion $prob(s) > \gamma_0$ is effective for QENN.

5   Conclusion

This paper proposes an effective termination criterion for QENN. The termination criterion is based on the probability of the best solution. Along with the evolution of QENN, the probability of the solutions with better fitness increases gradually. Eventually, the probability of the best solution converges to 1.

The experimental results demonstrate the convergence of QENN, the convergence of the probability of the best solution and the effectiveness of the termination criterion $prob(s) > \gamma_0$. The termination criterion $prob(s) > \gamma_0$ could give the information about the convergence of QENN accurately. QENN can be stopped when $prob(s)$ has been convergent with the termination criterion $prob(s) > \gamma_0$ and there is no need to set a large number to be the maximum generation of QENN.

The study of termination criterion for QENN can be useful in automatic controlling and save a mass of computing time by decreasing the number of generations of QENN.

Acknowledgments. This paper is supported by the National Natural Science Foundation of China under grant No. 61272175, the Research Foundation of Science & Technology Department of Sichuan province of China under grant No. 2012JY009 and the Key projects Foundation of the education department of Sichuan province of China under grant No. 2011ZA173. The authors would like to thank the anonymous reviewers for their careful reading of this paper and for their helpful and constructive comments.

References

1. Angeline, P.J., Saunders, G.M., Pollack, J.B.: An evolutionary algorithm that constructs recurrent neural networks. IEEE Transactions on Neural Networks 5(1), 54–65 (1994)
2. Bäck, T., Schwefel, H.P.: An overview of evolutionary algorithms for parameter optimization. Evolutionary Computation 1(1), 1–23 (1993)
3. Belew, R.K., McInerney, J., Schraudolph, N.N.: Evolving networks: Using the genetic algorithm with connectionist learning. Citeseer (1990)
4. Brown, M., Harris, C.J.: Neurofuzzy adaptive modelling and control. Prentice-Hall (1994)
5. Han, K.H., Kim, J.H.: Genetic quantum algorithm and its application to combinatorial optimization problem. In: Proceedings of the 2000 Congress on Evolutionary Computation, vol. 2, pp. 1354–1360. IEEE (2000)
6. Han, K.H., Kim, J.H.: Quantum-inspired evolutionary algorithm for a class of combinatorial optimization. IEEE Transactions on Evolutionary Computation 6(6), 580–593 (2002)
7. Han, K.H., Kim, J.H.: Quantum-inspired evolutionary algorithms with a new termination criterion, H_ϵ gate, and two-phase scheme. IEEE Transactions on Evolutionary Computation 8(2), 156–169 (2004)
8. Kitano, H.: Designing neural networks using genetic algorithms with graph generation system. Complex Systems Journal 4, 461–476 (1990)
9. Kwok, T.Y., Yeung, D.Y.: Constructive algorithms for structure learning in feedforward neural networks for regression problems. IEEE Transactions on Neural Networks 8(3), 630–645 (1997)
10. Leung, F.H.F., Lam, H.K., Ling, S.H., Tam, P.K.S.: Tuning of the structure and parameters of a neural network using an improved genetic algorithm. IEEE Transactions on Neural Networks 14(1), 79–88 (2003)
11. Lu, T.C., Yu, G.R., Juang, J.C.: Quantum-based algorithm for optimizing artificial neural networks. IEEE Transactions on Neural Networks 24(8), 1266–1278 (2013)

12. Oong, T.H., Isa, N.A.M.: Adaptive evolutionary artificial neural networks for pattern classification. IEEE Transactions on Neural Networks 22(11), 1823–1836 (2011)
13. Parekh, R., Yang, J., Honavar, V.: Constructive neural-network learning algorithms for pattern classification. IEEE Transactions on Neural Networks 11(2), 436–451 (2000)
14. Platel, M.D., Schliebs, S., Kasabov, N.: Quantum-inspired evolutionary algorithm: a multimodel eda. IEEE Transactions on Evolutionary Computation 13(6), 1218–1232 (2009)
15. Reed, R.: Pruning algorithms-a survey. IEEE Transactions on Neural Networks 4(5), 740–747 (1993)
16. Schaffer, J.D., Whitley, D., Eshelman, L.J.: Combinations of genetic algorithms and neural networks: A survey of the state of the art. In: International Workshop on Combinations of Genetic Algorithms and Neural Networks, COGANN 1992, pp. 1–37. IEEE (1992)
17. Sexton, R.S., Dorsey, R.E., Johnson, J.D.: Toward global optimization of neural networks: a comparison of the genetic algorithm and backpropagation. Decision Support Systems 22(2), 171–185 (1998)
18. Tsai, J.T., Chou, J.H., Liu, T.K.: Tuning the structure and parameters of a neural network by using hybrid taguchi-genetic algorithm. IEEE Transactions on Neural Networks 17(1), 69–80 (2006)
19. Yao, X.: A review of evolutionary artificial neural networks. International Journal of Intelligent Systems 8(4), 539–567 (1993)
20. Yao, X.: Evolving artificial neural networks. Proceedings of the IEEE 87(9), 1423–1447 (1999)
21. Yao, X., Liu, Y.: A new evolutionary system for evolving artificial neural networks. IEEE Transactions on Neural Networks 8(3), 694–713 (1997)
22. Zhang, R., Gao, H.: Improved quantum evolutionary algorithm for combinatorial optimization problem. In: 2007 International Conference on Machine Learning and Cybernetics, vol. 6, pp. 3501–3505. IEEE (2007)

On P_3-Convexity of Graphs with Bounded Degree

Lucia Draque Penso[1], Fábio Protti[2],
Dieter Rautenbach[1], and Uéverton S. Souza[2]

[1] Universität Ulm, Ulm, Germany
[2] IC – Universidade Federal Fluminense, Niterói, RJ, Brazil
{lucia.penso,dieter.rautenbach}@uni-ulm.de,
{fabio,usouza}@ic.uff.br

Abstract. Motivated by the large applicability as well as the hardness of P_3-convexity, we study new complexity aspects of such convexity restricted to graphs with bounded maximum degree. More specifically, we are interested in identifying either a minimum P_3-geodetic set or a minimum P_3-hull set of such graphs, from which the whole vertex set of G is obtained either after one or sufficiently many iterations, respectively. Each iteration adds to a set S all vertices of $V(G) \setminus S$ with at least two neighbors in S. We prove that: (i) a minimum P_3-hull set of a graph G can be found in polynomial time when $\delta(G) \geq \frac{n(G)}{c}$ (for some constant c); (ii) deciding if the size of a minimum P_3-hull set of a graph is at most k remains NP-complete even on planar graphs with maximum degree four; (iii) a minimum P_3-hull set of a cubic graph can be found in polynomial time; (iv) a minimum P_3-hull set can be found in polynomial time in graphs with minimum feedback vertex set of bounded size and no vertex of degree two; (v) deciding if the size of a minimum P_3-geodetic set of a planar graph with maximum degree three is at most k remains NP-complete.

Keywords: P_3-convexity, P_3-hull set, P_3-geodetic set, planar graphs, bounded degree, NP-hardness.

1 Introduction

Let $G = (V, E)$ be a graph. For $U \subseteq V$, let the interval $I[U]$ of U in G be the set $U \cup \{u \in V(G) \setminus U \mid |N_G(u) \cap U| \geq 2\}$. A set S of vertices of G is P_3-*geodetic* if $I[S]$ contains all vertices of G. The P_3-*geodetic number* $g_{P_3}(G)$ of a graph G is defined as the minimum cardinality of a P_3-geodetic set. The decision problem related to determining the P_3-geodetic number is known to be NP-complete for general graphs, and coincides with the well-studied 2-domination number [10,8,11,12,13].

A P_3-*hull* set U of G is a set of vertices such that:

- $U^0 = U$
- $U^k = I[U^{k-1}]$, for $k \geq 1$.
- $\exists\, k \geq 0 \mid U^k = V(G)$.

Q. Gu, P. Hell, and B. Yang (Eds.): AAIM 2014, LNCS 8546, pp. 263–274, 2014.
© Springer International Publishing Switzerland 2014

We define $H_G(S) \subseteq V(G)$ as $I[S]^{k+1}$ where the non-negative integer k is such that $I[S]^{k+1} = I[S]^k$, $k \geq 0$. The cardinality of a minimum P_3-hull set of G is the P_3-*hull number* of G, denoted by $h_{p3}(G)$. Again, the decision problem related to determining the P_3-hull number of a graph is still a well known NP-complete problem [4].

According to [5], as one of the most elementary models of the spreading of a property within a network – like sharing an idea or disseminating a virus – one can consider a graph G, a set U of vertices of G that initially possesses the property, and an iterative process whereby new vertices u are added to U whenever sufficiently many neighbors of u are already in U. The simplest non-trivial choice leads to the *irreversible 2-threshold processes* by Dreyer and Roberts [6]. Similar models were studied in various contexts, such as statistical physics, social networks, marketing, and distributed computing under different names such as bootstrap percolation, influence dynamics, local majority processes, irreversible dynamic monopolies, catastrophic fault patterns, and many others [1,2,3,4,5,6].

In the next sections, we analyze the complexity of these problems when some parameters related to the maximum and minimum degree of a graph are known. In the following subsection we review some results on planar satisfiability problems. In Section 2 we present some results on finding a minimum P_3-hull set of graphs with bounded degree. Finally, in Section 3 we analyze complexity aspects of finding a minimum P_3-geodetic set on planar graphs with bounded degree.

1.1 PLANAR SAT-AM3

SAT-AM3 [9]
Instance: A set $F = \{C_1, C_2, \ldots, C_m\}$ of clauses, built on a finite set $X = \{x_1, x_2, \ldots, x_n\}$ of boolean variables, such that each clause contains at most three literals, each variable appears at most three times, and each literal occurs at most twice.
Question: Is there a truth assignment to the variables in X that satisfies F?

SAT-AM3 is an NP-complete problem [9]. In [9] the problem was not defined with the restriction of each literal occurs at most twice, but without loss of generality, if a literal l occurs three times, the clauses containing l can be considered satisfied and removed from the formula F to be analyzed. Another variant of SAT is described below.

PLANAR 3-SAT [9]
Instance: A set $F = \{C_1, C_2, \ldots, C_m\}$ of clauses, built on a finite set $X = \{x_1, x_2, \ldots, x_n\}$ of boolean variables, where each clause contains at most three literals, and the bipartite graph $H_F = (V, E)$ such that $V = \{w_{c_1}, w_{c_2}, \ldots, w_{c_m}\} \cup \{v_{x_1}, v_{x_2}, \ldots, v_{x_n}\}$ and E contains exactly those pairs (w_{c_i}, v_{x_j}) such that either x_j or $\neg x_j$ belongs to the clause C_i, is planar.
Question: Is there a truth assignment to the variables in X that satisfies F?

Note that not every instance of SAT-AM3 is an instance of PLANAR 3-SAT. For example, $F = (\neg x_1 + x_2 + x_3)(x_2 + \neg x_3 + \neg x_5)(x_1 + \neg x_2 + x_4)(x_3 + \neg x_4)(\neg x_1 +$

x_5) is non-planar because it contains a subdivision of $K_{3,3}$. However, it is well known [9,14] that PLANAR 3-SAT is also an NP-complete problem.

At this point, we describe the intersection of these problems.

PLANAR SAT-AM3

Instance: A set $F = \{C_1, C_2, \ldots, C_m\}$ of clauses, built on a finite set $X = \{x_1, x_2, \ldots, x_n\}$ of boolean variables, where each clause contains at most three literals, each variable appears at most three times, each literal occurs at most twice, and the bipartite graph $H_F = (V, E)$ such that $V = \{w_{c_1}, w_{c_2}, \ldots, w_{c_m}\} \cup \{v_{x_1}, v_{x_2}, \ldots, v_{x_n}\}$ and E contains exactly those pairs (w_{c_i}, v_{x_j}) such that either x_j or $\neg x_j$ belongs to the clause C_i, is planar.

Question: Is there a truth assignment to the variables in X that satisfies F?

Lemma 1. PLANAR SAT-AM3 *is NP-complete.*

Proof. It is easy to see that the problem is in NP. To prove the hardness, we perform a reduction from PLANAR 3-SAT. Consider a general PLANAR 3-SAT expression F in which x_i appears k_i times. Assign $F' = F$, and for each x_i in F' replace the first occurrence of x_i by x_i^1, the second by x_i^2, and so on, where $x_i^1, x_i^2, \ldots, x_i^{k_i}$ are new variables. Add $(\neg x_i^1, x_i^2), (\neg x_i^2, x_i^3), \ldots, (\neg x_i^{k_i}, x_i^1)$ to F'. Clearly, F' is satisfiable if and only if F is satisfiable.

By Kuratowski's theorem a finite graph is planar if and only if it does not contain a subgraph that is a subdivision of K_5 or $K_{3,3}$. To show that $H_{F'}$ is planar, just observe that given a planar embedding of the bipartite graph corresponding to F, one can obtain a planar embedding of the graph corresponding to F' by replacing some vertices u of degree k with a cycle C of order k and a matching of k edges between $V(C)$ and the neighbors of u. It is easy to see that the constructed graph has a planar embedding.   □

2 P_3-Hull Set

In this section we consider both search and decision problems on P_3-hull sets.

P_3-HULL SET

Instance: A graph G.

Goal: Find a P_3-hull set of G with minimum cardinality.

P_3-HULL NUMBER

Instance: A graph G; an integer k.

Goal: Decide if G has a P_3-hull set with cardinality at most k.

Note that P_3-HULL NUMBER is clearly in NP. Moreover, it is easy to see that if P_3-HULL NUMBER is NP-complete then P_3-HULL SET is NP-hard.

Let $n(G)$ be the number of vertices of G, $N_G(x)$ the neighborhood of a vertex x in G, $d_G(x) = |N_G(x)|$ the degree of vertex x in G, and $\delta(G)$ and $\Delta(G)$ the minimum and maximum degree of a vertex in G, respectively.

Lemma 2. *Let k be a positive integer. If G is a graph, then*

$$\Delta_k(G) := \max\left\{\left|\bigcap_{x \in U} N_G(x)\right| \mid U \in \binom{V(G)}{k}\right\} \geq n(G)\frac{\binom{\delta(G)}{k}}{\binom{n(G)}{k}}.$$

Proof. Let $\mathcal{R} = \left\{(u, U) : u \in V(G), U \in \binom{V(G)}{k}, u \in \bigcap_{x \in U} N_G(x)\right\}$. Since for every vertex v of G there are $\binom{d_G(v)}{k} \geq \binom{\delta(G)}{k}$ pairs (u, U) in $\mathcal{R}$ with $u = v$, we have $|\mathcal{R}| \geq n(G)\binom{\delta(G)}{k}$. Conversely, by the definition of $\Delta_k(G)$, for every set $V \in \binom{V(G)}{k}$, there are at most $\Delta_k(G)$ pairs (u, U) in $\mathcal{R}$ with $U = V$, which implies $|\mathcal{R}| \leq \Delta_k(G)\binom{n(G)}{k}$. $\qquad\square$

Theorem 3. *Let c be a positive integer.*
If G is a graph with $\delta(G) \geq \frac{n(G)}{c}$, then

$$h_{P_3}(G) \leq 2\left\lceil\frac{\log(2c)}{\log\left(\frac{2c^2}{2c^2-1}\right)}\right\rceil + 2c^3.$$

Proof. In order to construct a small P_3-hull set of G we describe an inductive construction of a sequence $G_1, \ldots, G_k$ of induced subgraphs of G such that

- $G_i = G - H_G(S_{i-1})$ for a set S_{i-1} of at most $2(i-1)$ vertices of G,
- $n(G_i) \leq n(G)\left(1 - \frac{1}{2c^2}\right)^{i-1}$, and
- $\delta(G_i) \geq \frac{n(G_i)}{c}$

for $i \in [k]$.
   Let $G_1 = G$ and $S_0 = \emptyset$.
   Now let i be such that G_i and S_{i-1} are defined. If G_i is complete or $n(G_i) < 2c^3$, then terminate the construction of the sequence and set k to i. Since

$$h_{P_3}(G) \leq |S_{k-1}| + h_{P_3}(G_k) \leq 2(k-1) + 2c^3,$$

it suffices to bound k in order to complete the proof.
   Therefore, we may assume that G_i is not complete and that $n(G_i) \geq 2c^3$. By Lemma 2, there are two vertices u_i and v_i of G_i with at least

$$n(G_i)\frac{\binom{\delta(G_i)}{2}}{\binom{n(G_i)}{2}} \geq n(G_i)\frac{\binom{\frac{n(G_i)}{c}}{2}}{\binom{n(G_i)}{2}} = \frac{n(G_i)(n(G_i) - c)}{c^2(n(G_i) - 1)} \geq \frac{n(G_i)}{2c^2}$$

common neighbors. Let $S_i = S_{i-1} \cup \{u_i, v_i\}$ and $G_{i+1} = G - H_G(S_i)$. We obtain

$$
\begin{aligned}
n(G_{i+1}) &= n(G) - |H_G(S_i)| \\
&\leq n(G) - |H_G(S_{i-1}) \cup H_{G_i}(\{u_i, v_i\})| \\
&\leq n(G) - |H_G(S_{i-1})| - |H_{G_i}(\{u_i, v_i\})| \\
&= n(G_i) - |H_{G_i}(\{u_i, v_i\})| \\
&\leq n(G_i) - \frac{n(G_i)}{2c^2} \\
&= n(G_i) \left(1 - \frac{1}{2c^2}\right) \\
&\leq n(G) \left(1 - \frac{1}{2c^2}\right)^i .
\end{aligned}
$$

Since $G_{i+1} = G - H_G(S_i)$, we have $\delta(G_{i+1}) \geq \delta(G) - 1 \geq \frac{n(G)}{c} - 1$. Therefore,

$$
\frac{\delta(G_{i+1})}{n(G_{i+1})} > \frac{\frac{n(G)}{c} - 1}{n(G_i)\left(1 - \frac{1}{2c^2}\right)} \geq \frac{\frac{n(G_i)}{c} - 1}{n(G_i)\left(1 - \frac{1}{2c^2}\right)} \geq \frac{1}{c}.
$$

Since the minimum degree of all graphs G_i in the sequence is at least $\delta - 1$, the value of k is less than or equal to the smallest integer r with

$$
n(G) \left(1 - \frac{1}{2c^2}\right)^{r-1} \leq \frac{n(G)}{c} - 1.
$$

Since $\frac{n(G)}{c} - 1 \geq \frac{n(G)}{2c}$, we obtain

$$
k \leq \left\lceil \frac{\log(2c)}{\log\left(\frac{2c^2}{2c^2-1}\right)} \right\rceil + 1,
$$

which completes the proof.   □

Corollary 4. *A minimum P_3-hull set of a graph G with $\delta(G) \geq \frac{n(G)}{c}$ (for some constant c) can be found in polynomial time.*

Proof. The proof follows immediately from Theorem 3.   □

Theorem 5. P_3-HULL NUMBER *remains NP-complete on planar graphs with maximum degree four.*

Proof. To prove that deciding whether the P_3-hull number of a graph G is less than or equal k is NP-complete, we perform a reduction from PLANAR SAT-AM3, proved to be NP-complete in Lemma 1. Here *cross edges* are meant in the usual sense of a planar graph: edges crossing other edges in a specific embedding of a graph in the plane.

Given an instance F of PLANAR SAT-AM3, we construct an instance G of P_3-HULL SET as follows:

- For each variable x_i of F, create a gadget G_{x_i} composed of 62 vertices as illustrated in Figure 1. Note that G_{x_i} is composed of two subgadgets g_{x_i} and $g_{\bar{x}_i}$, which represent the literals x_i and $\bar{x}_i$, respectively.

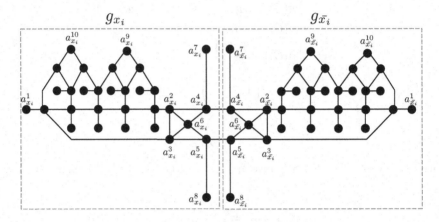

Fig. 1. Gadget G_{x_i}

- For each clause C_j of F, create a gadget G_{c_j} composed of the cycle $b_{c_j}^1$, $b_{c_j}^2$, $b_{c_j}^3$, $b_{c_j}^4$, $b_{c_j}^5$, $b_{c_j}^6$, $b_{c_j}^7$, $b_{c_j}^8$ plus the vertices $b_{c_j}^9$, $b_{c_j}^{10}$, $b_{c_j}^{11}$, $b_{c_j}^{12}$, $b_{c_j}^{13}$, $b_{c_j}^{14}$, $b_{c_j}^{15}$, $b_{c_j}^{16}$ and edges $(b_{c_j}^1, b_{c_j}^9), (b_{c_j}^2, b_{c_j}^{10}), (b_{c_j}^3, b_{c_j}^{11}), (b_{c_j}^4, b_{c_j}^{12}), (b_{c_j}^5, b_{c_j}^{13}), (b_{c_j}^6, b_{c_j}^{14}), (b_{c_j}^7, b_{c_j}^{15}), (b_{c_j}^8, b_{c_j}^{16})$. Figure 2 illustrates a gadget G_{c_j}.

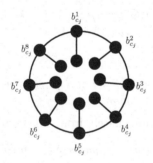

Fig. 2. Gadget G_{c_j}

- If the literal x_i occurs twice in F, then create the vertices $f_{x_i}^1$, $f_{x_i}^2$, and add edges $(f_{x_i}^1, a_{x_i}^7), (f_{x_i}^2, a_{x_i}^8)$. Otherwise, create only $f_{x_i}^1$ and add $(f_{x_i}^1, a_{x_i}^7)$.
- If the literal $\bar{x}_i$ occurs twice in F, then create the vertices $f_{\bar{x}_i}^1$, $f_{\bar{x}_i}^2$, and add edges $(f_{\bar{x}_i}^1, a_{\bar{x}_i}^7), (f_{\bar{x}_i}^2, a_{\bar{x}_i}^8)$. Otherwise, create only $f_{\bar{x}_i}^1$ and add $(f_{\bar{x}_i}^1, a_{\bar{x}_i}^7)$.
- For each clause C_j do:

1. if x_i is the first literal of C_j, then: if C_j contains the first occurrence of x_i then add edges $(a_{x_i}^7, b_{c_j}^1), (a_{x_i}^9, b_{c_j}^2)$; else add edges $(a_{x_i}^{10}, b_{c_j}^1), (a_{x_i}^8, b_{c_j}^2)$.
2. if x_i is the second literal of C_j, then: if C_j contains the first occurrence of x_i then add edges $(a_{x_i}^7, b_{c_j}^5), (a_{x_i}^9, b_{c_j}^6)$; else add edges $(a_{x_i}^{10}, b_{c_j}^5), (a_{x_i}^8, b_{c_j}^6)$.
3. if x_i is the third literal of C_j, then: if C_j contains the first occurrence of x_i then add edges $(a_{x_i}^7, b_{c_j}^7), (a_{x_i}^9, b_{c_j}^8)$; else add edges $(a_{x_i}^{10}, b_{c_j}^7), (a_{x_i}^8, b_{c_j}^8)$. If this step generates cross edges, remove the newly created edges, and repeat this step replacing $b_{c_j}^7$ and $b_{c_j}^8$ by $b_{c_j}^3$ and $b_{c_j}^4$, respectively. This operation keeps the graph planar, as one can check by verifying all possible configurations.
4. if $\bar{x}_i$ is the first literal of C_j, then: if C_j contains the first occurrence of $\bar{x}_i$ then add edges $(a_{\bar{x}_i}^7, b_{c_j}^2), (a_{\bar{x}_i}^9, b_{c_j}^1)$; else add edges $(a_{\bar{x}_i}^{10}, b_{c_j}^2), (a_{\bar{x}_i}^8, b_{c_j}^1)$.
5. if $\bar{x}_i$ is the second literal of C_j, then: if C_j contains the first occurrence of $\bar{x}_i$ then add edges $(a_{\bar{x}_i}^7, b_{c_j}^6), (a_{\bar{x}_i}^9, b_{c_j}^5)$; else add edges $(a_{\bar{x}_i}^{10}, b_{c_j}^6), (a_{\bar{x}_i}^8, b_{c_j}^5)$.
6. if $\bar{x}_i$ is the third literal of C_j, then: if C_j contains the first occurrence of $\bar{x}_i$ then add edges $(a_{\bar{x}_i}^7, b_{c_j}^8), (a_{\bar{x}_i}^9, b_{c_j}^7)$; else add edges $(a_{\bar{x}_i}^{10}, b_{c_j}^8), (a_{\bar{x}_i}^8, b_{c_j}^7)$. If this step generates cross edges, remove the newly created edges, and repeat this step replacing $b_{c_j}^7$ and $b_{c_j}^8$ by $b_{c_j}^3$ and $b_{c_j}^4$, respectively. As above, this operation keeps the graph planar, as one can check by verifying all possible configurations.

Let G be the graph obtained by the construction above from an instance F of PLANAR SAT-AM3. At this point, we will prove that F is satisfiable if and only if G has a hull set of size $8m + 23n$, where m is the number of clauses, and n is the number of variables of F.

If F is satisfiable, then we can obtain a P_3-hull set S of G by first adding all the pendant vertices of G to S. Note that G has $8m + 22n$ pendant vertices. Let A be a truth assignment of F. If $x_i = true$ in A we add $a_{x_i}^2$ to S, else we add $a_{\bar{x}_i}^2$ to S. As A is a truth assignment of F, each gadget G_{c_j} will be contaminated, i.e. in $H_G(S)$, and consequently all vertices of G will be contaminated. Hence S is a P_3-hull set of size $8m + 23n$.

Conversely, if G has a P_3-hull set S of size $8m + 23n$, S contains $8m + 22n$ pendant vertices and n non-pendant vertices of G. As we can observe in each gadget G_{x_i} of G, there is a subgraph B_{x_i} such that every vertex v of B_{x_i} is not a pendant vertex and either it is adjacent to only one leaf and has no non-pendant neighbor outside B_{x_i}, or v has only one neighbor outside B_{x_i}. Figure 3 illustrates a gadget G_{x_i} and its subgraph B_{x_i}. Consequently, each subgraph B_{x_i} must have exactly one vertex in S, which is not a pendant vertex. Otherwise either S is not a P_3-hull set or S has size greater than $8m + 23n$. At this point we can construct an assignment A of F by setting $x_i = true$ if and only if $S \cap V(g_{x_i}) \cap V(B_{x_i}) \neq \emptyset$. By construction, we can see that A is a truth assignment of F. □

A *feedback vertex set* of a graph is a set of vertices whose removal leaves a graph without cycles. In other words, each feedback vertex set contains at least one vertex of any cycle in the graph.

Lemma 6. *Let G be a cubic graph. $S \subseteq V(G)$ is a P_3-hull set of G if and only if S is also a feedback vertex set of G.*

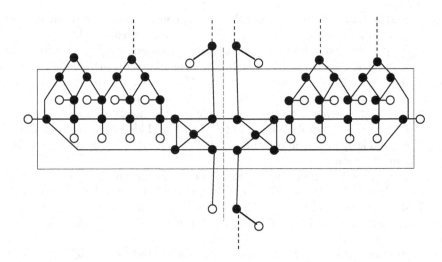

Fig. 3. Gadget G_{x_i} and its subgraph B_{x_i} inside the rectangle. The white vertices are pendant vertices in G and are not contained in B_{x_i}.

Proof. Let G be a cubic graph and S be a P_3-hull set of G. If $G[V \setminus S]$ has a cycle C, then each vertex $v \in C$ has at most one neighbor outside C, and consequently C is not in the hull of S, which is a contradiction because S is a P_3-hull set of G.

Conversely, let B be a feedback vertex set of G. As $G[V \setminus B]$ is a forest and G is cubic, all pendant vertices of $G[V \setminus B]$ are in $H_G(B)$; by removing these pendant vertices of $G[V \setminus B]$, we obtain a forest T where each leaf of T has two neighbors in $H_G(B)$. Applying this step recursively, we can see that all vertices of $G[V \setminus B]$ are in $H_G(B)$. $\qquad\square$

Proposition 1. [15] *A minimum feedback vertex set of a graph G with maximum degree at most three can be found in polynomial time.*

Corollary 7. *A minimum P_3-hull set of a cubic graph can be found in polynomial time.*

Proof. The proof follows immediately from Lemma 6 and Proposition 1. $\qquad\square$

Theorem 8. *Let $\mathscr{F}$ be the class of graphs with no vertex of degree two and with a minimum feedback vertex set of size bounded by a constant c. Then P_3-HULL SET on $\mathscr{F}$ can be solved in polynomial time.*

Proof. Let $G \in \mathscr{F}$. As G has a minimum feedback vertex set of size bounded by a constant c, we can find a minimum feedback vertex set B of G in polynomial time. Let L be the set of pendant vertices in G, and let $T = G \setminus \{B \cup L\}$. Since G has no vertex of degree two, each leaf of T has at least two neighbors in $\{B \cup L\}$ and just as in the proof of Lemma 6, $\{B \cup L\}$ is a hull set of G. As L is in any hull set of G, it is sufficient to examine all subsets of vertices in $V(G) \setminus L$ of size at most c to find a minimum P_3-hull set of G. $\qquad\square$

3 P_3-Geodetic Set

Now we consider the following decision problem:

P_3-GEODETIC NUMBER
Instance: A graph G; an integer k.
Goal: Decide if G has a P_3-geodetic set with cardinality at most k.

Note that P_3-GEODETIC NUMBER is clearly in NP.

As DOMINATING SET is NP-complete even restricted to planar graphs with maximum degree three [9], it is easy to see that P_3-GEODETIC NUMBER problem remains NP-complete on planar graphs with maximum degree four. Just take an instance G of such restricted DOMINATING SET problem and construct a graph G' by adding a new vertex w_v and a new edge (v, w_v) for each vertex v of G. Note that G has a dominating set of size k if and only if G' has a P_3-geodetic set of size $n + k$. As G is a planar graph with maximum degree 3, G' is a planar graph with maximum degree 4.

As P_3-GEODETIC NUMBER is NP-complete on planar graphs with maximum degree four, and trivially solvable in polynomial time on graphs with maximum degree two, it is natural to ask about the complexity of P_3-Geodetic Number on planar graphs with maximum degree 3.

Theorem 9. P_3-GEODETIC NUMBER *remains NP-complete on planar graphs with maximum degree three.*

Proof. Deciding whether the P_3-geodetic number of a graph G is less than or equal to k is clearly a problem in NP. To prove the NP-hardness we perform a reduction from PLANAR SAT-AM3, proved to be NP-complete in Lemma 1. Given an instance F of PLANAR SAT-AM3 we construct an instance G of P_3-GEODETIC SET as follows:

- for each variable x_i do: create in G a gadget g_{x_i} composed of a cycle $f_{x_i}^1, t_{x_i}^1, a_{x_i}^1, a_{x_i}^2, f_{x_i}^2, t_{x_i}^2, a_{x_i}^3, a_{x_i}^4$;
- for each clause C_i containing at most two literals do: create in G a gadget g_{c_i} composed of the vertices c_i^1, c_i^2 and edge (c_i^1, c_i^2);
- for each clause C_j containing exactly three literals do: create in G a gadget g_{c_j} composed of the vertices $c_j^1, c_j^2, c_j^3, l_j^1, l_j^2$ and the edges $(c_j^1, c_j^2), (c_j^1, c_j^3), (c_j^1, l_j^1), (c_j^3, l_j^2)$;
- for each clause C_j of F do:
    1. add an edge $(c_j^2, t_{x_i}^p)$ if x_i is the first or second literal of C_j and it is the p-th occurrence of x_i $(1 \leq p \leq 2)$;
    2. add an edge $(c_j^2, f_{x_i}^p)$ if $\neg x_i$ is the first or second literal of C_j and it is the p-th occurrence of $\neg x_i$ $(1 \leq p \leq 2)$;
    3. add an edge $(c_j^3, t_{x_i}^p)$ if x_i is the third literal of C_j and it is the p-th occurrence of x_i $(1 \leq p \leq 2)$;
    4. add an edge $(c_j^3, f_{x_i}^p)$ if $\neg x_i$ is the third literal of C_j and it is the p-th occurrence of $\neg x_i$ $(1 \leq p \leq 2)$.

At this point, we show that given an instance F of SAT-AM3, where n is the number of variables, m_1 the number of clauses with at most two literals, and m_2 the number of clauses with three literals, by the construction above we obtain a graph G such that: F is satisfiable if and only if G has a P_3-geodetic set S of size k, where $k = 4n + m_1 + 3m_2$.

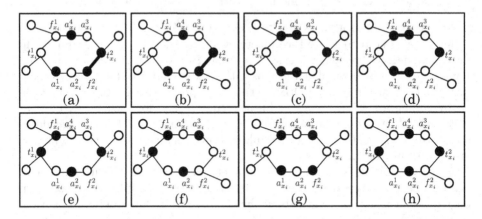

Fig. 4. $(a) - (d)$ Choices of vertices in S_A that imply in at least 5 vertices to be added to S_A; thicker edges mean that one of its endpoints must be added to S_A; $(e) - (h)$ Choices of vertices in S_A that imply in exactly 4 vertices to be added to S_A

Let F be a satisfiable formula and A be a truth assignment of F. We obtain a P_3-geodetic set S_A of G from A as follows: (i) every vertex with degree one is added to S_A; (ii) if $x_i = true$ in A then $t_{x_i}^1, t_{x_i}^2, a_{x_i}^2, a_{x_i}^4$ are added to S_A; (iii) if $x_i = false$ in A then $f_{x_i}^1, f_{x_i}^2, a_{x_i}^1, a_{x_i}^3$ are added to S_A; (iv) for each clause C_i with three literals, if c_i^3 has two neighbors in S_A then c_i^2 is added to S_A, otherwise c_i^1 is added to S_A. As A is a truth assignment of F, each gadget g_{c_i} of G has at least one neighbor in $S_A \cap \{\bigcup_1^n V(g_{x_i})\}$; consequently, S_A is a P_3-geodetic set of G of size $k = 4n + m_1 + 3m_2$.

Conversely, Let S_A be a P_3-geodetic set of G of size $k = 4n + m_1 + 3m_2$. We construct a truth assignment A for the variables $x_1, x_2, \ldots, x_n$ that satisfies all the clauses in F as follows. Any P_3-geodetic set of G contains: (i) at least one vertex of each gadget g_{c_i} if C_i has at most two literals; (ii) at least three vertices of each gadget g_{c_i} if C_i has three literals; (iii) at least four vertices of each gadget g_{x_i}. As S_A has size k, each gadget g_{x_i} has exactly four vertices in S_A, and at most two of these vertices has degree three in G: either $t_{x_i}^1$ and $t_{x_i}^2\}$, or $f_{x_i}^1$ and $f_{x_i}^2\}$. See Figure 4. At this point, we can construct a truth assignment A of F by assigning $x_i = true$ if and only if $t_{x_i}^1 \in S_A$ or $t_{x_i}^2 \in S_A$ and $t_{x_i}^2$ has degree three in G. By (i) and (ii), each gadget g_{c_i} must have at least one neighbor in S_A, otherwise either S_A would not be a P_3-geodetic set or we would have $|S_A| > k$. Consequently, by the construction of G and A, if S_A is a P_3-geodetic set of G of size k then A is a truth assignment of F.

Figure 5 illustrates a boolean formula F and the graph G obtained from F by the construction above. A possible P_3-geodetic set S_A is colored red.

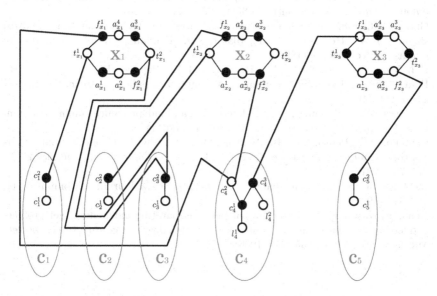

Fig. 5. (a) Satisfiable boolean formula $F = (x_1)(x_2)(x_1 + \neg x_2)(\neg x_1 + \neg x_2 + \neg x_3)(\neg x_3)$; (b) Graph G constructed from F

It is easy to see that G has maximum degree three. To show that G is planar, we can split G in two subgraphs $G_x = \{\bigcup_1^n g_{x_i}\}$ and $G_c = \{\bigcup_1^m g_{c_j}\}$. Note that G_x and G_c are both planar graphs. By contracting each graph g_{x_i} and each gadget g_{c_j} of G into a single vertex, we obtain the bipartite graph H_F which by assumption is planar. Hence, G is also a planar graph. □

References

1. Balister, P., Bollobás, B., Johnson, J.R., Walters, M.: Random majority percolation. Random Struct. Algorithms 36, 315–340 (2010)
2. Balogh, J., Bollobás, B.: Sharp thresholds in Bootstrap percolation. Physica A 326, 305–312 (2003)
3. Bermond, J.-C., Bond, J., Peleg, D., Perennes, S.: The power of small coalitions in graphs. Discrete Appl. Math. 127, 399–414 (2003)
4. Centeno, C.C., Dourado, M.C., Penso, L.D., Rautenbach, D., Szwarcfiter, J.L.: Irreversible conversion of graphs. Theor. Comput. Sci. 412, 3693–3700 (2011)
5. Centeno, C.C., Penso, L.D., Rautenbach, D., de Sá, V.G.P.: Immediate versus eventual conversion: comparing geodetic and hull numbers in P3-convexity. In: Golumbic, M.C., Stern, M., Levy, A., Morgenstern, G. (eds.) WG 2012. LNCS, vol. 7551, pp. 262–273. Springer, Heidelberg (2012)
6. Dreyer Jr., P.A., Roberts, F.S.: Irreversible k-threshold processes: Graph-theoretical threshold models of the spread of disease and of opinion. Discrete Appl. Math. 157, 1615–1627 (2009)

7. Cook, S.A.: The complexity of theorem-proving procedures. In: Proc. 3rd Ann. ACM Symp. on Theory of Computing Machinery, New York, pp. 151–158 (1971)

8. Hansberg, A., Volkmann, L.: On graphs with equal domination and 2-domination numbers. Discrete Mathematics 308(11), 2277–2281 (2008)

9. Garey, M.R., Johnson, D.S.: Computers and intractability. Freeman, N.York (1979)

10. Haynes, T.W., Hedetniemi, S.T., Slater, P.J.: Fundamentals of domination in graphs. Marcel Dekker (1998)

11. DeLaVina, E., Goddard, W., Henning, M.A., Pepper, R., Vaughan, E.R.: Bounds on the k-domination number of a graph. Applied Mathematics Letters 24(6), 996–998 (2011)

12. Hansberg, A., Volkmann, L.: On 2-domination and independence domination numbers of graphs. Ars Combinatoria 101, 405–415 (2011)

13. Centeno, C.C., Penso, L.D., Rautenbach, D., de Sà, V.G.P.: Geodetic Number versus Hull Number in P_3-Convexity. SIAM Journal on Discrete Mathematics 27(2), 717–731 (2013)

14. Lichtenstein, D.: Planar satisfiability and its uses. SIAM Journal on Computing 11, 329–343 (1982)

15. Ueno, S., Kajitani, Y., Gotoh, S.: On the nonseparating independent set problem and feedback set problem for graphs with no vertex degree exceeding three. Discrete Mathematics 72(1-3), 355–360 (1988)

The Competitive Diffusion Game
in Classes of Graphs

Elham Roshanbin

Department of Mathematics and Statistics, Dalhousie University,
Halifax, NS, B3H 4R2, Canada
e.roshanbin@dal.ca

Abstract. We study a game based on a model for the spread of influence through social networks. In game theory, a Nash-equilibrium is a strategy profile in which each player's strategy is optimized with respect to her opponents' strategies. Here we focus on a specific two player case of the game. We show that there always exists a Nash-equilibrium for paths, cycles, trees, and Cartesian grids. We use the centroid of trees to find a Nash-equilibrium for a tree with a novel approach, which is simpler compared to previous works. We also explore the existence of Nash-equilibriums for uni-cyclic graphs, and offer some open problems.

Keywords: Competitive information diffusion, Nash-equilibriums, Network game theory, Social networks.

1 Introduction

Social networks play an important role in society, and are actively studied in a number of different disciplines, including mathematics. Recent studies have concentrated on interactions and influence in a social network. Such studies can lead to better techniques for viral marketing. In viral marketing, different techniques are combined with the knowledge about the social network to achieve marketing objectives in a way which is analogous to the spread of viruses, where contagion occurs through the links of the network. Many of these studies try to find a model for the spread of an idea or innovation through a social network. Usually these models use a graph to show the structure of a network, in which every individual in the network is denoted by a vertex, and two vertices are adjacent if there exists a relation or link between them in the corresponding network.

In a very well studied point of view (look at [6] and [3]), the propagation process is modelled in a way that usually each node or vertex has two status, either active or inactive. The process starts by targeting (or setting active) a small subset of the nodes in the social network with the hope of getting a large number of the individuals at the end who become active, i.e., affected by the influence. These models are basically involved with optimization techniques. On the other hand, there are some other studies looking at the propagation process as a competition among the individuals in the network, see [5]. There exist also

Q. Gu, P. Hell, and B. Yang (Eds.): AAIM 2014, LNCS 8546, pp. 275–287, 2014.

some Voronoi game models involving a game among the parties or agents out of the network with representatives inside the network, where the objective is achieving the largest number of the users (see [4] and [7]).

In 2009, Alon et al [1], introduced a new model for the competitive diffusion process in social networks. Their approach was a novel way of modelling the spread of influence as a game, where the aim of this game is to influence users in the network through "infection" with a particular brand, spreading through the links of the network. In other words, suppose that we have a set of firms that want to advertise their products. Initially they target a small group of people, which they hope will extend into a larger group of society. Any individual, who has learned about a product brand from one of these firms first, either directly or through a social link, will be biased in favour of that brand. However, if a node is getting the influence from different products, she becomes confused and we cancel her out of the game. The gain of each firm is the total number of users that, at the end of the diffusion process, are biased towards its brand.

In the language of mathematics, we can model this competitive propagation process as a game on an undirected finite graph, in which our users form the vertex set of the graph, and the product of each firm is denoted by a distinct colour. A game $\Gamma = \langle G, N \rangle$ is induced by a graph G, representing the underlying social network, and a set of N players corresponding to the set of agents (we identify each player with a number i, $1 \leq i \leq N$). The strategy space of each player is the set of vertices V of G. That is, each player i, $1 \leq i \leq N$ selects a single node that is coloured in colour i at round 0, and every other vertex is uncoloured. If two or more agents select the same vertex at round 0, then, that vertex becomes gray, and those players automatically leave the game. If S_t is the set of the coloured vertices at round $t \geq 0$, then at round $t + 1$, every player i can colour an uncoloured vertex v in the neighbourhood of S_t by the following rule: If v has coloured neighbours only in colour i, then v gets colour i. If v has coloured neighbours with different colours, then it becomes gray. The players continue until no one can colour any uncoloured vertex. At the end, the pay-off of the i-th player is the number of the vertices in G which have colour i. Note that, in this game, after choosing the strategies of the players, every thing in the process is deterministic.

As an example, let G be a graph as shown in Figure 1, and take $N = 2$. If the first player with colour 1, and the second player with colour 2, choose the two

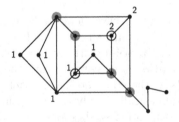

Fig. 1.

vertices with the circles around them at the beginning, then the pay-off of them will be the number of the vertices which are indexed by 1 and 2 in the figure, respectively. As we can see in Figure 1, there are four vertices that become gray by the rules of the game, and three vertices which are not reachable by any player and therefore, remain uncoloured at the end.

Note that, throughout this game, it is as if we delete all the gray vertices, so the metric of the graph is changing within the rounds of the game. This is unlike the Voronoi games [4], in which the gain of each agent is the number of individuals whose distance to the agent representative is less than the other agents.

In real networks finding a kind of stable situation in which every agent is satisfied is called a *Nash-equilibrium*. This is often of more interest than finding the winner of the game. A Nash-equilibrium is a strategy profile or a vector of strategies in the Cartesian product of the strategy sets of all players, such that the strategy of each player in such a vector is the best against the strategies of the others. In other words, in a Nash-equilibrium each player, by choosing that specific strategy, has maximized her pay-off with respect to the strategy of the other players. That is, no player can gain more by changing only her own strategy unilaterally. For further information about game theory concepts we refer the reader to [2]. Alon et al [1] in their paper, proved the existence of Nash-equilibriums for the game on graphs of diameter 2, and gave an example of a graph with diameter more than 2 which does not admit a Nash-equilibrium in the two-player case of the game. However, Takehara et al [10] provided a counter example with a graph of diameter 2 which does not admit a Nash-equilibrium, and presented a restatement of the theorem (about graphs with diameter at most 2) in [1] by putting some restrictions on the graph structure. Recently, Small and Mason [8] considered the existence of Nash-equilibriums for the two player game on trees, and also for the ILT model of online social networks [9], with focus on utility functions.

In this paper we will consider the special two player case of the above game for different families of graphs. However, we take a novel approach based on the graph properties of these families. In the second section, we prove the existence of Nash-equilibriums for trees, paths, cycles, and we consider the game for uni-cyclic graphs. Our proof for trees is much simpler and shorter compared to previous works [8]. In section 3, we show that Cartesian grids always admit a Nash-equilibrium, and we end by suggesting some open problems. In the paper we assume that the graphs are connected. We denote the vertex set and the edge set of a graph G by $V = V(G)$ and $E = E(G)$, respectively. For two vertices like $u, v \in G$, we call the length of the shortest path between u and v in G the *distance* between u and v, and we denote it by $d(u, v)$. If S is a subset of the vertices in G, by $G[S]$ we mean the subgraph induced by S. We denote a path and a cycle on n vertices by P_n and C_n, respectively. For two graphs G and H, we show the Cartesian product of G and H by $G \square H$. We refer the reader to [11] for graph-theoretic notation and terminology.

2   Trees, Paths, Cycles, and Uni-Cyclic Graphs

In this section, we consider some simple facts about the game, and use them to find Nash-equilibriums for different known families of graphs. The following definition help us to have a simpler language to describe the obtained results.

Assume that we are playing the game on a graph G, and, u and v are two distinct vertices of G. Suppose that in some round of the game there is a shortest path $P : u, v_1, \ldots, v_{n-1}, v$, and the vertices u and v have been coloured by two different players, such that no other vertex of P has been coloured yet. Then, we call path P a *blocked path* induced by the vertices u and v, or simply, a blocked path if $\min\{d(v_i, u), d(v_i, v)\} < d(v_i, w)$, for every $1 \leq i \leq n - 1$, and for all vertices w ($w \neq u, v$) which have been coloured so far throughout the game.

We need the following lemma to find a better understanding of the dynamic of a path between a pair of vertices with different colours throughout the game. We omit the proof, which follows immediately from the definition.

Lemma 1. *Suppose that we have a game on graph G. If P is a blocked path of length n induced by vertices v_1 and v_2 in G, by the end of the game, each player wins the first $\lfloor (n + 1)/2 \rfloor$ nearest vertices in path P, and in the case that the length of P is even, one vertex in the middle becomes gray.*

A vertex v of a graph G is called a *cut vertex* if removing v from G results in a graph which is not connected. An edge uv is a *cut edge* if deletion of uv from G is a disconnected graph. The following lemma is quite useful for some of the results as we will see later on.

Lemma 2. *Assume that graph $G = G_1 \cup G_2$ is the union of two induced subgraphs G_1 and G_2 such that, for some cut vertex like v, $G_1 \cap G_2 = \{v\}$. Then, any possible Nash-equilibrium of the two player game on G consists of either two vertices in G_1 or two vertices in G_2.*

Proof. Assume that $\{u_1, u_2\}$ form a Nash-equilibrium such that $u_1 \in G_1 - G_2$ and $u_2 \in G_2 - G_1$. Then, each player changing her strategy to v can increase her pay-off. Because, this way, she can reach on the vertices in the other side earlier than before.

We now state and prove our first result on the competitive diffusion game for paths.

Theorem 1. *In a two-player game on a path of length n, the set of possible Nash-equilibriums is determined as below.*

(i) If n is odd, then the two adjacent vertices in the middle form the only possible Nash-equilibrium, and the equilibrium pay-offs are equal to $(n + 1)/2$.

(ii) If n is even, then any two vertices in the middle (i.e., we have two possibilities, the central vertex and one of its neighbours) form a Nash-equilibrium, and the equilibrium pay-offs are both equal to $n/2$.

Proof. With a simple discussion using Lemma 1, we can show that if vertex v is the strategy of her opponent, then the best strategy for any of the players is to choose a neighbour of v which separates v from a larger number of the vertices in P. So in a possible Nash-equilibrium, the strategies of the players must be adjacent. However, if the players choose two adjacent vertices as their strategies which are not selected as in (i) or (ii), then the player who is closer to one of the end points can improve her pay-off by changing her strategy to another neighbour of her opponent. So, such a case is not a Nash-equilibrium. Finally, if they both have taken their strategies as in (i) or (ii), then no one can improve her pay-off by changing her strategy. Therefore, (i) and (ii) form the only possible Nash-equilibriums of this game.

Theorem 2. *In a two player game on cycle C_n of length n we have the following statements.*

(i) If n is odd, then every two vertices on C_n selected by the players as their strategies, form a Nash-equilibrium, and the pay-offs are equal to $(n-1)/2$.

(ii) If n is even, then two vertices on C_n form a Nash-equilibrium if and only if they are of odd distance, and the equilibrium pay-offs are equal to $n/2$.

Proof. When we have a two player game on a cycle C_n, the strategies of the players divide the cycle into two blocked paths. If n is odd, then one of the blocked paths is always of odd length and the other one is of even length. Obviously, by Lemma 1, every player wins $(n-1)/2$ vertices, and one vertex in the middle of the even path becomes gray. Since this happens for any selection of the vertices, any two vertices form a Nash-equilibrium when n is odd.

If n is even, then the two blocked paths are both even or odd. If they are both of odd length, then by Lemma 1, each player wins exactly half of the vertices on C_n, and no one can improve this. If the blocked paths are both even, then every player wins $(n/2)-1$, and one of the vertices in each blocked path becomes gray. Thus, each player can improve her pay-off by changing her strategy to an adjacent vertex. Hence, two vertices of C_n form a Nash-equilibrium if and only if they are of odd distance.

A maximal sub-tree which contains a vertex v of a tree T as a leaf is called a *branch* of T at v. The *weight* of a vertex v of T, denoted by $wt(v)$ is the maximum number of vertices in a branch at v (not including v). A vertex u is a *centroid vertex* of T if it has the minimum weight among all vertices. The *centroid* of T is the set of all centroid vertices of T.

Theorem 3. *[12] If $C = C(T)$ is the centroid of a tree T of order n, then we have,*

(i) C consists of either a single vertex or two adjacent vertices.

(ii) If $C = \{c_1, c_2\}$, then $wt(c_1) = wt(c_2) = n/2$.

(iii) $C = \{c\}$ if and only if, $wt(c) \leq (n-1)/2$.

Note that, according to the above theorem, in both possible cases for the centroid of a tree T, if $v \notin C(T)$, then $wt(v) > n/2$.

The following theorem, using the centroid of a tree shows that there exists a Nash-equilibrium for any tree.

Theorem 4. *In a two-player game on a tree T of order n with centroid C, we have the following statements.*

   (i) If $C = \{c_1, c_2\}$, then C is the unique Nash-equilibrium, and the equilibrium pay-offs are equal to $n/2$.

   (ii) If $C = \{c\}$, then $\{c, v\}$ is an equilibrium, in which v is a neighbour of c in a branch with maximum weight attached at c, and any equilibrium for this game consists of such two vertices.

Proof. Assume that v_1 and v_2 are the strategies of the players, and g_1 and g_2 are their pay-offs, respectively. Since there exists a unique path between any two vertices in a tree, then we conclude that the gain of v_1 is a subset of a branch attached at v_2 like B_2 which contains this unique path. Thus, we have $g_1 \leq |B_2| \leq wt(v_2)$. Similarly, $g_2 \leq wt(v_1)$.

Now, if v_1 and v_2 are not adjacent, then the path between v_1 and v_2 is a blocked path of length more than one and therefore, by Lemma 1, either every player wins half of the vertices on it, or there is a gray vertex in the middle of this path which no one gains. Thus, in such a case, we have the strict inequalities $g_1 < wt(v_2)$ and $g_2 < wt(v_1)$.

On the other hand, we know that always one of the branches attached at v_1 (similarly v_2) has the maximum weight, and if the second player chooses the neighbour of v_1 on such a branch, then she gains exactly $wt(v_1)$ vertices. Similarly, the first player can gain $wt(v_2)$. Hence, for the first player we have $g_1 \leq wt(v_2)$, and the equality achieved if and only if she chooses a vertex adjacent to v_2 from a maximum branch attached at v_2 (we have a similar result for the second player). In other words, fixing the strategy of a player on a vertex like v, the best strategy for the other player is to select a neighbour of v on a maximum branch attached at v. Therefore, in a possible Nash-equilibrium v_1 and v_2 must be adjacent.

Now, assume that v_1 and v_2 are adjacent, and for example (without loss of generality), $g_1 = wt(v_2)$. We know $g_1 + g_2 \leq n$. Also, by Theorem 3, if v_1 and v_2 are not in C then, $wt(v_1) > \frac{n}{2}$, and $wt(v_2) > \frac{n}{2}$. Consequently, $g_2 < \frac{n}{2} < wt(v_1)$, and therefore, the second player can move to a vertex adjacent to v_1 which achieves the maximum weight and increases her pay-off. Hence, such a case is not a Nash-equilibrium. Therefore, in a possible Nash-equilibrium at least one of the players' strategies must be in C. Now, by the above discussion and by Theorem 3, we can easily see that, the best strategy for the other player is to choose the strategy in (i) or in (ii), depending on the structure of C.

Suppose that G is a *uni-cyclic* graph, that is, G has only one cycle C. We can easily see that, $G - C$ is a forest, such that each tree component of this forest is adjacent to exactly one vertex on C. For each vertex $v \in C$, if there are $t = d(v) - 2$ different tree components in $G - C$ that are connected to v, we denote the union of each of these trees together with v (which is like adding a leaf to a tree and making a new tree) by T_{iv}, for $1 \leq i \leq t$; that is, all T_{iv}s share v.

Suppose that we have a two-player game on a uni-cyclic graph G with cycle C. By the above definition, we can assume that every vertex v on C has a weight $wt_C(v) := | \cup_{i=1}^{d(v)-2} T_{iv}|$. As we will see, sometimes we play the game on the weighted cycle C (instead of G) by the regular rules. The only difference here is that, the gain of each player after taking vertex v is increased by the weight of v. In such cases, we denote C by C_W (when we are playing the game only on C with weighted vertices).

We use the above notations for the results on uni-cyclic graphs. We use the following lemma, which is an immediate result of Theorem 3, to prove the next theorem (we omit the proof here).

Lemma 3. *Assume that T is a tree with centroid C. Then, for any vertex v which is not in C the maximum branch attached at v is the one that contains C (which is the only branch attached at v with weight more than $\frac{n}{2}$).*

In general, we have two possibilities for a uni-cyclic graph G with cycle C; either there is a vertex v on C with $|T_{iv}| \geq \frac{n}{2} + 1$, for some i, $1 \leq i \leq d(v) - 2$, or $|T_{iv}| \leq \frac{n}{2}$ for all $v \in C$, and $1 \leq i \leq d(v) - 2$. So, we have the following theorem.

Theorem 5. *Suppose that G is a uni-cyclic graph with cycle C. If there is a vertex v on C with $|T_{iv}| \geq \frac{n}{2} + 1$, for some $1 \leq i \leq d(v) - 2$, then there exists a Nash-equilibrium by playing on T_{iv}. Otherwise, if there exists a Nash-equilibrium for this game, then it must consist of a set of two vertices either on C_W or on a T_{iv}, for some $v \in C$ and $1 \leq i \leq d(v) - 2$.*

Proof. If there is a vertex v on C with $|T_{iv}| \geq \frac{n}{2} + 1$, for some i, then the players' strategies must be somewhere on T_{iv}. Because, first, if no one selected her strategy on T_{iv}, then the player with the smaller gain by moving to v can improve her pay-off (because this way she wins more than half of the vertices in G). So, in a possible Nash-equilibrium, at least one of the players must choose her strategy on a vertex in T_{iv}. Moreover, since v is a cut vertex, by Lemma 2, both of them should choose their equilibrium strategies in T_{iv}.

Now, we show that in such a case, we always have a Nash-equilibrium. In fact, in this case, we can replace $G - T_{iv}$ by a path P consisting of $|G - T_{iv}|$ vertices. If we take $T = T_{iv} \cup P$ (obviously, T is a tree) and $C(T)$ to be the centroid of T, then for the neighbour of v on P, called u, we have,

$$wt_T(u) = |T_{iv}| \geq \frac{n}{2} + 1 > \frac{n}{2}.$$

Thus, by Lemma 3, the centroid of T is in T_{iv}. Moreover, can easily see that, playing in a Nash-equilibrium of T is like playing in a Nash-equilibrium of G. Because, no one can increase her pay-off unilaterally. Therefore, by Theorem 4, we know that T always has a Nash-equilibrium.

Now, assume that for every $v \in C$ and each $1 \leq i \leq d(v) - 2$, $|T_{iv}| \leq \frac{n}{2}$, and there exists a Nash-equilibrium for this game. If the equilibrium vertices both are not included simultaneously in any T_{iv}, for a vertex v on C, and some $1 \leq i \leq d(v) - 2$, then, since every vertex in C of weight greater than one is a cut vertex, by Lemma 2, the strategies must be selected on C_W.

If for every vertex $v \in C$, $wt_C(v) \leq \frac{n}{2}$, then the following lemma could be helpful.

Lemma 4. *Assume that G is a uni-cyclic graph with weighted cycle C_W such that $wt(v) \leq \frac{n}{2}$ for all $v \in C$. Then, in a two player game on C_W with Nash-equilibrium $\{u, v\}$ we have,*

(i) *either $\{u, v\}$ is a Nash-equilibrium for the regular game on G, or*

(ii) *one of the neighbours of u (or v) together with v (or u) form a Nash-equilibrium for G.*

Proof. Assume that u and v are the strategies of the first and the second player in a Nash-equilibrium for the game on C_W. Also, suppose that g_x denotes the pay-off of a player who takes a vertex like x as her strategy. By definition, we know that no one can increase her pay-off by changing her strategy to another vertex on C_W. So, we have $g_u \geq g_z$, for any $z \in C_W$. Now, we consider the changes in the pay-off of the first player after moving to any vertex like w on a tree attached at a vertex like z on C, with $w \neq z$. We can easily see that, if the first player changes her strategy to vertex $w \neq z$, then $g_w < g_z$. Because, this way she gains the vertices on C_W at a later time ($d(w, v) > d(z, v)$). Therefore, she loses at least one of the vertices that she was able to take by choosing z. Thus, we have, $g_u \geq g_z > g_w$. Hence, the only way for the first player to increase her pay-off is to move to a vertex on T_{iv}, for some $1 \leq i \leq d(v) - 2$. Now, we can take a path P of length $|G - \cup_{i=1}^{d(v)-2} T_{iv}|$ and let T be the tree obtained from connecting P to $\cup_{i=1}^{d(v)-2} T_{iv}$ via v. We can see that, finding the best strategy with respect to v among the vertices in $\cup_{i=1}^{d(v)-2} T_{iv}$ in the game on G, is equivalent to finding such a strategy in the game on T. By proof of Theorem 4, in a game on a tree always the best strategy against an opponent is to play in her neighbourhood. So, if for any neighbour of v, like $w \in T_{iv}$, $g_w > g_u$ and g_w is the maximum over such neighbours of v, then the best strategy for the first player (against v) is to move to w. Moreover, in this case, $\{w, v\}$ forms a Nash-equilibrium for G. Because, in one side, w is the best strategy against v, and in the other side, the second player, moving to a vertex $z \neq w$ in T_{iv}, will gain $g_z < |T_{iv}| \leq \frac{n}{2} \leq |G - T_{iv}| = g_v$. Also, moving to a vertex in $G - T_{iv}$, she will lose some of the vertices (by getting further with respect to her opponent). Thus, v is also the best strategy against w.

However, if for every neighbour of v, like w, $g_w \leq g_u$, then, u is the best strategy against v in G. We can do the same discussion for the second player, and conclude that, either u together with one of its neighbours form a Nash-equilibrium for G, or otherwise, v is the best strategy against u. Therefore, $\{u, v\}$ forms a Nash-equilibrium for G.

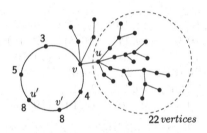

Fig. 2.

In reverse, if we have a Nash-equilibrium for G on cycle C, then it is also a Nash-equilibrium for C_W. But, if we find a Nash-equilibrium for G such that one of the strategies chosen by the players is out of C, then this case does not necessarily help to find a Nash-equilibrium for C_W. In Figure 2 we see a uni-cyclic graph G, in which $\{u, v\}$ form a Nash-equilibrium (nobody can increase her pay-off moving to another vertex). But, if we try to play the game on the weighted cycle C_W, then the only possible Nash-equilibrium is $\{u', v'\}$, which has no intersection with $\{u, v\}$ and is obtained independently.

As a consequence of Lemma 4, if we find a Nash-equilibrium for the game on C_W, then we can find a Nash-equilibrium for the game on G. This conclusion shows the importance of the following theorem as the last result of this section. We omit the proof here which is a long technical one, and will be published in a future paper.

Theorem 6. *In a two player game on a uni-cyclic graph G with cycle C (or weighted cycle C_W) of lengths 3, 4, and 5 always there exists a Nash-equilibrium.*

The uni-cyclic graph G in Figure 3 is an example of a weighted 6-cycle that does not admit any Nash-equilibrium. First, the weight of each tree attached at a vertex on the cycle C is less than half of the whole number of the vertices. Hence, using Lemma 2, we can easily consider all different possibilities to conclude that there can not be any Nash-equilibrium in which one of the players chooses a strategy out of C. So, by Theorem 5, it is enough to consider the game on C_W. Now, we have the following bimatrix as the pay-off matrix (see [2]) of the players in C_W (note that it is a symmetric game and the columns are corresponding to vertices $v_1, v_2, v_3, v_4, v_5,$ and v_6, as well as the rows, respectively):

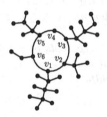

Fig. 3.

$$\begin{bmatrix}
(0,0) & (21^*,15) & (14,10) & (19,17^*) & (16,8) & (25^*,11) \\
(15,21^*) & (0,0) & (19,17) & (16,8) & (25^*,11) & (14,10) \\
(10,14) & (17,19) & (0,0) & (25^*,11) & (14,10) & (15,21^*) \\
(17^*,19) & (8,16) & (11,25^*) & (0,0) & (15,21) & (10,14) \\
(8,16) & (11,25^*) & (10,14) & (21,15) & (0,0) & (17,19) \\
(11,25^*) & (10,14) & (21^*,15) & (14,10) & (19,17) & (0,0)
\end{bmatrix}$$

From game theory (see [2]), we know that a possible Nash-equilibrium for such a game is determined by an entry of this matrix, in which the first component is the largest in the same column and the second component is the largest in the row. Here, for each column and each row we determine such components with a star. As we can see, there is no entry with a star on both components. Thus, there is no Nash-equilibrium for this game.

As another example, assume that G is a uni-cyclic graph with trees of equal order attached at the vertices of the cycle. Then, the two-player game on G is like playing on a weighted cycle with equal weight on all vertices. So, we can easily see that the set of Nash-equilibriums is determined exactly as for a regular cycle. The only difference is that here the pay-off of the players is a multiple (a constant multiple, which is equal to the weight of the vertices on C) of the pay-off in the regular game on a cycle without weights.

3   Cartesian Grids

In this section we investigate the existence of Nash-equilibriums for the *Cartesian grids*. In graph theory, a *grid* (or Cartesian grid) is the Cartesian product of two paths. If $G = P_n \square P_m$, then we call such a grid a $m \times n$ *grid* [11]. We call a subgraph of G which is also a grid by itself, a *subgrid* of G. If A and B are two vertices of a grid G, then G_{AB} is the maximal subgrid of G which contains A and B as the corner points and consists of all the shortest paths between A and B in G.

We need the following concepts to reach the result on grids. Assume that G is a graph and v is a vertex of G. Then, the *eccentricity* of v is defined to be $\max\{d(v,u) : u \in G\}$. The *center* of G is the set of the vertices in G which have the minimum eccentricity [11]. We have the following fact about the center of a grid, which is quite easy to prove only using the definition.

Theorem 7. *Assume that G is a $m \times n$ grid with center C, in which m and n are positive integers. Then, depending on the parity of m and n, we have the following possibilities for C.*

(i) If m and n are odd, then C consists of a single point in the middle.

(ii) If one of m and n is odd and the other one is even, then C consists of two adjacent vertices in the middle.

(iii) If m and n are even, then C consists of a 1×1 subgrid in the middle.

Using the center of a grid, we can always find a Nash-equilibrium for the two player competitive diffusion game on grids.

Theorem 8. *Assume that we have a two player game on a $m \times n$ grid G, in which m and n are positive integers, and $m \leq n$. Let C be the center of G. Then,*

(i) If m and n are odd, then the single vertex in $C = \{c\}$ together with one of the neighbours of c like v which is placed in the same row as c, form a Nash-equilibrium.

(ii) If one of m and n is odd and the other one is even, then the two vertices in $C = \{c_1, c_2\}$ form the unique Nash-equilibrium.

(iii) If m and n are even, then any pair of adjacent vertices in C form a Nash-equilibrium.

Proof. Assume that A and B are the strategies of the players, with g_1 and g_2 as their pay-offs, respectively. Then, there is a vertical as well as a horizontal line

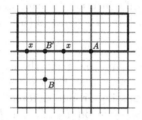

Fig. 4.

which passes through point A in the grid plane, and forms part of the perimeter of some rectangles created by A (in total, there are at most four possible such rectangles as we see in Figure 4, depending on the position of A). We observe that B is always inside of one of those rectangular regions created by A. Now, if G_{AB} is a square, then the distance between A and B is even. Thus, by Lemma 1, there must be some gray vertices appeared on the diagonal points of G_{AB} through out the game, and obviously, one of the players by changing her position and making these gray vertices vanish can gain more. So, this can not be a Nash-equilibrium.

If G_{AB} is not a square, then B is further with respect to one of these rectangles like R_{Ai} than the others. Thus, assuming that B' is the closest point of R_{Ai} with respect to B, for any point like x on the perimeter of R_{Ai}, we have,

$$d(x, A) \leq d(x, B') + d(B', B) = d(x, B).$$

Therefore, through the rounds of the game, the first player (choosing A) gets x before the second player. Thus, the first player wins at least all the vertices in R_{Ai}. Hence, in a possible Nash-equilibrium we have,

$$g_2 \leq mn - |R_{Ai}| \leq mn - \min\{|R_{Aj}| : R_{Aj} \text{ is a rectangle created by A}\}. \quad (1)$$

But, this bound can be achieved only when B is the neighbour of A opposite to the smallest rectangle created by A. Otherwise, the first player wins all the vertices in the smallest rectangle created by A, plus at least the neighbour of A opposite to this rectangle. Thus, the best strategy for the second player is to achieve this bound as discussed. Similarly, we can consider the rectangles created by B, and again we have,

$$g_1 \leq mn - |R_{Bi}| \leq mn - \min\{|R_{Bj}| : R_{Bj} \text{ is a rectangle created by B}\}, \quad (2)$$

which can be achieved only when A is the neighbour of B opposite to the smallest rectangle created by B. Hence, in a possible Nash-equilibrium, the strategies of the players should be adjacent. If the players do not choose their strategies as in (i), (ii), or (iii), then using the above discussion and inequities (1) and (2), we can see that one of the players can increase her pay-off. Thus, such a case is not a Nash-equilibrium.

Now, assume that players choose their strategies like in (i), (ii), or (iii). Then, no one can increase her pay-off, since no one can enlarge the smallest rectangle created by her strategy. Therefore, (i), (ii), and (iii) form the Nash-equilibriums of this game.

Although for the two player game on grids there exists a Nash-equilibrium, it seems that for the three player case the existence of Nash-equilibriums is not certain. For example, discussing around different possibilities, we can easily see that for the three player game on $P_2 \square P_n$, or $P_3 \square P_n$, there is no Nash-equilibrium. In general, we have the following conjecture.

Conjecture 1. There exist no Nash-equilibrium for a three player game on a Cartesian grid.

Another family of graphs that we often consider for a graph theoretic problem are bipartite graphs. We can simply discuss that for a complete bipartite graph, a Nash-equilibrium is to choose a vertex as the strategy of the first player from the first part and a vertex for the second player from the second part. This way, each player wins all the vertices in the opposite part except for the strategy of her opponent. But, finding a Nash-equilibrium for an arbitrary bipartite graph in general seems challenging.

Acknowledgments. I would like to sincerely thank my supervisors Anthony Bonato and Jeannette Janssen for their support and encouragement, and for opening new directions for this research project.

References

1. Alon, N., Feldman, M., Procaccia, A.D., Tennenholtz, M.: A note on competitive diffusion through social networks. Information Processing Letters 110, 221–225 (2010)

2. Barron, E.N.: Game Theory, An Introduction, 2nd edn. Wiley-Inter science, John Wiley and Sons, Hoboken, NJ (2008)
3. Bhagat, S., Goyal, A., Lakshmanan, L.V.S.: Maximizing Product Adoption in Social Networks. In: Proceedings of the Fifth ACM International Conference on Web Search and Data Mining, pp. 603–612 (2012)
4. Dürr, C., Thang, N.K.: Nash equilibria in Voronoi games on graphs. In: Arge, L., Hoffmann, M., Welzl, E. (eds.) ESA 2007. LNCS, vol. 4698, pp. 17–28. Springer, Heidelberg (2007)
5. Immorlica, N., Kleinberg, J.M., Mahdian, M., Wexler, T.: The role of compatibility in the diffusion of technologies through social networks. In: Proceedings of the 8th ACM Conference on Electronic Commerce, pp. 75–83 (2007)
6. Kempe, D., Kleinberg, J.M., Tardos, É.: Maximizing the spread of influence through a social network. In: Proceedings of the 9th International Conference on Knowledge Discovery and Data Mining (KDD), pp. 137–146 (2003)
7. Mavronicolas, M., Monien, B., Papadopoulou, V.G., Schoppmann, F.: Voronoi games on cycle graphs. In: Ochmański, E., Tyszkiewicz, J. (eds.) MFCS 2008. LNCS, vol. 5162, pp. 503–514. Springer, Heidelberg (2008)
8. Small, L., Mason, O.: Nash Equilibria for Competitive Information Diffusion on Trees. Information Processing Letters 113, 217–219 (2013)
9. Small, L., Mason, O.: Information diffusion on the iterated local transitivity model of online social networks. Discrete Applied Mathematics 161, 1338–1344 (2013)
10. Takehara, R., Hachimori, M., Shigeno, M.: A comment on pure-strategy Nash equilibria in competitive diffusion games. Information Processing Letters 112, 59–60 (2012)
11. West, D.B.: Introduction to Graph Theory, 2nd edn. Prentice Hall Inc., Upper Saddle River (2001)
12. Wilf, H.S.: The uniform selection of free trees. Journal of Algorithms 2, 204–207 (1981)

A New Linear Kernel for Undirected Planar Feedback Vertex Set: Smaller and Simpler

Mingyu Xiao⋆

School of Computer Science and Engineering,
University of Electronic Science and Technology of China, China
myxiao@gmail.com

Abstract. We show that any instance I of the FEEDBACK VERTEX SET problem in undirected planar graphs can be reduced to an equivalent instance I' such that (i) the size of the instance and the size of the minimum feedback vertex set do not increase, (ii) and the size of the minimum feedback vertex set in I' is at least $\frac{1}{29}$ of the number of vertices in I'. This implies a $29k$ kernel for this problem with parameter k being the size of the feedback vertex set. Our result improves the previous results of $97k$ and $112k$.

Keywords: Kernelization, Feedback Vertex Set, Planar Graphs.

1 Introduction

A *feedback vertex set* of a directed or undirected graph is a subset of vertices intersecting all cycles in the graph. The FEEDBACK VERTEX SET problem asks us to find a feedback vertex set of minimum size in a given directed or undirected graph. This problem has applications in operating system, computer architecture communities, database system, rank aggregation and so on [18,16].

It is known that FEEDBACK VERTEX SET is NP-hard even in undirected planar graphs [12]. Due to the importance of this problem, FEEDBACK VERTEX SET has been extensively studied in exact and parameterized algorithms. Exact algorithms with running time better than the trivial bound of 2^n have been developed recently for both of FEEDBACK VERTEX SET in undirected graphs (UFVS) and FEEDBACK VERTEX SET in directed graphs (DFVS) [10,14,15], where n stands for the number of vertices in the graph. The running time bound for UFVS has been improved to $1.7266^n n^{O(1)}$ [18] but the best result for DFVS is still $1.9977^n n^{O(1)}$ by Razgon [15]. As for parameterized algorithms, we consider the parameterized problems with parameter k being the size of the feedback vertex set. Let k-UFVS (resp. k-DFVS) denote the problem of checking whether or not a given undirected (resp. directed) graph has a feedback vertex set of size at most k. A parameterized problem is *fixed-parameter tractable* (FPT) if it can be solved in $f(k)poly(n)$ time, where $poly(n)$ is an arbitrary polynomial function

⋆ Supported by NFSC of China under the Grant 61370071 and Fundamental Research Funds for the Central Universities under the Grant ZYGX2012J069.

Q. Gu, P. Hell, and B. Yang (Eds.): AAIM 2014, LNCS 8546, pp. 288–298, 2014.

and $f(k)$ is an arbitrary computable function. There is a long list of contributions to fast FPT algorithms for k-UFVS [13,9,7,6]. Now it can be solved in $3.83^k n^{O(1)}$ time [6]. Whether k-DFVS is FPT or not had been a big open problem in parameterized algorithms for many years. Finally it was solved affirmatively by Chen *et. al.* [8].

Kernelization is one of the most active topics in parameterized algorithms. In this area, we aim to find a polynomial-time algorithm that reduces any instance (I, k) of a parameterized problem to an instance (I', k') of this problem such that $k' \leq k$ and the size of I' is bounded by a computable function $g(k')$ of k', where (I', k') is called a *kernel* of this problem and $g(k')$ is called the *size* of the kernel. We are interested in finding kernels of polynomial size. Whether or not k-DFVS has a polynomial kernel is still an open problem. As for k-UFVS, there is a long list of contributions to kernelizations. The first polynomial kernel for k-UFVS in general graphs was developed in [5], which was improved to a cubic kernel [4]. The current best result is the $4k^2$ kernel by Thomassé [17]. Linear kernels have been obtained for k-UFVS in some graph classes, such as bounded-genus graphs and H-minor free graphs [2,11]. For k-UFVS in planar graphs, Bodlaender and Pennikx [3] gave the first linear kernel of size $112k$, which was improved to $97k$ by Abu-Khzam and Khuzam [1]. In this paper, we gave a kernel of size $29k$, greatly improving previous results. In fact, all reduction rules in our algorithm are parameter-independent. It means that even if the parameter k is not part of the input, our algorithm can still reduce an instance I to an equivalent instance I' such that the size of I' is at most 29 times of the size of the solution to I'.

2 Preliminaries

Let $G = (V, E)$ be an undirected graph with possible parallel edges and self-loops. The vertex set and edge set of a graph G' are denoted by $V(G')$ and $E(G')$ respectively. If there is at least one edge between two vertices, we say that the two vertices are *adjacent*. For two adjacent vertices, any one is a *neighbor* of the other one. For a vertex subset or a subgraph V', the set of vertices in $V \setminus V'$ (or $V \setminus V(V')$ for a subgraph V'') adjacent to at least one vertex in V' is denoted by $N(V')$. Furthermore, we use $N_H(V')$ to denote the set $N(V') \cap H$ for a vertex subset H (or $N(V') \cap V(H)$ for a subgraph H). We may denote a singleton set $\{a\}$ by a. For a vertex v, the *degree* of v is defined to be $d(v) = |N(v)|$ and the number of edges incident on v is denoted by $e(v)$. Then $e(v) \geq d(v)$ since the graph may contain parallel edges. A degree-2 vertex is called a *strong degree-2 vertex* if there are parallel edges between it and any of the two neighbors of it. A vertex is called *non-trivial* if it is a strong degree-2 vertex or a vertex of degree ≥ 3. We use (A, B, E) to denote a bipartite graph with edges between two vertex sets A and B.

The subgraph induced by a vertex subset $V' \subseteq V$ is denoted by $G[V']$. A path $v_1 v_2 \cdots v_r$ in the graph is called an *induced path* if there is exactly one edge between vertices v_i and v_{i+1} for $i \in \{1, 2, \cdots, r - 1\}$ and no edge between vertices v_{i_1} and v_{i_2} with $|i_1 - i_2| \geq 2$. Note that we do not allow parallel edges in induced paths.

A graph without any cycle is called a *forest*. A *tree* is a connected graph without any cycle. In a forest, a degree-1 vertex is called a *leaf-vertex*, a vertex of degree ≥ 2 is called an *inner-vertex*, and a vertex of degree ≥ 3 is called a *branch-vertex*. A path in a tree is called a *branch* if it is a maximal path not containing any branch-vertex. Then after deleting all branch-vertices from a tree, each component is a branch of the tree. A branch is called a *leaf-branch* if it contains at least one leaf-vertex as its endpoint and a branch is called an *inner-branch* if all of its vertices are degree-2 vertices in the tree.

Contracting a vertex subset or a subgraph V' means deleting V', introducing a new vertex v, for any vertex $u \in V \setminus V'$ (or $u \in V \setminus V(V')$ for a subgraph V') adding x edges between v and u if there are x edges between V' and u before deleting V', and adding a self-loop incident on v if there is a cycle in the induced graph $G[V']$ (or subgraph V'). *Contracting an edge* means contracting the two endpoints of it. In our algorithm, we assume that the initial graph is a connected graph, because if the graph has more than one component we can simply take each component as the input to solve it.

2.1   Some Properties of Planar Graphs

Here we give some properties of planar graphs. One of them will be used to get a bound of the vertex number in our analysis and one of them will be used to show that after executing each of our operations on a planar graph the resulting graph is still a planar graph.

The famous Euler's formula gives a relation among the number of vertices, the number of edges and the number of faces of a planar graph. Let f be the number of faces of a planar graph $G = (V, E)$. It holds that

$$|V| - |E| + f = 2.$$

This formula can be used to get some upper bound of the edge number in a planar graph. For a planar graph, each edge belongs to 2 faces and each face contains at least 3 edges. For a bipartite planar graph, each edge belongs to 2 faces and each face contains at least 4 edges (since the bipartite graph has no odd cycle). Then by Euler's formula, we can get

Proposition 1. *For any planar graph, it holds that*

$$|E| \leq 3|V| - 6 \tag{1}$$

and, for any bipartite planar graph, it holds that

$$|E| \leq 2|V| - 4. \tag{2}$$

It is also easy to observe the following.

Proposition 2. *After applying any one of the following operations on a planar graph, the resulting graph is still planar*

(i) *deleting a vertex or an edge,*
(ii) *contracting an edge,*
(iii) *adding an edge between two adjacent vertices or two neighbors of a degree-2 or degree-3 vertex, and*
(iv) *adding a degree-2 vertex adjacent to a pair of adjacent vertices.*

3 Reduction Rules

Since that all the reduction operations in the paper are parameter-independent, we do not include the parameter when we discuss our reduction operations. A *reduction rule* is a procedure that takes a planar graph G as the input and outputs a planar graph G' and a set $S_0 \subseteq V$ such that for any minimum feedback vertex set S' of G', the union $S_0 \cup S'$ is a minimum feedback vertex set of G. In what follows, when introducing a reduction rule, we assume that all previous reduction rules cannot be applied on the current instance anymore.

It is easy to observe the following reduction rules to deal with vertices having self-loops and some vertices of degree at most 2 in the graph (except strong degree-2 vertices).

Rule 1. *If there is a cycle contains only one vertex (a self-loop incident on the vertex), delete the vertex from the graph and put it to S_0.*

Rule 2. *For a degree-1 vertex v,*
(i) *delete v from the graph if there is only one edge between v and its unique neighbor, and*
(ii) *delete v and its unique neighbor u and put u to S_0 if there are parallel edges between v and u.*

Rule 3. *For a degree-2 vertex v with two neighbors u_1 and u_2,*
(i) *delete v and add an edge between u_1 and u_2 if there are only two edges incident on v, and*
(ii) *delete v and u_{i_1} and put u_{i_1} to S_0 if there are parallel edges between v and u_{i_1} but only one edge between v and u_{i_2}, where $\{i_1, i_2\} = \{1, 2\}$.*

Note that after applying the above reduction rules all the vertices in the graph are non-trivial vertices. This property will be used in our analysis. In order to get a kernel of the problem, we need to reduce some local structures where a large number of vertices are only adjacent to a few vertices. Next, we will design some rules for this kind of local structures.

Lemma 1. *Let $G = (V, E)$ be a graph such that all above reduction rules can not be applied anymore. If there is a vertex subset $V' \subset V$ such that $|V'| \geq 3$, the induced graph $G[V']$ has no cycle, and $|N(V')| \leq 2$, then there is a minimum feedback vertex set of G containing all vertices in $N(V')$.*

Proof. We know that $|N(V')| = 2$ (because if $|N(V')| = 1$ then either Rule 2 or Rule 3 can be applied on any leaf-vertex in $G[V']$). Let $N(V') = \{w_1, w_2\}$.

There is no degree-0 vertex in $G[V']$, any degree-1 vertex in $G[V']$ is adjacent to both of w_1 and w_2 in G, and any degree-2 vertex in $G[V']$ is adjacent to at least one of w_1 and w_2 in G.

If there are at least three degree-1 vertices a_1, a_2 and a_3 in $G[V']$, then any minimum feedback vertex set S of G contains at least two vertices in $\{a_1, a_2, a_3, w_1, w_2\}$. Note that after deleting w_1 and w_2 from G, no vertex in V' is contained in a cycle. We know that $\{w_1, w_2\}$ intersects any cycle containing at least one vertex in $V' \cup \{w_1, w_2\}$. Thus, $S' = S \setminus \{a_1, a_2, a_3, w_1, w_2\} \cup \{w_1, w_2\}$ is still a minimum feedback vertex set of G. If there are at most two degree-1 vertices in $G[V']$, then $G[V']$ can only be a path P. There are at least three vertices in P since $|V'| \geq 3$. Any no-endpoint vertex in P is adjacent to at least one vertex in $\{w_1, w_2\}$ in G. It is easy to see that we need at least two vertices to intersect all cycles in the subgraph $G[V' \cup \{w_1, w_2\}]$. Then any minimum feedback vertex set S of G contains at least two vertices in $V' \cup \{w_1, w_2\}$. If S does not contain both of w_1 and w_2, then we replace the vertices in $(V' \cup \{w_1, w_2\}) \cap S$ with $\{w_1, w_2\}$ in S to get another minimum feedback vertex set of G. So there is always a minimum feedback vertex set containing both vertices in $N(V')$. $\square$

This lemma can be used to get reduction rules to deal with some vertex-cuts of size at most 2. However, our algorithm only uses a very special case where $|V'| = 3$.

Rule 4. *Let $P = v_1 v_2 v_3$ be an induced path in G such that $|N(P)| \leq 2$. Delete $N(P)$ from the graph and put all vertices in $N(P)$ to S_0. (See Figure 1 (a))*

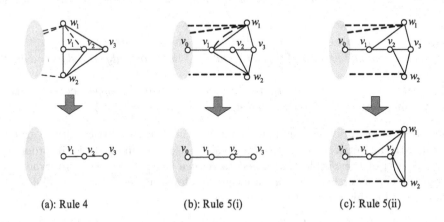

(a): Rule 4   (b): Rule 5(i)   (c): Rule 5(ii)

Fig. 1. Illustrations for Rules 4 and 5

The next three rules are also used to deal with some induced paths with small number of neighbors.

Rule 5. *Let $v_0 v_1 v_2 v_3$ be an induced path in G where each vertex in $V' = \{v_1, v_2, v_3\}$ is a non-trivial vertex and $|N(V')| = 3$. Let $N(V') = \{v_0, w_1, w_2\}$ and $V'' = \{v_1, v_2, v_3, w_1, w_2\}$.*
(i) If the size of the minimum feedback vertex set of the induced subgraph $G[V'']$ is at least 2, delete w_1 and w_2, and put the two vertices to S_0; and
(ii) If the size of the minimum feedback vertex set of the induced subgraph $G[V'']$ is 1, add a new edge between w_1 and w_2, and contract edge $v_2 v_3$ into a single vertex v_2. (See Figure 1 for an illustration)

Proof. Each vertex in V' is adjacent to at least one vertex in $\{w_1, w_2\}$ and v_3 is adjacent to both of w_1 and w_2, because any vertex in V' is non-trivial and none of v_2 and v_3 can be adjacent to v_0.

(i): When the size of the minimum feedback vertex set of $G[V'']$ is at least 2, any minimum feedback vertex set S of G contains at least two vertices in V''. However, w_1 and w_2 intersect all cycles containing some vertex in V''. Then $S \backslash V'' \cup \{w_1, w_2\}$ is still a minimum feedback vertex set of G. There is a minimum feedback vertex set containing w_1 and w_2 and then we can include them to the solution set directly.

(ii): When the size of the minimum feedback set of $G[V'']$ is 1, the vertex intersecting all cycles in $G[V'']$ can only be v_2 or v_3. For this case, there are no parallel edges between v_3 and a neighbor of it. Let G' be the resulting graph after executing the operation in (ii) on G and S' be a minimum feedback vertex set of G'. We show that S' is also a minimum feedback vertex set of G.

First, we show that S' is a feedback vertex set of G. Case 1. $v_2 \in S'$: Note there are no parallel edges incident on v_3 in G. The two graphs $G \backslash \{v_2\}$ and $G' \backslash \{v_2\}$ are almost the same except the degree-2 vertex v_3 in $G \backslash \{v_2\}$ is replaced with an edge in $G' \backslash \{v_2\}$. Then there is no cycle in $G \backslash S'$ if there is no cycle in $G' \backslash S'$. Case 2. $v_2 \notin S'$: Now S' contains at least two vertices in $\{w_1, w_2, v_1\}$. Assume to the contrary that there is a cycle C in $G \backslash S'$. The cycle C must contain some edges in the induced subgraph $G[V'']$. Furthermore, C is contained in $G[V'']$, because if C also contains an edge not in $G[V'']$ then C should pass through at least two vertices in $\{w_1, w_2, v_1\}$ in $G \backslash S'$. Then the cycle C is either $v_2 w_{i_0}$ or $v_2 v_3 w_{i_0}$, where $w_{i_0} \in \{w_1, w_2\}$. Both cases imply a cycle $v_2 w_{i_0}$ in $G' \backslash S'$, a contradiction. So S' is a feedback vertex set of G.

To show that S' is also a minimum feedback vertex set of G, we only need to prove that the size of a minimum feedback vertex of G is not greater than the size of a minimum feedback vertex of G'. Let S be a minimum feedback vertex set in G. If S contains only one vertex v^* in V'', then v^* can only be v_2 or v_3. For this case, we can see that $S \backslash \{v^*\} \cup \{v_2\}$ is a feedback vertex set of G'. If S contains at least two vertices in V'', then $S' = S \backslash V'' \cup \{w_1, w_2\}$ is still a minimum feedback vertex set of G. Furthermore, S' is also a feedback vertex set of G'. Thus, the size of a minimum feedback vertex of G' will not be larger than the size of a minimum feedback vertex of G. □

The next two rules are firstly used in [1] to get a kernel for the problem. We also need to use them. Figure 2 gives illustrations for these two rules.

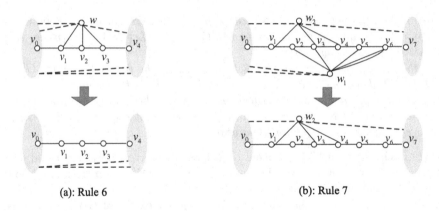

(a): Rule 6   (b): Rule 7

Fig. 2. Illustrations for Rules 6 and 7

Rule 6. *Let $v_0v_1v_2v_3v_4$ be an induced path in G where v_1, v_2 and v_3 are three degree-3 vertices having a common neighbor w. Delete w from the graph and put w to S_0.*

To prove the correctness of this reduction rule, we only need to show that there is at least one minimum feedback vertex set of G containing w. Let S be a minimum feedback vertex set of G not containing w. Since w is not in S, we know that at least two vertices v_{i_1} and v_{i_2} in $\{v_1, v_2, v_3\}$ are in S. It is easy to see that any cycle that contains at least one of v_{i_1} and v_{i_2} also contains at least one of w and v_1. Then $S' = S \setminus \{v_{i_1}, v_{i_2}\} \cup \{w, v_1\}$ is a minimum feedback vertex set of G containing w.

Rule 7. *Let $v_0v_1v_2v_3v_4v_5v_6v_7$ be an induced path in G where each vertex in $V' = \{v_1, v_2, v_3, v_4, v_5, v_6\}$ is a vertex of degree ≥ 3 and $|N(V')| = 4$. Let $N(V') = \{v_0, v_7, w_1, w_2\}$. Assume that $|N_{V'}(w_1)| \geq |N_{V'}(w_2)|$. Delete w_1 from the graph and put w_1 to S_0.*

We can verify that there is a minimum feedback vertex set of G containing w_1 in the above rule. We omit the full proof here since it can be obtained in [1].

Our algorithm takes a planar graph as the input, iteratively execute the above reduction rules in order until none of them can be applied anymore, and then return the resulting planar graph. It is easy to verify that after executing each of our reduction rules the resulting graph is still planar by Proposition 2. The algorithm runs in polynomial time, since each reduction rule can be applied in polynomial time and reduces at least one vertex in the graph. We are only interested in the size of the resulting graph, so we omit a more detailed analysis of the complexity of the algorithm. In fact, all of our reduction rules can be applied in general graphs. We only analyze a kernel for planar graphs. We call a graph *reduced* if none of the above reduction rules can be applied.

4   The Analysis

In this section, we assume that G is a reduced planar graph. Let S be an arbitrary feedback vertex set of G, $k = |S|$ and $n = |V(G)|$. We will prove that $n \leq 29k - 57$.

Let F be the remaining graph after deleting the solution set S from the graph G, i.e., $F = G[V \setminus S]$. Then F is a forest. Note that each degree-1 vertex in F is adjacent to at least 2 vertices in S in G and each degree-2 vertex in F is adjacent to at least one vertex in S in G since all vertices in G are non-trivial. We analyze the number of vertices in the forest F. Assume that F has l leaf-vertices and c connected components.

Lemma 2. *It holds that*

$$l \leq 2k + 2c - 4. \tag{3}$$

Proof. For each component of F that is not a single edge, we contract all inner-vertices in it into a single inner-vertex. For each component of F that is a single edge e, we introduce a degree-2 vertex adjacent to the two endpoints of e and deleting e. Then the new added vertex will become an inner vertex in this component. Let F' be the resulting forest. Then F' has c inner-vertices. We consider the bipartite graph $H = (A = L, B = I \cup S, E)$, where L is the set of leaf-vertices in F' and I is the set of inner-vertices in F'. There is an edge between $a \in L$ and $b \in I$ in H if there is an edge between a and b in F'. There is an edge between $a \in L$ and $s \in S$ in H if there is an edge between a and s in G. The bipartite graph H is still a planar graph since it can be obtained from G by contracting some edges, deleting some edges or adding a degree-2 vertex adjacent to two endpoints of an edge. By (2), the number of edges in H is at most $2|V(H)| - 4 \leq 2(l + c + k) - 4$. On the other hand, each vertex $a \in L$ is adjacent to at least two vertices in S (since G has only no-trivial vertices and then each leaf-vertex in F is adjacent to at least two vertices in S) and adjacent to one inner-vertex in I. Then there are at least $3|L|$ edges between L and $I \cup S$. We get $3l \leq 2(l + c + k) - 4$, which is (3). $\qquad\square$

Lemma 3. *The number of branch-vertices in F is at most $l - 2c$ and the number of inner-branches in F is at most $l - 3c$.*

Proof. Assume that the c trees in F have $l_1, l_2, \cdots, l_c$ leaf-vertices respectively. It is easy to see that a tree with l_i leaf-vertices has at most $l_i - 2$ branch-vertices and at most $l_i - 3$ inner-branches. Then F has at most $\sum_{i=1}^{c}(l_i - 2) = l - 2c$ branch-vertices and at most $\sum_{i=1}^{c}(l_i - 3) = l - 3c$ inner-branches. $\qquad\square$

A sub-path P of a branch is *good* if $|N_S(P)| \geq 3$, i.e., a good sub-path is adjacent to at least 3 vertices in S. We use the following method to partition each branch of F into several sub-paths, called *chains*, such that each vertex in the branch is contained in exactly one of the sub-paths. Let $Q = v_1 v_2 \cdots v_q$ be a branch, where we assume that v_q is a leaf-vertex in F if Q is a leaf-branch.

If Q is not a good path, we take Q as a single chain. Otherwise we partition Q into chains $P_1 = v_1 v_2 \cdots v_{i_1}, P_2 = v_{i_1+1} v_{i_1+2} \cdots v_{i_2}, \cdots, P_r = v_{i_{r-1}+1} v_{i_{r-1}+2} \cdots v_q$ such that for any chain $P_j = v_{i_{j-1}+1} v_{i_{j-1}+2} \cdots v_{i_j}$, either (i) P_j is good and $v_{i_{j-1}+1} v_{i_{j-1}+2} \cdots v_{i_j-1}$ is not good or (ii) P_j is not good and it holds $i_j = q$.

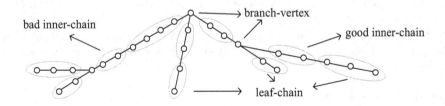

Fig. 3. Partitioning each branch in a tree into chains

We fix such a partition $\mathcal{P}$ of F and analyze the number of vertices in it. See Figure 3 for an illustration for the partition. We can see that each vertex of degree ≤ 2 in F is contained in a chain. A chain is *good* if it is a good path and *bad* otherwise. A chain is called a *leaf-chain* if it contains at least one leaf-vertex in F and an *inner-chain* if it does not contain any leaf-vertex in F. We will analyze the size and number of the chains. According to the way we partition a branch into chains, we know that

Lemma 4. *The number of bad inner-chains in $\mathcal{P}$ is at most $l - 3c$ and each bad inner-chain contains at most 5 vertices.*

Proof. In the partition $\mathcal{P}$ of F, each inner-branch can contain at most one bad inner-chain and each leaf-branch can not contain any bad inner-chain. Then we know that the number of bad inner-chains in $\mathcal{P}$ is bounded by the number of inner-branches in F. By Lemma 3, the number of inner-branches in F is at most $l - 3c$, which implies the first claim in the lemma. Next, we consider the second claim. Assume to the contrary that a bad inner-chain P contains at least 6 vertices. Since G is a reduced graph and all vertices in it are non-trivial, we know that each vertex in the bad inner-chain P is adjacent to at least one vertex in S. Then the condition of either Rule 6 or Rule 7 will hold, which implies a contradiction. □

Lemma 5. *The number of bad leaf-chains in $\mathcal{P}$ is at most $2k + 2c - 4$ and each bad leaf-chain contains at most 2 vertices.*

Proof. Since each leaf-chain contains at least one leaf-vertex, we know that the number of bad leaf-chains in $\mathcal{P}$ is bounded by the number of leaf-vertices in F. By Lemma 2, the number of leaves in F is at most $2k + 2c - 4$. We get the first claim. For the second claim, we can see that if a bad leaf-chain contains at least 3 vertices, then either Rule 4 or Rule 5 can be applied. □

Lemma 6. *The number of good chains in $\mathcal{P}$ is at most $2k - 4$ and each good chain contains at most 6 vertices.*

Proof. We consider a bipartite planar graph $H = (A, B, E)$, where the number of vertices in A equals to the number of good chains in $\mathcal{P}$, each vertex $a \in A$ is corresponding a good chain Q_a in $\mathcal{P}$, and $B = S$. The edge set E is defined by this: there is an edge between vertices $a \in A$ and $s \in B$ if and only if there is an edge between s and a vertex in Q_a. Note that H can be obtained from G by contracting all edges in good chains and then deleting some vertices and edges. By Proposition 2, we know that H is still a planar graph. According to the definition of good chains, we know that each vertex $a \in A$ is adjacent to at least 3 vertices in B. Then there are at least $3|A|$ edges between A and B. By (2), we have that $3|A| \leq 2(|A| + |B|) - 4 = 2|A| + 2k - 4$. Then we get

$$|A| \leq 2k - 4.$$

For the second claim, we first assume to the contrary that a good chain Q contains at least 7 vertices. We look at the sub-path Q' containing only the first 6 vertices in Q. According to the definition of chains, we know that Q' should be a bad path. Then either Rule 6 or Rule 7 can be applied on Q', a contradiction to the fact that the graph is a reduced graph. $\qquad\square$

Lemma 7. *The number of vertices in F is at most $28k - 57$.*

Proof. Let n_2 denote the number of degree-1 and degree-2 vertices in F and n_3 denote the number of vertices of degree ≥ 3, which are branch-vertices in F. Then $n_3 \leq l - 2c$ by Lemma 3. Each vertex of degree ≤ 2 is contained in a chain in $\mathcal{P}$. Each chain is either a good chain or a bad chain. Each bad chain is either a bad inner-chain or a bad leaf-chains. By Lemma 4, Lemma 5 and Lemma 6, we get

$$
\begin{aligned}
|V(F)| &= n_2 + n_3 \\
&\leq n_2 + (l - 2c) \\
&\leq 5(l - 3c) + 2(2k + 2c - 4) + 6(2k - 4) + (l - 2c) \\
&= 6l + 16k - 13c - 32 \\
&\leq 6(2k + 2c - 4) + 16k - 13c - 32 \qquad \text{by}(3) \\
&= 28k - c - 56 \\
&\leq 28k - 57.
\end{aligned}
$$

$\qquad\square$

The number of vertices in G is $|V(F)| + |S| \leq 28k - 57 + k = 29k - 57$. Therefore, we obtain the following theorem

Theorem 1. *Let G be a planar graph such that none of the above reduction rules can be applied on it. Then the size of a minimum feedback vertex set of G is more than $\frac{|V(G)|}{29}$.*

References

1. Abu-Khzam, F.N., Bou Khuzam, M.: An improved kernel for the undirected planar feedback vertex set problem. In: Thilikos, D.M., Woeginger, G.J. (eds.) IPEC 2012. LNCS, vol. 7535, pp. 264–273. Springer, Heidelberg (2012)
2. Bodlaender, H.L., Fomin, F.V., Lokshtanov, D., Penninkx, E., Saurabh, S., Thilikos, D.M.: (Meta) kernelization. In: FOCS 2009, pp. 629–638. IEEE Computer Society, Washington, DC (2009)
3. Bodlaender, H.L., Penninkx, E.: A Linear Kernel for Planar Feedback Vertex Set. In: Grohe, M., Niedermeier, R. (eds.) IWPEC 2008. LNCS, vol. 5018, pp. 160–171. Springer, Heidelberg (2008)
4. Bodlaender, H.L., van Dijk, T.C.: A Cubic Kernel for Feedback Vertex Set and Loop Cutset. Theory Comput. Syst. 46(3), 566–597 (2010)
5. Burrage, K., Estivill-Castro, V., Fellows, M.R., Langston, M.A., Mac, S., Rosamond, F.A.: The undirected feedback vertex set problem has a poly(k) kernel. In: Bodlaender, H.L., Langston, M.A. (eds.) IWPEC 2006. LNCS, vol. 4169, pp. 192–202. Springer, Heidelberg (2006)
6. Cao, Y., Chen, J., Liu, Y.: On feedback vertex set new measure and new structures. In: Kaplan, H. (ed.) SWAT 2010. LNCS, vol. 6139, pp. 93–104. Springer, Heidelberg (2010)
7. Chen, J., Fomin, F., Liu, Y., Lu, S., Villanger, Y.: Improved algorithms for feedback vertex set problems. J. Comput. Syst. Sci. 74, 1188–1198 (2008)
8. Chen, J., Liu, Y., Lu, S., O'Sullivan, B., Razgon, I.: A fixed-parameter algorithm for the directed feedback vertex set problem. J. ACM 55, 1–19 (2008)
9. Dehne, F., Fellows, M.R., Langston, M.A., Rosamond, F.A., Stevens, K.: An $O(2^{O(k)}n^3)$ FPT algorithm for the undirected feedback vertex set problem. In: Wang, L. (ed.) COCOON 2005. LNCS, vol. 3595, pp. 859–869. Springer, Heidelberg (2005)
10. Fomin, F.V., Gaspers, S., Pyatkin, A.V., Razgon, I.: On the minimum feedback vertex set problem: Exact and enumeration algorithms. Algorithmica 52(2), 293–307 (2008)
11. Fomin, F.V., Lokshtanov, D., Saurabh, S., Thilikos, D.M.: Bidimensionality and kernels. In: SODA 2010, Philadelphia, PA, USA, pp. 503–510 (2010)
12. Garey, M.R., Johnson, D.S.: Computers and intractability: A guide to the theory of NP-completeness. Freeman, San Francisco (1979)
13. Guo, J., Gramm, J., Huffner, F., Niedermeier, R., Wernicke, S.: Compression-based fixed-parameter algorithms for feedback vertex set and edge bipartization. J. Comput. Syst. Sci. 72(8), 1386–1396 (2006)
14. Razgon, I.: Exact computation of maximum induced forest. In: Arge, L., Freivalds, R. (eds.) SWAT 2006. LNCS, vol. 4059, pp. 160–171. Springer, Heidelberg (2006)
15. Razgon, I.: Computing minimum directed feedback vertex set in $O(1.9977^n)$. In: 10th Italian Conference on Theoretical Computer Science, ICTCS 2007, Rome, Italy, pp. 70–81 (2007)
16. Silberschatz, A., Galvin, P.: Operating System Concepts, 4th edn. Addison-Wesley (1994)
17. Thomassé, S.: A $4k^2$ kernel for feedback vertex set. ACM Transactions on Algorithms 6(2) (2010)
18. Xiao, M., Nagamochi, H.: An Improved Exact Algorithm for Undirected Feedback Vertex Set. Journal of Combinatorial Optimization (2014), doi: 10.1007/s10878-014-9737-x; A preliminary version appears as: Xiao, M., Nagamochi, H.: An Improved Exact Algorithm for Undirected Feedback Vertex Set. In: Widmayer, P., Xu, Y., Zhu, B. (eds.) COCOA 2013. LNCS, vol. 8287, pp. 153–164. Springer, Heidelberg (2013)

Clustering Performance
of 3-Dimensional Hilbert Curves

H.K. Dai[1] and H.C. Su[2]

[1] Computer Science Department, Oklahoma State University,
Stillwater, Oklahoma 74078, U.S.A.
dai@cs.okstate.edu
[2] Department of Computer Science, Arkansas State University,
Jonesboro, Arkansas 72401, U.S.A.
suh@astate.edu

Abstract. A discrete space-filling curve provides a linear traversal or indexing of a multi-dimensional grid space. This paper presents an analytical study of the clustering performance of the 3-dimensional Hilbert curve family. The underlying measure is the mean number of clusters over all identically shaped cubic subgrids. We derive an exact formula for the statistics for the Hilbert curve family, and have verified all exact formulas (intermediate and final) involved in the derivations in the analytical study with computer programs over various grid- and subgrid-orders.

Keywords: space-filling curve, Hilbert curve, z-order curve, clustering.

1 Preliminaries

Discrete space-filling curves have many applications in databases, parallel computation, algorithms, in which linearization techniques of multi-dimensional arrays or grids are needed. Sample applications include heuristics for Hamiltonian traversals, multi-dimensional space-filling indexing methods, image compression, and dynamic unstructured mesh partitioning.

For a positive integer n, denote $[n] = \{1, 2, \ldots, n\}$. An m-dimensional (discrete) space-filling curve of length n^m is a bijective mapping $C : [n^m] \to [n]^m$, thus providing a linear indexing/traversal or total ordering of the grid points in $[n]^m$. An m-dimensional grid is said to be of order k if it has side-length $n = 2^k$; a space-filling curve has order k if its codomain is a grid of order k. The generation of a sequence of multi-dimensional space-filling curves of successive orders usually follows a recursive framework (on the dimensionality and order), which results in a few classical families, such as Gray-coded curves, Hilbert curves, Peano curves, and z-order curves (see, for examples, [1] and [6]).

Denote by H_k^m and Z_k^m an m-dimensional Hilbert and z-order, respectively, space-filling curve of order k. Figure 1 illustrates the recursive constructions of H_k^m and Z_k^m for $m = 2$ and $k = 1, 2$, and $m = 3$ and $k = 1$.

We measure the applicability of a family of space-filling curves based on their common structural characteristics, which are informally described as follows.

Q. Gu, P. Hell, and B. Yang (Eds.): AAIM 2014, LNCS 8546, pp. 299–311, 2014.

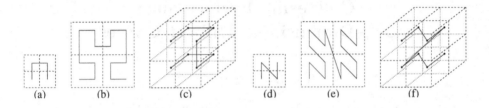

Fig. 1. Recursive constructions of Hilbert and z-order curves of higher order (respectively, H_k^m and Z_k^m) by interconnecting symmetric subcurves, via reflection and/or rotation, of lower order (respectively, H_{k-1}^m and Z_{k-1}^m) along an order-1 subcurve (respectively, H_1^m and Z_1^m): (a) H_1^2; (b) H_2^2; (c) H_1^3; (d) Z_1^2; (e) Z_2^2; (f) Z_1^3

Clustering performance measures the distribution of continuous runs of grid points (clusters) over identically shaped subspaces of $[n]^m$, which can be characterized by the mean number of clusters and the average inter-cluster distance (in $[n^m]$) within a subspace. The studies of clustering and inter-clustering performances for space-filling curves are motivated by the applicability of multi-dimensional space-filling indexing methods, in which an m-dimensional data space is mapped onto a 1-dimensional data space (external storage structure) by adopting a 1-dimensional indexing method based on an m-dimensional space-filling curve. Locality preservation reflects proximity between the grid points of $[n]^m$, that is, close-by points in $[n]^m$ are mapped to close-by indices/numbers in $[n^m]$, or vice versa.

For an m-dimensional space-filling curve $C : [n^m] \rightarrow [n]^m$ and a subgrid G of $[n]^m$, a cluster of G induced by C is a maximal (contiguous) subinterval I of $[n^m]$ such that $C(I) \subseteq G$. We can partition and order $C^{-1}(G)$ into a disjoint union of clusters. An inter-cluster gap of G is a subinterval of $[n^m]$ delimited by two consecutive clusters of G, and the corresponding inter-cluster distance is the length of the inter-cluster gap. Empirical and analytical studies of clustering and inter-clustering performances of various low-dimensional space-filling curves have been reported in the literature (see [6] and [2] for details). Generally, the Hilbert and z-order curve families exhibit good performance in this respect. A few locality measures have been proposed and analyzed for space-filling curves in the literature (see for example studies in [4], [1], and [3]).

Moon, Jagadish, Faloutsos, and Saltz [6] prove that in a sufficiently large m-dimensional H_k^m-structural grid space, the asymptotic mean number of clusters over all rectilinear polyhedral queries with common surface area $S_{m,k}$ approaches $\frac{1}{2}\frac{S_{m,k}}{m}$ as k approaches ∞. They also extend the work by Jagadish [5] to obtain an exact formula for the mean number of clusters over all rectangular $2^q \times 2^q$ subgrids of an H_k^2-structural grid space. Xu and Tirthapura [7] generalize the above asymptotic mean number of clusters over all rectilinear polyhedral queries with common surface area from m-dimensional Hilbert curves to arbitrary continuous space-filling curves. Note that rectangular queries with common volume yield the optimal asymptotic mean number of clusters for a continuous space-filling curve.

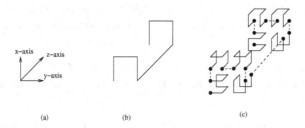

Fig. 2. The canonical 3-dimensional-Hilbert curves of orders 1 and 2: (a) coordinate system; (b) canonical H_1^3; (c) canonical H_2^3

Note that we present the skeletons for proving the main results for the Hilbert curve family without lengthy details in the abstract. The computation of the mean-clustering statistics proceeds to establishing many systems of recurrences and their closed-form solutions over three geometric regions of 3-dimensional Hilbert curves (boundary face-, edge-, and vertex-regions). Closed-form solutions for most systems of recurrences in our study are computed via the mathematical and analytical software Maple. Complete results: proofs, derivations, closed-form solutions, and exact formulas, and verifying computer programs for the Hilbert curve family are available from the authors.

2   Clustering Performance of Hilbert Curve Family

For a mathematical formalism of discrete Hilbert curves that facilitates combinatorial studies of multi-dimensional Hilbert indexing, see [1] for details. One of the salient characteristics of Hilbert curves is their "self-similarity" — a Hilbert curve can be generated by interconnecting identical subcurves via reflection and rotation (see Figure 2 for an example curve in the 3-dimensional case). For m-dimensional Hilbert curves, this self-similar structural property guides us to decompose H_k^m into 2^m identical H_{k-1}^m-subcurves (via reflection and/or rotation), which are amalgamated together by an H_1^m-curve. For $m = 2$, the "orientation" of H_k^2 uniquely determines those of the four H_{k-1}^2-subcurves and thus only one H_k^2 exists modulo symmetry, whereas for $m = 3$, there are 1536 structurally different 3-dimensional Hilbert curves [1].

Our study is based on the canonical H_k^3-curve whose recursive construction (on order k) employs a canonical H_1^3 depicted in Figure 2(b) as the basis and for the amalgamating H_1^3-curve. Figure 2(c) illustrates the recursive construction of the canonical 3-dimensional order-2 Hilbert curve H_2^3.

Remark 1. For most self-similar m-dimensional order-k space-filling curve C_k^m indexing the grid $[2^k]^m$, we can view C_k^m as a C_{k-q}^m-curve interconnecting $2^{m(k-q)}$ number of C_q^m-subcurves for all $q \in [k]$.

Remark 1 above motivates our analytical study of clustering performances to be based on query subgrids of size $2^q \times 2^q \times 2^q$.

For a canonical 3-dimensional order-k Hilbert curve H_k^3, let $\mathcal{S}_q(H_k^3)$ denote the total number of clusters over all $2^q \times 2^q \times 2^q$ query subgrids of the H_k^3-structural grid space $[2^k]^3$.

Remark 2. For an m-dimensional space-filling curve $C : [n^m] \to [n]^m$ and a subgrid G of $[n]^m$, the number of clusters within G is $\frac{1}{2}$ of the number of edges of C cut by G.

For a canonical 3-dimensional order-k Hilbert curve H_k^3, $\mathcal{E}_q(H_k^3)$ denote the total number of edge-cuts by all $2^q \times 2^q \times 2^q$ query subgrids. As suggested by Remark 2, it suffices to compute $\mathcal{E}_q(H_k^3)$, which yields $\mathcal{S}_q(H_k^3)$ ($= \frac{1}{2}\mathcal{E}_q(H_k^3)$) and the mean number of clusters over all $2^q \times 2^q \times 2^q$ query subgrids.

We express $\mathcal{E}_q(H_k^3)$ from three sources of edge-cut contribution as follows:

1. The H_k^3-structural grid space $[2^k]^3$ has $(2^k)^3 - 1$ edges (of H_k^3), each of which is cut by two opposite boundary faces (of face-size $(2^q)^2$) of a $2^q \times 2^q \times 2^q$ query grid — not subject to any boundary-constraint of the grid space $[2^k]^3$; therefore, the inclusion of all $2^q \times 2^q \times 2^q$ unconstrained query grids results in the edge-cut contribution of $(2^{3k} - 1)2 \cdot 2^{2q}$,

2. The contribution in item 1 should exclude the total number of edge-cuts by all $2^q \times 2^q \times 2^q$ constrained query grids that overlap with some boundary (faces, edges, vertices) of the grid space $[2^k]^3$ — denoted by $\epsilon_q(H_k^3)$, and

3. An additive adjustment of "$+2$" for the entrance and exit edge-cuts of H_k^3.

Thus, we have:
$$\mathcal{E}_q(H_k^3) = (2^{3k} - 1)2 \cdot 2^{2q} - \epsilon_q(H_k^3) + 2.$$

With respect to the canonical orientation of H_k^3 shown in Figure 2(a) and (b), we cover the 3-dimensional order-k grid space with three orthogonal systems of 2^k layers with normal vectors: z-, y-, and x-axes, respectively: for each $\alpha \in [2^k]$,

$$L_{k,\alpha}^{(xy)} = \{v \in [2^k]^3 \mid \text{z-coordinate of } v \text{ is } \alpha\} \quad L_{k,\alpha}^{(xz)} = \{v \in [2^k]^3 \mid \text{y-coordinate of } v \text{ is } \alpha\}$$
$$L_{k,\alpha}^{(yz)} = \{v \in [2^k]^3 \mid \text{x-coordinate of } v \text{ is } \alpha\}$$

We label the left-, right-, bottom-, top-, front-, and back-side boundary faces of a canonical H_k^3 (see Figure 3(b)) by L, R, B, T, F, and F', respectively. Denote by $\mathcal{S}_\mathcal{F} = \{L, R, B, T, F\}$, and by $\Lambda_{k,\lambda}^\Gamma$ the Γ-side boundary face-region of λ layers, where $\Gamma \in \mathcal{S}_\mathcal{F}$, i.e., $\Lambda_{k,\lambda}^L, \Lambda_{k,\lambda}^R, \Lambda_{k,\lambda}^B, \Lambda_{k,\lambda}^T, \Lambda_{k,\lambda}^F$ the left-, right-, bottom-, top-, front-side λ-layer boundary face-regions, respectively (note that $\Lambda_{k,\lambda}^{F'}$ is a reflexive $\Lambda_{k,\lambda}^F$ with z-axis as their common normal vector and any computation on $\Lambda_{k,\lambda}^{F'}$ will be treated as on $\Lambda_{k,\lambda}^F$):

$$\Lambda_{k,\lambda}^L = \cup_{\beta=1}^\lambda L_{k,\beta}^{(xz)} \qquad \Lambda_{k,\lambda}^R = \cup_{\beta=2^k-\lambda+1}^{2^k} L_{k,\beta}^{(xz)} \qquad \Lambda_{k,\lambda}^B = \cup_{\alpha=1}^\lambda L_{k,\alpha}^{(yz)}$$
$$\Lambda_{k,\lambda}^T = \cup_{\alpha=2^k-\lambda+1}^{2^k} L_{k,\alpha}^{(yz)} \qquad \Lambda_{k,\lambda}^F = \cup_{\gamma=1}^\lambda L_{k,\gamma}^{(xy)} \qquad \Lambda_{k,\lambda}^{F'} = \cup_{\gamma=2^k-\lambda+1}^{2^k} L_{k,\gamma}^{(xy)}$$

The intersection of two (respectively, three) adjacent boundary face-regions results in a boundary edge-region (respectively, a boundary vertex-region). Accordingly, the underlying constraint required for a query grid refers to simultaneously overlapping with the two (respectively, three) boundary faces.

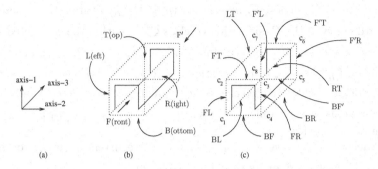

Fig. 3. For a canonical H_k^3: (a) coordinate system; (b) labelings of faces; (c) labelings of edges and vertices

For computing the total number of edge-cuts in $\epsilon_q(H_k^3)$ by all $2^q \times 2^q \times 2^q$ query grids G (constrained to overlapping with a boundary face, that is, $G \cap H_k^3 \subseteq \Lambda_{k,2^q-1}^\Gamma$ for some $\Gamma \in \mathcal{S}_\mathcal{F}$), we apply the inclusion-exclusion principle to the membership of the constrained query grid G in successive boundary face-, edge-, and vertex-regions. A uniform approach in computing the number of edge-cuts by all $2^q \times 2^q \times 2^q$ constrained query grids in the three types of boundary regions employ the following strategy:

1. Invoke Remark 1 that expresses a canonical H_k^3 as a H_{k-q}^3-curve interconnecting $2^{3(k-q)}$ number of H_q^3-subcurves, each of which is isomorphic to a canonical H_q^3 via reflection and/or rotation;

2. Complete the underlying boundary region of $2^q - 1$ layers to one of 2^q layers, and the completed boundary region is saturated with canonical H_q^3-subcurves from different oriented classes — each of which is characterized by the (simultaneous) boundary face(s) of its member-H_q^3-subcurves embedded as subface(s) of the boundary face(s) of the underlying boundary region, and compute the cardinality of each of these oriented classes; and

3. Compute two sets of statistics of edge-cut contribution due to all $2^q \times 2^q \times 2^q$ constrained query grids: (1) for each oriented class in item 2, the edge-set of its representative canonical H_q^3-subcurve, and (2) the set of all interconnecting edges for adjacent canonical H_q^3-subcurves in the boundary region.

2.1 Edge-Cuts of H_q^3-Subcurves in Boundary Face-Regions

For each Γ-side $(2^q - 1)$-layer boundary face-region $\Lambda_{k,2^q-1}^\Gamma$, where $\Gamma \in \{L, R, B, T, F, F'(=F)\}$ (F'-side treated as F-side), its completion $\Lambda_{k,2^q}^\Gamma$ is saturated with canonical H_q^3-subcurves, which are categorized into five (face) oriented classes $\mathcal{Q}_{k,q}^{\Gamma,\Gamma'}$, where $\Gamma' \in \mathcal{S}_\mathcal{F}$ ($\mathcal{Q}_{k,q}^{\Gamma,F'}$ treated as $\mathcal{Q}_{k,q}^{\Gamma,F}$), such that the Γ'-side oriented class consists of all the canonical Γ'-side H_q^3-subcurves whose Γ'-side face is embedded as a subface of the Γ-side boundary face.

For each $\Gamma \in \mathcal{S}_{\mathcal{F}}$, denote by $\eta_{k,q}^{\Gamma}$ the set $(\eta_{k,q}^{\Gamma,\Gamma'})_{\Gamma'\in\mathcal{S}_{\mathcal{F}}}$ of statistics, where $\eta_{k,q}^{\Gamma,\Gamma'}$ is the cardinality of the Γ'-side oriented class $\mathcal{Q}_{k,q}^{\Gamma,\Gamma'}$ within the completed Γ-side boundary face-region $\Lambda_{k,2^q}^{\Gamma}$ of a canonical H_k^3, i.e., $\eta_{k,q}^{\Gamma,\Gamma'} = |\mathcal{Q}_{k,q}^{\Gamma,\Gamma'}|$; and denote by $\eta_{k,q}^{\text{total}}$ the set $(\eta_{k,q}^{\text{total},\Gamma'})_{\Gamma'\in\mathcal{S}_{\mathcal{F}}}$ of statistics, where $\eta_{k,q}^{\text{total},\Gamma'}$ is the summation of cardinalities of all Γ'-side oriented classes $\mathcal{Q}_q^{\Gamma,\Gamma'}$ in all (six) completed Γ-side boundary face-regions of a canonical H_k^3, i.e., $\eta_{k,q}^{\text{total},\Gamma'} = \sum_{\Gamma\in\{L,R,B,T,F,F'(=F)\}} \eta_{k,q}^{\Gamma,\Gamma'}$ (F'-side treated as F-side).

For each $\Gamma \in \mathcal{S}_{\mathcal{F}}$, we develop a system of recurrences (in k) for $\eta_{k,q}^{\Gamma} = (\eta_{k,q}^{\Gamma,\Gamma'})_{\Gamma'\in\mathcal{S}_{\mathcal{F}}}$ and solve for its closed-form solution.

Lemma 1. *For a canonical H_k^3 structured as an H_{k-q}^3-curve interconnecting $2^{3(k-q)}$ number of canonical H_q^3-subcurves, and for every $\Gamma' \in \mathcal{S}_{\mathcal{F}}$,*

$$\eta_{k,q}^{L,\Gamma'} = 2\eta_{k-1,q}^{B,\Gamma'} + 2\eta_{k-1,q}^{F,\Gamma'} \quad \eta_{k,q}^{R,\Gamma'} = 2\eta_{k-1,q}^{L,\Gamma'} + 2\eta_{k-1,q}^{F,\Gamma'} \quad \eta_{k,q}^{B,\Gamma'} = 2\eta_{k-1,q}^{T,\Gamma'} + 2\eta_{k-1,q}^{F,\Gamma'}$$

$$\eta_{k,q}^{T,\Gamma'} = 4\eta_{k-1,q}^{R,\Gamma'} \qquad \eta_{k,q}^{F,\Gamma'} = 2\eta_{k-1,q}^{B,\Gamma'} + \eta_{k-1,q}^{F,\Gamma'} + \eta_{k-1,q}^{L,\Gamma'} \quad \eta_{q,q}^{\Gamma,\Gamma'} = \begin{cases} 1 & \text{if } \Gamma = \Gamma' \\ 0 & \text{if } \Gamma \neq \Gamma' \end{cases}$$

The closed-form solution for the system of recurrences $(\eta_{k,q}^{\Gamma,\Gamma'})_{\Gamma,\Gamma'\in\mathcal{S}_{\mathcal{F}}}$ is employed to yield one for $\eta_{k,q}^{\text{total}} = (\eta_{k,q}^{\text{total},\Gamma'})_{\Gamma'\in\mathcal{S}_{\mathcal{F}}}$.

The two sets of statistics obtained in Lemma 1 allow us to focus on computing the edge-cut contribution from a representative canonical H_q^3 of $\mathcal{Q}_{k,q}^{\Gamma,\Gamma'}$ for all $\Gamma,\Gamma' \in \mathcal{S}_{\mathcal{F}}$, which are in turn inferred by the following three sets of statistics when considering the edge-cut contribution from the x-, y-, and z-oriented edges in a canonical H_q^3, by all $2^q \times 2^q \times 2^q$ Γ-side constrained query grids (overlapping with the Γ-side boundary face of the H_q^3).

For an axis-orientation $\alpha \in \{x, y, z\}$, denote $\overline{\Pi}_q = (\overline{\Pi}_{q,\alpha})_{\alpha\in\{x,y,z\}}$ where $\overline{\Pi}_{q,\alpha}$ is the number of all α-oriented edges in a canonical H_q^3. For all $\alpha \in \{x, y, z\}$ and $\Gamma \in \mathcal{S}_{\mathcal{F}}$, denote $D_q = (D_{q,\alpha}^{\Gamma})_{\alpha\in\{x,y,z\},\Gamma\in\mathcal{S}_{\mathcal{F}}}$ where $D_{q,\alpha}^{\Gamma}$ is the total number of the α-oriented edge-cuts of a canonical H_q^3 by all $2^q \times 2^q \times 2^q$ Γ-side constrained query grids (overlapping with the Γ-side boundary face of the H_q^3), and the completion of D_q, $\overline{D}_q = (\overline{D}_{q,\alpha}^{\Gamma})_{\alpha\in\{x,y,z\},\Gamma\in\mathcal{S}_{\mathcal{F}}}$, where $\overline{D}_{q,\alpha}^{\Gamma}$ is the total number of the α-oriented edge-cuts of a canonical H_q^3 by all $2^q \times 2^q \times 2^q$ Γ-side query grids overlapping or abutting with the Γ-side boundary face of the H_q^3.

Lemma 2. *For a canonical H_q^3, the following system of recurrences (in q) for $\overline{\Pi}_q = (\overline{\Pi}_{q,\alpha})_{\alpha\in\{x,y,z\}}$:*

$$\overline{\Pi}_{q,x} = \begin{cases} 2\overline{\Pi}_{q-1,x} + 4\overline{\Pi}_{q-1,y} + 2\overline{\Pi}_{q-1,z} + 4 & \text{if } q > 1 \\ 4 & \text{if } q = 1 \end{cases}$$

$$\overline{\Pi}_{q,y} = \begin{cases} 2\overline{\Pi}_{q-1,y} + 4\overline{\Pi}_{q-1,z} + 2\overline{\Pi}_{q-1,x} + 2 & \text{if } q > 1 \\ 2 & \text{if } q = 1 \end{cases}$$

$$\overline{\Pi}_{q,z} = \begin{cases} 2\overline{\Pi}_{q-1,z} + 4\overline{\Pi}_{q-1,x} + 2\overline{\Pi}_{q-1,y} + 1 & \text{if } q > 1 \\ 1 & \text{if } q = 1 \end{cases}$$

has a closed-form solution.

Lemma 3. *For a canonical* H_q^3, *the following system of recurrences (in q) for*
$\overline{D}_q = (\overline{D}_{q,\alpha}^\Gamma)_{\alpha \in \{x,y,z\}, \Gamma \in \mathcal{S}_{\mathcal{F}}}$:

$$\overline{D}_{q,x}^L = 2(\overline{D}_{q-1,z}^B + \overline{D}_{q-1,y}^F + 2^{q-1}\overline{\Pi}_{q-1,z} + \overline{D}_{q-1,x}^R + 2^{q-1}\overline{\Pi}_{q-1,y} + \overline{D}_{q-1,y}^F + 2^q + 1)$$

$$\overline{D}_{q,z}^L = 2(\overline{D}_{q-1,y}^B + \overline{D}_{q-1,x}^F + 2^{q-1}\overline{\Pi}_{q-1,y} + \overline{D}_{q-1,z}^R + 2^{q-1}\overline{\Pi}_{q-1,x} + \overline{D}_{q-1,x}^F) + 1$$

$$\overline{D}_{q,x}^R = (2^q + 1)\overline{\Pi}_{q,x} - \overline{D}_{q,x}^L \qquad \overline{D}_{q,z}^R = (2^q + 1)\overline{\Pi}_{q,y} - \overline{D}_{q,y}^L$$

$$\overline{D}_{q,y}^B = 2(\overline{D}_{q-1,x}^F + \overline{D}_{q-1,y}^T + 2^{q-1}\overline{\Pi}_{q-1,x} + \overline{D}_{q-1,z}^L + 2^{q-1}\overline{\Pi}_{q-1,y} + \overline{D}_{q-1,z}^L + 2^{q-1})$$

$$\overline{D}_{q,z}^B = 2(\overline{D}_{q-1,y}^F + \overline{D}_{q-1,z}^T + 2^{q-1}\overline{\Pi}_{q-1,y} + \overline{D}_{q-1,x}^L + 2^{q-1}\overline{\Pi}_{q-1,z} + \overline{D}_{q-1,x}^L) + 2^{q-1} + 1$$

$$\overline{D}_{q,y}^T = (2^q + 1)\overline{\Pi}_{q,y} - \overline{D}_{q,y}^B \qquad \overline{D}_{q,z}^T = (2^q + 1)\overline{\Pi}_{q,z} - \overline{D}_{q,z}^B$$

$$\overline{D}_{q,x}^F = (2^q + 1)\overline{\Pi}_{q,x}/2 - reflexive \qquad \overline{D}_{q,y}^F = (2^q + 1)\overline{\Pi}_{q,y}/2 - reflexive$$

has a closed-form solution. For all $\alpha \in \{x, y, z\}$ and $\Gamma \in \mathcal{S}_{\mathcal{F}}$, $D_{q,\alpha}^\Gamma = \overline{D}_{q,\alpha}^\Gamma - \overline{\Pi}_{q,\alpha}$, and the closed-form solutions for $\overline{\Pi}_q$ and $\overline{D}_q$ yield one for D_q.

For each $\Gamma \in \mathcal{S}_{\mathcal{F}}$, the Γ-side $(2^q - 1)$-layer boundary face-region $\Lambda_{k,2^q-1}^\Gamma$ consists of some edges of H_{k-q}^3-curve that interconnects $2^{3(k-q)}$ number of H_q^3-subcurves in $\Lambda_{k,2^q}^\Gamma$. Observe that the end-vertices of an interconnecting edge are vertices (where three adjacent boundary faces intersect) of the two adjacent H_q^3-subcurves, hence the interconnecting edge lies in the outermost layer of the boundary face-region. Note that the edge-cut contribution of an interconnecting edge in $\Lambda_{k,2^q-1}^\Gamma$ is $2 \cdot 2^q$ by all $2^q \times 2^q \times 2^q$ query subgrids contained in H_k^3 and $2((2^q)^2 - 2^q)$ by all $2^q \times 2^q \times 2^q$ constrained query grids overlapping with the Γ-side boundary face.

Denote by $\tilde{\eta}_{k,q}$ the set $(\tilde{\eta}_{k,q}^\Gamma)_{\Gamma \in \mathcal{S}_{\mathcal{F}}}$ of statistics, where $\tilde{\eta}_{k,q}^\Gamma$ is the number of all interconnecting edges in the Γ-side $(2^q - 1)$-layer boundary face-region $\Lambda_{k,2^q-1}^\Gamma$.

Lemma 4. *For a canonical H_k^3 structured as an H_{k-q}^3-curve interconnecting $2^{3(k-q)}$ number of canonical H_q^3-subcurves, the following system of recurrences (in k) for $\tilde{\eta}_{k,q} = (\tilde{\eta}_{k,q}^\Gamma)_{\Gamma \in \mathcal{S}_{\mathcal{F}}}$:*

$$\tilde{\eta}_{k,q}^L = 2\tilde{\eta}_{k-1,q}^F + 2\tilde{\eta}_{k-1,q}^B + 2 \qquad \tilde{\eta}_{k,q}^R = 2\tilde{\eta}_{k-1,q}^F + 2\tilde{\eta}_{k-1,q}^L + 3 \qquad \tilde{\eta}_{k,q}^B = 2\tilde{\eta}_{k-1,q}^F + 2\tilde{\eta}_{k-1,q}^T$$

$$\tilde{\eta}_{k,q}^T = 4\tilde{\eta}_{k-1,q}^R \qquad \tilde{\eta}_{k,q}^F = \tilde{\eta}_{k-1,q}^F + \tilde{\eta}_{k-1,q}^L + 2\tilde{\eta}_{k-1,q}^B + 3$$

$$\tilde{\eta}_{q+1,q}^L = 2 \qquad \tilde{\eta}_{q+1,q}^R = 3 \qquad \tilde{\eta}_{q+1,q}^B = 0$$

$$\tilde{\eta}_{q+1,q}^T = 0 \qquad \tilde{\eta}_{q+1,q}^F = 3$$

has a closed-form solution.

Theorem 1. *For a canonical H_k^3 structured as an H_{k-q}^3-curve interconnecting $2^{3(k-q)}$ number of canonical H_q^3-subcurves, the number of edge-cuts in all boundary face-regions of the H_k^3 by $2^q \times 2^q \times 2^q$ constrained query grids overlapping with some boundary face(s) is:*

$$\sum_{\Gamma' \in \mathcal{S}_{\mathcal{F}}} \eta_{k,q}^{total,\Gamma'} \left((D_{q,\alpha}^{\Gamma'} + D_{q,\beta}^\Gamma) 2^q \cdot 2 + \overline{\Pi}_{q,\gamma} 2^{2q} \right) + \sum_{\Gamma \in \mathcal{S}_{\mathcal{F}}} \tilde{\eta}_{k,q}^\Gamma (2^{2q+1} - 2^{q+1})$$

$$= 3 \cdot 2^{2k+3q+1} - 2^{2k+2q+2} - \frac{117}{35} \cdot 2^{2k-q-1} - \frac{93}{5} \cdot 2^{2k-3} - 3 \cdot 2^{2q+1} + 3 \cdot 2^{q+1}$$

$$+ many\ lower\text{-}order\ terms.$$

2.2 Edge-Cuts of H_q^3-Subcurves in Boundary Edge-Regions

We label an edge by its two incident boundary faces (their intersection), that is, an edge formed by Γ- and Γ'-side boundary faces is labeled by $\Gamma\Gamma' \in$ $\{\text{BF}, \text{BL}, \text{BR}, \text{FL}, \text{FT}, \text{FR}, \text{LT}, \text{RT}\}$ as illustrated in Figure 3(c). Denote by $\mathcal{S}_{\mathcal{E}} = \{\text{BF}, \text{BL}, \text{BR}, \text{FL}, \text{FT}, \text{FR}, \text{LT}, \text{RT}\}$.

The derivation of the edge-cut contribution by all $2^q \times 2^q \times 2^q$ constrained query grids overlapping with an edge (at least two boundary faces) is more complicated. The added complexity comes from two sources: (1) Most underlying geometrical entities and their associated statistics are characterized by edges $\Gamma\Gamma' \in \mathcal{S}_{\mathcal{E}}$ (rather than by faces $\Gamma \in \mathcal{S}_{\mathcal{F}}$), and (2) When considering the edge-cut contribution from a representative canonical H_q^3 of an oriented class (in a boundary edge-region of H_k^3) by a constrained query grid overlapping with a boundary edge of the H_q^3, a finer categorization of edge-cuts based on the parallelity/orthogonality of the edge-cutting face of a query grid versus the edge-forming boundary faces of the H_q^3 is employed with additional sets of statistics.

For each $XX' \in \mathcal{S}_{\mathcal{E}}$, denote by $\hat{\eta}_{k,q}^{XX'}$ the set $(\hat{\eta}_{k,q}^{XX',YY'})_{YY'\in\mathcal{S}_{\mathcal{E}}}$ of statistics, where $\hat{\eta}_{k,q}^{XX',YY'}$ is the number of all canonical H_q^3-subcurves whose YY'-edge is embedded as a subedge of the XX'-edge of the H_k^3, and by $\hat{\eta}_{k,q}^{\text{total}}$ the set $(\hat{\eta}_{k,q}^{\text{total},YY'})_{YY'\in\mathcal{S}_{\mathcal{E}}}$ of statistics, where $\hat{\eta}_{k,q}^{\text{total},YY'} = \sum_{XX'\in\mathcal{S}_{\mathcal{E}}} \hat{\eta}_{k,q}^{XX',YY'}$. We develop a system of recurrences (in k) for the set of statistics $\hat{\eta}_{k,q}^{XX'} = (\hat{\eta}_{k,q}^{XX',YY'})_{YY'\in\mathcal{S}_{\mathcal{E}}}$ and solve for its closed-form solution.

Lemma 5. *For a canonical H_k^3 structured as an H_{k-q}^3-curve interconnecting $2^{3(k-q)}$ number of canonical H_q^3-subcurves, and for every $YY' \in \mathcal{S}_{\mathcal{E}}$,*

$$
\begin{array}{lll}
\hat{\eta}_{k,q}^{\text{FL},YY'} = \hat{\eta}_{k-1,q}^{\text{BL},YY'} + \hat{\eta}_q^{\text{BF},YY'} & \hat{\eta}_{k,q}^{\text{FT},YY'} = 2\hat{\eta}_{k-1,q}^{\text{BR},YY'} & \hat{\eta}_{k,q}^{\text{FR},YY'} = \hat{\eta}_{k-1,q}^{\text{FL},YY'} + \hat{\eta}_{k-1,q}^{\text{BF},YY'} \\[4pt]
\hat{\eta}_{k,q}^{\text{BF},YY'} = \hat{\eta}_{k-1,q}^{\text{FL},YY'} + \hat{\eta}_{k-1,q}^{\text{FT},YY'} & \hat{\eta}_{k,q}^{\text{BR},YY'} = 2\hat{\eta}_{k-1,q}^{\text{RT},YY'} & \hat{\eta}_{k,q}^{\text{RT},YY'} = 2\hat{\eta}_{k-1,q}^{\text{FR},YY'} \\[4pt]
\hat{\eta}_{k,q}^{\text{LT},YY'} = 2\hat{\eta}_{k-1,q}^{\text{FR},YY'} & \hat{\eta}_{k,q}^{\text{BL},YY'} = 2\hat{\eta}_{k-1,q}^{\text{BF},YY'} & \hat{\eta}_{q,q}^{XX',YY'} = \begin{cases} 1 & \text{if } XX' = YY' \\ 0 & \text{if } XX' \neq YY' \end{cases}
\end{array}
$$

The closed-form solution for the system of recurrences $(\hat{\eta}_{k,q}^{XX',YY'})_{XX',YY'\in\mathcal{S}_{\mathcal{E}}}$ is employed to yield one for $\hat{\eta}_{k,q}^{\text{total}} = (\hat{\eta}_{k,q}^{\text{total},YY'})_{YY'\in\mathcal{S}_{\mathcal{E}}}$.

Consider the edge-cut contribution from a representative canonical H_q^3 of an oriented class in a boundary edge-region of H_k^3 by a $2^q \times 2^q \times 2^q$ constrained query grid G overlapping with a boundary $\Gamma\Gamma'$-edge of the H_q^3, where $\Gamma\Gamma' \in \mathcal{S}_{\mathcal{E}}$. Edge-cuts of H_q^3 by four boundary faces of G are grouped into three categories based on the parallelity/orthogonality of the boundary faces of G: (1) Two parallel boundary faces of G orthogonal to both Γ- and Γ'-side boundary faces of the H_q^3 (that is, the $\Gamma\Gamma'$-edge is the common normal vector of the two parallel boundary faces of G), (2) One boundary face of G parallel and orthogonal to the Γ- and Γ'-side boundary faces respectively, and (3) One boundary face of G parallel and orthogonal to the Γ'- and Γ-side boundary faces respectively. The number of edge-cuts of categories 2 and 3 is computed via the statistics

$D_q = (D_{q,\alpha}^{\Gamma})_{\alpha \in \{x,y,z\}, \Gamma \in \mathcal{S}_{\mathcal{F}}}$. For the edge-cuts of category 1, we consider the following two sets of statistics.

Denote $E_q = (E_q^{\Gamma\Gamma'})_{\Gamma\Gamma' \in \mathcal{S}_{\mathcal{E}}}$ where $E_q^{\Gamma\Gamma'}$ is the total number of the edge-cuts of a canonical H_q^3 by all the parallel boundary faces with $\Gamma\Gamma'$-edge as their common normal vector of $2^q \times 2^q \times 2^q$ constrained query grids overlapping with the boundary $\Gamma\Gamma'$-edge of the H_q^3, and the completion of E_q, $\overline{E}_q = (\overline{E}_q^{\Gamma\Gamma'})_{\Gamma\Gamma' \in \mathcal{S}_{\mathcal{E}}}$ where $\overline{E}_q^{\Gamma\Gamma'}$ is the total number of the edge-cuts of a canonical H_q^3 by all the parallel boundary faces with $\Gamma\Gamma'$-edge as their common normal vector of $2^q \times 2^q \times 2^q$ query grids overlapping or abutting with the boundary $\Gamma\Gamma'$-edge of the H_q^3.

Lemma 6. *For a canonical H_q^3, the following system of recurrences (in q) for* $\overline{E}_q = (\overline{E}_q^{\Gamma\Gamma'})_{\Gamma\Gamma' \in \mathcal{S}_{\mathcal{E}}}$:

$$\overline{E}_q^{BF} = 2\overline{E}_{q-1}^{BL} + \overline{E}_{q-1}^{FL} + \overline{E}_{q-1}^{FR} + 2\overline{E}_{q-1}^{FT} + 2\overline{E}_{q-1}^{LT} + 2^{q-1}(\overline{D}_{q-1,1}^{F} + 2\overline{D}_{q-1,2}^{F} + \overline{D}_{q-1,1}^{L})$$
$$+2\overline{D}_{q-1,3}^{L} + \overline{D}_{q-1,1}^{R} + \overline{D}_{q-1,2}^{T}) + (\overline{\Pi}_{q-1,1} + \overline{\Pi}_{q-1,2})2^{2q-2} + 2^{2q-1} + 2^{q-1}$$

$$\overline{E}_q^{BL} = 2\overline{E}_{q-1}^{BF} + 4\overline{E}_{q-1}^{FL} + 2\overline{E}_{q-1}^{RT} + 2^{q-1}(2\overline{D}_{q-1,2}^{B} + 2\overline{D}_{q-1,2}^{F} + 2\overline{D}_{q-1,1}^{L} + 2\overline{D}_{q-1,3}^{R})$$
$$+(2\overline{\Pi}_{q-1,2})2^{2q-2} + 2^{q-1}$$

$$\overline{E}_q^{BR} = 4\overline{E}_{q-1}^{FL} + 2\overline{E}_{q-1}^{FT} + 2\overline{E}_{q-1}^{LT} + 2^{q-1}(2\overline{D}_{q-1,1}^{L} + 2\overline{D}_{q-1,3}^{L} + 2\overline{D}_{q-1,2}^{T} + 2\overline{D}_{q-1,3}^{T})$$
$$+(2\overline{\Pi}_{q-1,3})2^{2q-2} + 2^{2q-1} + 2^{q}$$

$$\overline{E}_q^{FL} = \overline{E}_{q-1}^{BL} + \overline{E}_{q-1}^{BR} + 2\overline{E}_{q-1}^{BF} + 2\overline{E}_{q-1}^{FR} + 2\overline{E}_{q-1}^{FT} + 2^{q-1}(\overline{D}_{q-1,2}^{B} + \overline{D}_{q-1,3}^{B} + 2\overline{D}_{q-1,2}^{F})$$
$$+\overline{D}_{q-1,3}^{L} + \overline{D}_{q-1,1}^{R} + \overline{D}_{q-1,3}^{R} + \overline{D}_{q-1,2}^{T}) + (\overline{\Pi}_{q-1,2} + \overline{\Pi}_{q-1,3})2^{2q-2} + 2^{2q} + 2^{q+1} + 1$$

$$\overline{E}_q^{FR} = 2\overline{E}_{q-1}^{FR} + 2\overline{E}_{q-1}^{FL} + 2\overline{E}_{q-1}^{BF} + \overline{E}_{q-1}^{LT} + \overline{E}_{q-1}^{RT} + 2^{q-1}(\overline{D}_{q-1,2}^{B} + 2\overline{D}_{q-1,1}^{F} + 2\overline{D}_{q-1,2}^{F})$$
$$+\overline{D}_{q-1,1}^{L} + \overline{D}_{q-1,2}^{L} + \overline{D}_{q-1,3}^{L}) + (\overline{\Pi}_{q-1,1} + \overline{\Pi}_{q-1,2})2^{2q-2} + 2^{2q} + 2^{q+1} + 1$$

$$\overline{E}_q^{FT} = 2\overline{E}_{q-1}^{BF} + 2\overline{E}_{q-1}^{BR} + \overline{E}_{q-1}^{FL} + \overline{E}_{q-1}^{FR} + 2\overline{E}_{q-1}^{RT} + 2^{q-1}(\overline{D}_{q-1,2}^{B} + 2\overline{D}_{q-1,3}^{B} + \overline{D}_{q-1,1}^{F})$$
$$+2\overline{D}_{q-1,3}^{R} + 2\overline{D}_{q-1,3}^{T}) + (2\overline{\Pi}_{q-1,3})2^{2q-2} + 2^{2q-1} + 3 \cdot 2^{q-1} + 1$$

$$\overline{E}_q^{LT} = 2\overline{E}_{q-1}^{BF} + 2\overline{E}_{q-1}^{FR} + 4\overline{E}_{q-1}^{FR} + 2^{q-1}(4\overline{D}_{q-1,1}^{F} + 2\overline{D}_{q-1,2}^{F} + 2\overline{D}_{q-1,1}^{R})$$
$$+(2\overline{\Pi}_{q-1,1})2^{2q-2} + 2^{q-1}$$

$$\overline{E}_q^{RT} = 2\overline{E}_{q-1}^{BL} + 4\overline{E}_{q-1}^{FR} + 2\overline{E}_{q-1}^{FT} + 2^{q-1}(2\overline{D}_{q-1,3}^{B} + 4\overline{D}_{q-1,1}^{F} + 2\overline{D}_{q-1,1}^{R})$$
$$+(2\overline{\Pi}_{q-1,1})2^{2q-2} + 2^{2q-1}$$

$$\overline{E}_0^{\Gamma\Gamma'} = 0 \text{ for every } \Gamma\Gamma' \in \mathcal{S}_{\mathcal{E}}$$

has a closed-form solution. For $E_q = (E_q^{\Gamma\Gamma'})_{\Gamma\Gamma' \in \mathcal{S}_{\mathcal{E}}}$,

$$E_q^{FL} = \overline{E}_q^{FL} - \overline{D}_{q,x}^{F} - \overline{D}_{q,x}^{L} + \overline{\Pi}_{q,x} \qquad E_q^{FT} = \overline{E}_q^{FT} - \overline{D}_{q,y}^{F} - \overline{D}_{q,y}^{T} + \overline{\Pi}_{q,y}$$
$$E_q^{FR} = \overline{E}_q^{FR} - \overline{D}_{q,x}^{F} - \overline{D}_{q,x}^{R} + \overline{\Pi}_{q,x} \qquad E_q^{BF} = \overline{E}_q^{BF} - \overline{D}_{q,y}^{B} - \overline{D}_{q,y}^{F} + \overline{\Pi}_{q,y}$$
$$E_q^{BR} = \overline{E}_q^{BR} - \overline{D}_{q,z}^{B} - \overline{D}_{q,z}^{R} + \overline{\Pi}_{q,z} \qquad E_q^{RT} = \overline{E}_q^{RT} - \overline{D}_{q,z}^{R} - \overline{D}_{q,z}^{T} + \overline{\Pi}_{q,z}$$
$$E_q^{LT} = \overline{E}_q^{LT} - \overline{D}_{q,z}^{L} - \overline{D}_{q,z}^{T} + \overline{\Pi}_{q,z} \qquad E_q^{BL} = \overline{E}_q^{BL} - \overline{D}_{q,z}^{B} - \overline{D}_{q,z}^{L} + \overline{\Pi}_{q,z}$$

and the closed-form solutions for $\overline{\Pi}_q$, $\overline{D}_q$, *and* $\overline{E}_q$ *yield one for* E_q.

Analogous to the case of the boundary face-regions, an interconnecting edge of two adjacent canonical H_q^3-subcurves in a boundary edge-region lies in the outermost layer of the boundary edge-region; that is, all such interconnecting edges correspond to the edges of an amalgamating H_{k-q}^3-curve that are in the outermost layers of the boundary edge-regions. The edge-cut contribution of an interconnecting edge is $2(2^q - 1)^2$ by all $2^q \times 2^q \times 2^q$ constrained query grids overlapping with a boundary edge-region.

Denote by $\check{\eta}_{k,q}$ the set $(\check{\eta}_{k,q}^{\Gamma\Gamma'})_{\Gamma\Gamma'\in\mathcal{S}_{\mathcal{E}}}$ of statistics, where $\check{\eta}_{k,q}^{\Gamma\Gamma'}$ is the number of all interconnecting edges in the boundary $\Gamma\Gamma'$-edge-region.

Lemma 7. *For a canonical H_k^3 structured as an H_{k-q}^3-curve interconnecting $2^{3(k-q)}$ number of canonical H_q^3-subcurves, the following system of recurrences (in k) for $\check{\eta}_{k,q} = (\check{\eta}_{k,q}^{\Gamma\Gamma'})_{\Gamma\Gamma'\in\mathcal{S}_{\mathcal{E}}}$:*

$$\check{\eta}_{k,q}^{\mathrm{FL}} = \check{\eta}_{k-1,q}^{\mathrm{BL}} + \check{\eta}_{k-1,q}^{\mathrm{BF}} + 1 \qquad \check{\eta}_{k,q}^{\mathrm{FT}} = 2\check{\eta}_{k-1,q}^{\mathrm{BR}} \qquad \check{\eta}_{k,q}^{\mathrm{FR}} = \check{\eta}_{k-1,q}^{\mathrm{FL}} + \check{\eta}_{k-1,q}^{\mathrm{BF}} + 1$$
$$\check{\eta}_{k,q}^{\mathrm{BF}} = \check{\eta}_{k-1,q}^{\mathrm{FL}} + \check{\eta}_{k-1,q}^{\mathrm{FT}} \qquad \check{\eta}_{k,q}^{\mathrm{BR}} = 2\check{\eta}_{k-1,q}^{\mathrm{RT}} \qquad \check{\eta}_{k,q}^{\mathrm{RT}} = 2\check{\eta}_{k-1,q}^{\mathrm{FR}}$$
$$\check{\eta}_{k,q}^{\mathrm{LT}} = 2\check{\eta}_{k-1,q}^{\mathrm{FR}} \qquad \check{\eta}_{k,q}^{\mathrm{BL}} = 2\check{\eta}_{k-1,q}^{\mathrm{BF}} \qquad \check{\eta}_{q,q}^{\Gamma\Gamma'} = 0 \text{ for every } \Gamma\Gamma' \in \mathcal{S}_{\mathcal{E}}$$

has a closed-form solution.

Theorem 2. *For a canonical H_k^3 structured as an H_{k-q}^3-curve interconnecting $2^{3(k-q)}$ number of canonical H_q^3-subcurves, the number of edge-cuts in all boundary edge-regions of the H_k^3 by $2^q \times 2^q \times 2^q$ constrained query grids overlapping with some boundary edge(s) is:*

$$\sum_{\Gamma\Gamma'\in\mathcal{S}_{\mathcal{E}}} (\hat{\eta}_{k,q}^{\mathrm{total},\Gamma\Gamma'}(2E_q^{\Gamma\Gamma'} + 2^q D_{q,\alpha}^{\Gamma} + 2^q D_{q,\alpha}^{\Gamma'}))$$

(where α is the axis-orientation parallel to both Γ and Γ')

$$+ \sum_{\Gamma\Gamma'\in\{\mathrm{BF,BF,BL,BR,FL,FL,FR,FR,FT,FT,LT,RT}\}} \check{\eta}_{k-q}^{\Gamma\Gamma'}(2^q - 1)^2 2$$

$$= 3 \cdot 2^{k+4q+1} - 2^{k+3q+3} + 2^{k+2q+1} - \frac{271}{77} \cdot 2^{k+q} - \frac{947}{231} \cdot 2^{k+1} + \frac{27}{77} \cdot 2^{k-q+4}$$

$$- \frac{5}{11} \cdot 2^{k+1} q - 3 \cdot 2^{2q+1} + 3 \cdot 2^{q+2} - 6 + \text{ many lower-order terms.}$$

2.3 Edge-Cuts of H_q^3-Subcurves in Boundary Vertex-Regions

Note that we consider $k > q$ only. Denote by $c_{1,k}, c_{2,k}, c_{3,k}, c_{4,k}$ (see Figure 3(c)) the four differently oriented vertices (traversal order of the interconnecting H_1^3-curve) of a canonical H_k^3, and by $c_{i,k,q}$ the H_q^3 embedded in the i-th vertex of a canonical H_k^3, where $i \in \{1, 2, 3, 4\}$. Note that $c_{i,k,k} = c_{i,k}$. By zooming in the vertices of a canonical H_k^3, observe that (when $k > q$): $c_{1,k,q} = c_{1,k-1,q}$, $c_{2,k,q} = c_{4,k-1,q}$, $c_{3,k,q} = c_{4,k-1,q}$, and $c_{4,k,q} = c_{2,k-1,q}$. Hence, the vertices in a canonical H_k^3 are: two $c_{1,q}$, two $c_{2,q}$, two $c_{4,q}$, and two $c_{2,q}$ if $k - q$ is even or two $c_{4,q}$ otherwise.

Consider the edge-cut contribution from a representative canonical H_q^3 of an oriented class in a boundary vertex-region of H_k^3 by a $2^q \times 2^q \times 2^q$ constrained query grid G overlapping with a boundary vertex of the H_k^3. The edge-cut contribution of an edge within the boundary vertex-region is 1 — by one (edge-)orthogonal boundary face of G.

Denote by Δ the operator for the total number of edge-cuts in a boundary vertex-region (operand) by all identically shaped $2^q \times 2^q \times 2^q$ constrained query grids overlapping with the boundary vertex. The total number of edge-cuts within the eight boundary vertex-regions by all identically shaped $2^q \times 2^q \times 2^q$ constrained query grids is:

$$\begin{cases} 2\Delta(c_{1,q,q}) + 2\Delta(c_{2,q,q}) + 4\Delta(c_{4,q,q}) & \text{if } k - q \text{ is even,} \\ 2\Delta(c_{1,q,q}) + 4\Delta(c_{2,q,q}) + 2\Delta(c_{4,q,q}) & \text{otherwise.} \end{cases}$$

Lemma 8. *For a canonical H_k^3 structured as an H_{k-q}^3-curve interconnecting $2^{3(k-q)}$ number of canonical H_q^3-subcurves, the number of edge-cuts within a boundary vertex-region by all identically shaped $2^q \times 2^q \times 2^q$ constrained query grids overlapping with the boundary vertex is as follows:*

$$\Delta(c_{1,q,q}) = E_q^{FL} + E_q^{BF} + E_q^{BL} \qquad \Delta(c_{2,q,q}) = E_q^{FL} + E_q^{FT} + E_q^{LT}$$
$$\Delta(c_{3,q,q}) = E_q^{FR} + E_q^{FT} + E_q^{RT} \qquad \Delta(c_{4,q,q}) = E_q^{FR} + E_q^{BF} + E_q^{BR}$$

Theorem 3. *For a canonical H_k^3 structured as an H_{k-q}^3-curve interconnecting $2^{3(k-q)}$ number of canonical H_q^3-subcurves, the number of edge-cuts in all boundary vertex-regions of the H_k^3 by $2^q \times 2^q \times 2^q$ constrained query grids overlapping with some boundary vertex is:*

$$\begin{cases} 2\Delta(c_{1,q,q}) + 2\Delta(c_{2,q,q}) + 4\Delta(c_{4,q,q}) & \text{if } k-q \text{ is even} \\ 2\Delta(c_{1,q,q}) + 4\Delta(c_{2,q,q}) + 2\Delta(c_{4,q,q}) & \text{otherwise} \end{cases}$$

$$= \begin{cases} 4E_q^{FL} + 4E_q^{FR} + 6E_q^{BF} + 2E_q^{FT} + 2E_q^{BL} + 2E_q^{LT} + 4E_q^{BR} & \text{if } k-q \text{ is even} \\ 6E_q^{FL} + 2E_q^{FR} + 4E_q^{BF} + 4E_q^{FT} + 2E_q^{BL} + 4E_q^{LT} + 2E_q^{BR} & \text{otherwise} \end{cases}$$

$$= 4E_q^{FL} + 2E_q^{FR} + 4E_q^{BF} + 2E_q^{FT} + 2E_q^{BL} + 2E_q^{LT} + 2E_q^{BR}$$

$$+ (1 + (-1)^{k-q+1})(E_q^{FR} + E_q^{BF} + E_q^{BR}) + (1 + (-1)^{k-q})(E_q^{FL} + E_q^{FT} + E_q^{LT})$$

$$= 2^{5q+1} - 2^{4q+2} + 2^{3q+1} - \frac{69}{7} \cdot 2^{2q-2} + \frac{11}{7} \cdot 2^{2q-2}(-1)^{k-q} - \frac{1}{21} \cdot 2^{q-2} - q \cdot 2^q$$

$$+ \frac{11}{7} \cdot 2^q (-1)^{k-q+1} - 2 + \text{ many lower-order terms.}$$

3 Total and Mean Numbers of Clusters for Canonical H_k^3-Curves, and Verification

The results in Section 2 yield the total number of edge-cuts in all boundary (face-, edge-, or vertex-)regions of a canonical H_k^3 by $2^q \times 2^q \times 2^q$ constrained query grids overlapping with some boundary (faces, edges, or vertices) of the grid space $[2^k]^3$:

$$\epsilon_q(H_k^3) = \text{number of edge-cuts in Theorem 1} - \text{number of edge-cuts in Theorem 2}$$
$$+ \text{number of edge-cuts in Theorem 3.}$$

Hence, an exact formula for the total number of the edge-cuts of a canonical H_k^3 by all $2^q \times 2^q \times 2^q$ query subgrids is obtained via:

$$\mathcal{E}_q(H_k^3) = (2^{3k} - 1)2 \cdot 2^{2q} - \epsilon_q(H_k^3) + 2$$

$$= 2^{3k+2q+1} - 3 \cdot 2^{2k+3q+1} + 2^{2k+2q+2} + \frac{117}{35} \cdot 2^{2k-q-1} + \frac{93}{5} \cdot 2^{2k-3} + 3 \cdot 2^{k+4q+1}$$

$$- 2^{k+3q+3} + 2^{k+2q+1} - \frac{271}{77} \cdot 2^{k+q} - \frac{947}{231} \cdot 2^{k+1} + \frac{27}{77} \cdot 2^{k-q+4} - \frac{5}{11} \cdot 2^{k+1}q - 2^{5q+1}$$

$$+ 2^{4q+2} - 2^{3q+1} + \frac{13}{7} \cdot 2^{2q-2} - \frac{11}{7} \cdot 2^{2q-2}(-1)^{k-q} + q2^q + \frac{505}{21} \cdot 2^{q-2}$$

$$- \frac{11}{7} \cdot 2^q(-1)^{k-q+1} - 2 + \text{ many lower-order terms.}$$

By Remark 2, an exact formula for the total number of clusters over all $2^q \times 2^q \times 2^q$ query subgrids of a canonical H_k^3-curve is obtained via:

$$S_k(H_k^3) = \frac{1}{2}\mathcal{E}_q(H_k^3),$$

and that for the mean number of clusters over all $2^q \times 2^q \times 2^q$ query subgrids of a canonical H_k^3-curve is obtained via:

$$\frac{S_q(H_k^3)}{(2^k - 2^q + 1)^3} \left(= \frac{\mathcal{E}_q(H_k^3)}{2(2^k - 2^q + 1)^3} \right).$$

We have verified all the exact formulas (intermediate and final) involved in the derivations in the analytical study with computer programs over various grid- and subgrid-orders: $k \in \{3, 4, \ldots, 10\}$ and $q \in \{2, 3, \ldots, k\}$.

4   Conclusion

Our analytical study of clustering performance of the 3-dimensional order-k continuous Hilbert curve family is based on the mean-clustering statistics — the mean number of clusters over all subgrids of size $2^q \times 2^q \times 2^q$. For sufficiently large k ($\gg q$), the mean number of clusters is approximated via:

$$\frac{S_q(H_k^3)}{(2^k - 2^q + 1)^3} \approx \frac{2^{3k+2q+1}}{2(2^k - 2^q + 1)^3} \approx 2^{2q},$$

which is consistent with the asymptotic results in [6] and [7]: $\frac{1}{2}\frac{S_{m,k}}{m} = \frac{1}{2}\frac{6(2^q)^2}{3} = 2^{2q}$. Our work in progress includes: (1) the completion of the mean-clustering analytical study for the discontinuous z-order curve family — the exact results allow us to compare the relative performances of Hilbert and z-order curve families with respect to the measure, and (2) the formulation of a multi-dimensional random walk to study the clustering performance of continuous multi-dimensional space-filling curves (such as Hilbert curves) and obtain a closed-form approximation to the mean-clustering statistics for the Hilbert curve family — which may shed some light on if the random walk may furnish an effective model to develop approximations to clustering and locality statistics for space-filling curves.

References

1. Alber, J., Niedermeier, R.: On multi-dimensional curves with Hilbert property. Theory of Computing Systems 33(4), 295–312 (2000)
2. Dai, H.K., Su, H.C.: Approximation and analytical studies of inter-clustering performances of space-filling curves. In: Proceedings of the International Conference on Discrete Random Walks. Discrete Mathematics and Theoretical Computer Science, vol. AC, pp. 53–68 (September 2003)
3. Dai, H.K., Su, H.C.: On p-norm based locality measures of space-filling curves. In: Fleischer, R., Trippen, G. (eds.) ISAAC 2004. LNCS, vol. 3341, pp. 364–376. Springer, Heidelberg (2004)

4. Gotsman, C., Lindenbaum, M.: On the metric properties of discrete space-filling curves. IEEE Transactions on Image Processing 5(5), 794–797 (1996)
5. Jagadish, H.V.: Analysis of the Hilbert curve for representing two-dimensional space. Information Processing Letters 62(1), 17–22 (1997)
6. Moon, B., Jagadish, H.V., Faloutsos, C., Saltz, J.H.: Analysis of the clustering properties of the Hilbert space-filling curve. IEEE Transactions on Knowledge and Data Engineering 13(1), 124–141 (2001)
7. Xu, P., Tirthapura, S.: On the optimality of clustering through a space filling curve. In: Proceedings of the 31st Symposium on Principles of Database Systems, PODS 2012, pp. 215–224 (May 2012)

Broadcast Networks with Near Optimal Cost

Hovhannes A. Harutyunyan

Department of Computer Science and Software Engineering, Concordia University,
Montreal, Quebec, H3G 1M8, Canada
haruty@cs.concordia.ca

Abstract. Broadcasting is a basic problem of communication in usual
networks. Many papers have investigated the construction of minimum
broadcast networks, the cheapest possible network architecture (having
the fewest communication lines), in which broadcasting can be accomplished as fast as theoretically possible from any vertex. Other papers
considered the problem of determining the minimum broadcast time of
a given vertex in an arbitrary network. In this paper, for given n we construct optimal networks on n vertices which we define to be the product
of the broadcast time and the number of edges of the network. On the
way we start the study of an interesting problem, the problem of minimum time broadcasting in networks with given number of vertices and
edges.

Keywords: Broadcast, minimum broadcast graph, optimal networks.

1 Introduction

Computer networks have become essential in several aspects of modern society.
The performance of information dissemination in networks often determines their
overall efficiency. One of the fundamental information dissemination problems
is broadcasting. Broadcasting is a process in which a single message is sent
from one member of a network to all other members. Inefficient broadcasting
could degrade the performance of a network seriously. Therefore, it is of major
interest to improve the performance of a network by using efficient broadcasting
algorithm.

Broadcasting is an information dissemination problem in a connected network, in which one node, called the *originator*, must distribute a message to all
other nodes by placing a series of calls along the communication lines of the network. Once informed, the informed nodes aid the originator in distributing the
message. This is assumed to take place in discrete time units. The broadcasting
is to be completed as quickly as possible, subject to the following constraints:

- Each call involves only one informed node and one of its uninformed neigh-
  bors.
- Each call requires one unit of time.
- A node can participate in only one call per unit of time.
- In one unit of time, many calls can be performed in parallel.

Q. Gu, P. Hell, and B. Yang (Eds.): AAIM 2014, LNCS 8546, pp. 312–322, 2014.

Formally, any network can be modelled as a connected graph $G = (V, E)$, where V is the set of vertices (or nodes) and E is the set of edges (or communication lines).

Given a vertex u as the originator, we define the *broadcast time* of vertex u, $t(u)$, as the minimum number of time units required to complete broadcasting from vertex u. Note that for any vertex u in a connected graph G on n vertices, $t(u) \geq \lceil \log n \rceil$, since during each time unit the number of informed vertices can at most double. All logarithms in this paper are base 2.

The broadcast time $t(G)$ of the graph G is defined as $\max\{t(u)|u \in V\}$. Another obvious lower bound on the broadcast time of a vertex in graph G is the diameter of graph G, $t(u) \geq D(G)$, where $D(G)$ is the diameter of G.

G is called a *broadcast graph* or *broadcast network* if $t(G) = \lceil \log n \rceil$. The *broadcast function* $B(n)$ is the minimum number of edges in any broadcast graph (network) on n vertices. A *minimum broadcast graph* or *mbg* is a broadcast graph on n vertices with only $B(n)$ edges. Therefore, an *mbg* is the cheapest possible broadcast network architecture (having the fewest possible edges), in which broadcasting can be accomplished as fast as theoretically possible from any originator vertex.

The problem of determining $t(u)$ for a vertex u in an arbitrary graph is NP-complete [19]. The problem remains NP-complete even for 3-regular planar graphs [25]. Research in [5] has showed that the broadcast time cannot be approximated within a factor 3.

The literature on this subject can be divided primarily into two major areas: one on designing approximation algorithms and heuristics to determine $t(u)$ for a vertex u in an arbitrary graph, the other on designing minimum broadcast graphs. For the first problem several approximation algorithms with a polylogarithmic ratio are suggested in [5], [26], [6]. The best approximation algorithm is presented in [5]. The second problem is also very difficult. The values of $B(n)$ and constructions of *mbgs* are known only for some small values of n, $n \leq 63$, $n = 2^p$ [8] and $n = 2^p - 2$ [4], [20].

Since *mbgs* seem to be extremely difficult to construct, a long sequence of papers presented techniques to construct broadcast graphs, and to obtain upper bounds on $B(n)$ (see [1], [3], [4], [8], [11], [12], [13], [15], [18], [20], [22] and [23]).

To guarantee minimum time broadcasting we have to design dense graphs. For example, the minimum time broadcasting in a graph on $n = 2^p$ vertices requires at least degree $\log n = p$ for every vertex. Thus, $B(2^p) = p2^{p-1}$. However, $B(2^p + 1) \leq 2^{p+1}$ (see for example [13]). So, the density of broadcast graphs depend on the value of n.

In many applications the broadcast time can be relaxed which will allow to use sparser graphs. This may decrease the overall cost of the network. In this paper we will consider the problem of optimizing the total cost of the network which is defined to be the product of the broadcast time and the number of edges of the network.

The two main parameters in communication networks are the broadcast time of the network and its total number of links. The problem of minimizing both parameters at the same time, of course, are contradictory tasks. The sparse networks have small number of communication lines but a larger broadcast time while dense networks may have relatively small broadcast time but are costly because of many communication lines. The choice of using sparse or dense networks depends on the network applications and also on the upper bound on the broadcast time. In this paper, we consider the problem of minimizing the product of the broadcast time and the number of edges of the network.

In many network applications, all communication links of the network should be active during the entire communication process. For such applications, the total cost of the network is defined as the product of the number of edges and the broadcast time of the network. Using our notation $total\ cost = m \times t(G)$ for network G with m edges and broadcast time $t(G)$. In this paper we construct networks on n vertices with near optimal cost for any n.

To study this problem we will consider broadcasting in graphs with fixed number of vertices and edges. The latter problem is interesting problem of its own which was not studied in the literature.

The paper is organized as follows. In section 2 we give an upper bound on the minimum broadcast time for any graph with fixed number of vertices and edges. In section 3 for any n we present networks on n vertices with small broadcast time and show that these graphs have near optimal cost.

2 $G_{m,n}$ Graphs with Minimum Broadcast Time

In this section we consider the problem of finding the minimum possible broadcast time of any graph from a class of graphs with given Number of vertices and edges.

In our construction of networks with optimal cost we will use some classes of graphs that are known for their good broadcasting properties. Below we give the definitions of these classes of graphs.

Definition 1. *The binomial tree B^k of dimension k has 2^k vertices and can be constructed recursively. The tree B^0 is a single vertex which is its root. The tree B^{k+1} is obtained from two copies of B^k by connecting their roots. Each of the two roots can be the root of B^{k+1}.*

It is easy to see that the broadcast time of B^k originated at the root is equal to k. Fig. 1 illustrates B^4.

Definition 2. *([1], [2]) Knödel graph on n vertices (where n is even), denoted $KG(n)$, is $KG(n) = (V, E)$ where $V = \{0, 1, \ldots, n-1\}$ and $(x, y) \in E$ iff $x + y \equiv 2^k - 1 \pmod{n}$ for $k = 1, 2, \ldots, \lfloor \log n \rfloor$.*

The Knödel graph is well known for its good communication and graph theoretic properties. See [9], [1], [2], [16], [17]. It is proved ([22], [1]) that $KG(n)$ is a broadcast graph on n vertices for any even n, and so $t(KG(n)) = \lceil \log n \rceil$. Fig. 2 presents $KG(20)$.

B^4

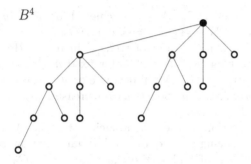

Fig. 1. Binomial tree B^4

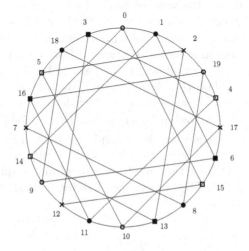

Fig. 2. $KG(20)$ with broadcast time 5

Definition 3. *Let $G_{m,n}$ be the class of all connected simple graphs with n vertices and m edges, for some $n-1 \leq m \leq \frac{n(n-1)}{2}$. Let $\tau(G_{m,n})$ be the minimum broadcast time over all graphs from $G_{m,n}$.*

First, we consider the problem of determining $\tau(G_{m,n})$, and we give lower and upper bounds on $\tau(G_{m,n})$.

Lemma 1. $\tau(G_{m,n})$ *is monotonically decreasing as a function of m.*

Proof. If $\tau(G_{m,n}) = t(G)$ for some graph G with n vertices and m edges, then we construct a graph H by adding one edge (which is not in G) to the set of edges of G. Then graph H belongs to the set $G_{m+1,n}$. Since G is a spanning subgraph of H then broadcasting in H will be completed no later than broadcasting in G. Thus, $t(G) \geq t(H)$ and $\tau(G_{m,n}) = t(G) \geq t(H) \geq \tau(G_{m+1,n})$.

The monotonicity of $\tau(G_{m,n})$ as function of n is a difficult problem. For example, when $n = 2^k$ then $\tau(G_{m,n}) = k$ and this is achieved only for $m \geq \frac{n \log n}{2}$, but when $n = 2^{k-1} + 1$ then $\tau(G_{n,m}) = k$ is achieved for $m < 3n$ (see for example [13]). In particular, $\tau(G_{2^{k+\log 3},2^k+1}) = k$ but $\tau(G_{2^{k+\log 3},2^k}) = k + 1$. So, if we fix m and decrease n then the value of $\tau(G_{m,n})$ increases. However if we do the same comparison picking $n = 2^k + 2^{k-1}$ for fixed $m = 2n$ and decrease n up to the value $2^k + 1$ then $\tau(G_{m,n})$ will decrease or stay the same. This is due to the fact that the broadcast function $B(n)$ is increasing as a function of n where $2^k + 1 \leq n \leq 2^k + 2^{k-1}$ ([14]).

Therefore, the question of the monotonicity of $\tau(G_{m,n})$ as a function of n heavily depends on n being a power of 2 or how far it is from a power of 2. This also makes the problem of finding the exact value of $\tau(G_{m,n})$ for all m from interval $n - 1 \leq m \leq \frac{n\lceil \log n \rceil}{2}$ extremely difficult. So, in the text below we will try to accurately estimate the value of $\tau(G_{m,n})$ for large values of n and m.

The two boundary cases should be considered here are $m = n - 1$ and $m = \frac{n\lceil \log n \rceil}{2}$. The first boundary case follows from Lemma 1 and the fact that any connected simple graph on n vertices contains at least $n - 1$ edges. It is obvious that $\tau(G_{m,n}) \leq \tau(G_{n-1,n})$ for all $m \geq n$. The second boundary follows from the well known upper bound on $B(n)$ for all n. In [7] the author presents a recursive construction of broadcast graphs (graphs which have broadcast time $\lceil \log n \rceil$) for all n and proves the following upper bound, $B(n) \leq \frac{n\lceil \log n \rceil}{2}$. This actually means that for any n there is a broadcast graph G with at most $\frac{n\lceil \log n \rceil}{2}$ edges and $t(G) = \lceil \log n \rceil$. Thus, using our notation $\tau(G_{m,n}) = \tau(G_{n\lceil \log n \rceil/2,n}) = \lceil \log n \rceil$ for all $m \geq n\lceil \log n \rceil/2$ since the broadcast time of any graph on n vertices is lower bounded by $\lceil \log n \rceil$. Then we get the following bounds, $\tau(G_{n\lceil \log n \rceil/2,n}) \leq \tau(G_{m,n}) \leq \tau(G_{n-1,n})$.

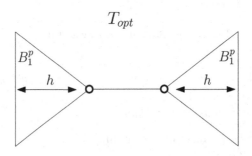

T_{opt}

Fig. 3. Optimal tree T_{opt}

In [21] and [23] the authors present the construction of optimal trees T_{opt} (in [21] it is called minimal broadcast trees), the trees on n vertices with minimum possible broadcast time. Fig. 3 illustrates the general structure of the optimal tree. It is proved in [21] that for all $n > 8$ the optimal tree has the structure presented in Fig. 3. In particular, B_1^p is a subtree of binomial tree B^p containing only the vertices at distance at most h from the root of B^p. The exact values of p and h are calculated by complicated combinatorial arguments, but for large n it

is proved that the optimal tree on n vertices has broadcast time $\log n / \log \frac{\sqrt{5}+1}{2}$,
$t(T_{opt}) = \log n / \log \frac{\sqrt{5}+1}{2} \approx 1.43 \log n$.

Figure 4 illustrates the optimal tree on 22 vertices with broadcast time 7.

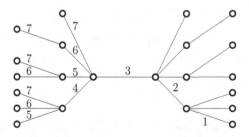

Fig. 4. The optimal tree on 22 vertices with broadcast time 7

The above result on optimal trees actually gives $\tau(G_{n-1,n}) = \log n / \log \frac{\sqrt{5}+1}{2}$
$\approx 1.43 \log n$ for large values of n.

From our discussion above we actually have lower and upper bounds on
$\tau(G_{m,n})$.

Lemma 2. $\lceil \log n \rceil \leq \tau(G_{m,n}) \leq \log n / \log \frac{\sqrt{5}+1}{2} \approx 1.43 \log n$ *for large* n.

Below we will give upper bounds on $\tau(G_{m,n})$ for some m, where $n - 1 \leq m \leq \frac{n \lceil \log n \rceil}{2}$.

Lemma 3. $\tau(G_{n,n}) \leq \log n / \log \frac{\sqrt{5}+1}{2} - \left(\frac{1}{\log \frac{\sqrt{5}+1}{2}} - 1\right) \approx 1.43 \log n - 0.43$ *for
large* n.

Proof. Consider two copies of the optimal tree T_{opt} on $n/2$ vertices (assuming n
is even). Connect the two central vertices of these optimal trees with 2 additional
edges (see Fig. 5). The obtained graph G clearly has n vertices and n edges. The
broadcast procedure in G is exactly the same as the broadcast procedure in
the optimal tree except when the two central vertices of one optimal tree are
informed they both send the message to the other central vertices of the second
optimal tree. Then the broadcasting in all four subtrees will follow the broadcast
procedure as in the optimal tree T on $n/2$ vertices. Thus, broadcasting in graph
G takes one additional time unit compared to T_{opt}. Therefore, $\tau(G_{n,n}) \leq t(G) =$
$\tau(G_{n/2-1,n/2}) + 1 = \log \frac{n}{2} / \log \frac{\sqrt{5}+1}{2} + 1 = \log n / \log \frac{\sqrt{5}+1}{2} - \left(\frac{1}{\log \frac{\sqrt{5}+1}{2}} - 1\right)$.

Before turning to the general case we give a tight upper bound on $\tau(G_{n+1,n})$.
Fig. 5 presents graph H from $G_{n+1,n}$ with $t(H) \approx 1.43 \log n - 1.72$ for large
n. The construction and the broadcast procedure in graph H is the same as in
Lemma 3, except we replace the 4-cycle (which is a minimum broadcast graph
on 4 vertices) with a minimum broadcast graph on 7 vertices presented in [8].

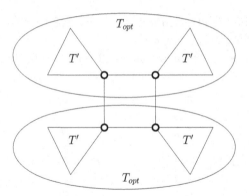

Fig. 5. The graph with n vertices and n edges from Lemma 3

Lemma 4. $\tau(G_{n+1,n}) \leq \log n / \log \frac{\sqrt{5}+1}{2} - \left(\frac{\log 3.5}{\log \frac{\sqrt{5}+1}{2}} - 2\right) \approx 1.43 \log n - 1.72$ *for large n.*

Proof. The proof is essentially the same as of Lemma 3, except we use 7 copies of only one subtree of the optimal tree on $n/3.5$ vertices. The broadcast procedure is also the same except it will take 3 time units for any central vertex to inform all the other central vertices of the minimum broadcast graph on 7 vertices. Thus, $\tau(G_{n+1,n}) \leq t(G) = \tau(G_{n/3.5-1,n/3.5}) + 2 = \log \frac{n}{3.5} / \log \frac{\sqrt{5}+1}{2} + 2 = \log n / \log \frac{\sqrt{5}+1}{2} - \left(\frac{\log 3.5}{\log \frac{\sqrt{5}+1}{2}} - 2\right) \approx 1.43 \log n - 1.72$.

The results of the last two Lemmas can be generalized by replacing the 4-cycle or the minimum broadcast graph on 7 vertices with any minimum broadcast graph. We will use such a construction in section 3 to give a tight upper bound on the networks with optimal cost.

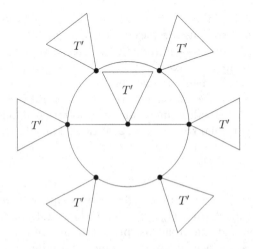

Fig. 6. The graph with n vertices and $n+1$ edges from Lemma 4

Several papers considered time-relaxed broadcast graphs (see [14] for example). r-relaxed broadcast graph is a graph on n vertices with broadcast time at most $\lceil \log n \rceil + r$. Minimum r-relaxed broadcast graph is a r-relaxed broadcast graph with minimum possible number of edges, denoted by $B_r(n)$. The best upper bound on $B_r(n)$ for general n is presented in [14]. They construct r-relaxed broadcast graphs on n vertices from $B^{\lceil \log n \rceil}$, the binomial tree of dimension $\lceil \log n \rceil$, by first deleting $2^{\lceil \log n \rceil} - n$ farthest leaves from $B^{\lceil \log n \rceil}$ and then connecting the root of $B^{\lceil \log n \rceil}$ with the vertices at distance d for all $d \equiv i (\mod r)$ from the root for each $0 \le i \le r - 1$. Graph G constructed above has at most $n + n/r$ edges and broadcast time $\lceil \log n \rceil + r$.

Lemma 5. $\tau(G_{n+n/r,n}) \le \lceil \log n \rceil + r$ for any $1 \le r \le \log n / \log \frac{\sqrt{5}+1}{2}$.

3   Networks with Optimal Cost

In this section we will design networks with near optimal cost for any n.

Recall that the total cost of the broadcast network (graph) is the product of its broadcast time and the number of edges. Denote the cost of a network on n vertices with minimum possible cost by $c(n)$. Following the notation from section 2, the optimal (minimum) cost of a network on n vertices will be $c(n) = \min\{m \times \tau(G_{m,n}) : n - 1 \le m \le \frac{n(n-1)}{2}\}$.

Lemma 6. $c(n) \ge n \log n + \Omega(n)$.

Proof. Note that since $\tau(G_{m,n}) \ge \lceil \log n \rceil$ and $m \ge n-1$ then $c(n) \ge \lceil \log n \rceil (n-1) = n\lceil \log n \rceil - \lceil \log n \rceil$. However, as mentioned above, when $m = n - 1$ then $\tau(G_{m,n}) \approx 1.43 \log n$ which gives the total cost of $1.43 \log n(n - 1) > n \log n$ for large n. For all other values of m, $m = n + a$, where $a \ge 0$ and $\tau(G_{m,n}) = \lceil \log n \rceil + c$, where $c \ge 0$. Thus, $c(n) = n\lceil \log n \rceil + cn + a\lceil \log n \rceil + ac \ge n \log n + \Omega(n)$ for large n.

The main result of this section is a construction of a graph on n vertices with $c(n) = n \log n + \Theta(n \log \log n)$, thus, giving an asymptotic optimal value of the minimum cost of any network on n vertices.

From the above and discussion in section 2 it follows that an optimal network on n vertices should have broadcast time close to $\lceil \log n \rceil$. Moreover, to achieve the lower bound from Lemma 6, c must be a constant and $a \log n = O(n)$. This would be possible to claim if there exist a c-relaxed broadcast graph (with broadcast time $\lceil \log n \rceil + r$), where c is a constant, with $n + n/\log n$ edges, $B_c(n) \le n + n/\log n$. However, there is no such c-relaxed broadcast graph construction known.

The best know r-relaxed broadcast graph constructions are presented in [14]. We used this construction in Lemma 5 and denoted it by G. Recall that G has $n+r$ edges and $t(G) = \log n + r$. So, the cost of G will be $(n+n/r)(\log n + r)$. The last expression is minimized for $r = \sqrt{\log n}$, and the cost of G will be at least $(n+n/\sqrt{\log n})(\log n + \sqrt{\log n}) = n \log n + 2n\sqrt{\log n} + n = n \log n + \Theta(n\sqrt{\log n})$.

Now we will construct another graph with slightly smaller cost. From Lemma 6 it follows that the leading term cannot be smaller than $n \log n$, but we will improve the second term.

Our construction of $G_{m,n}$ graphs with "good" broadcast time is a combination of Knödel graph $KG(k)$ on k vertices and the optimal broadcast tree T' on n/k vertices (see Fig.7).

We construct graph $H = (V, E)$ on n vertices and $m = n - k + \frac{k \log k}{2}$ edges as follows. Consider the graph obtained from k copies of the optimal tree T' on $\frac{n}{k}$ vertices where the roots of the trees form a Knödel graph on k vertices (see Fig. 7).

Theorem 1. *For any n there exists a graph on n vertices with $c(n) \leq n \log n + \Theta(n \log(\log n))$ for large values of n.*

Proof. Consider graph H described above, which is a combination of the Knödel graph and binomial tree. To broadcast from a farthest leaf in subtree T' (which is a subtree of a binomial tree) we first send the message to the root of the same subtree T' via shortest path then the root vertex broadcasts the message in the Knödel graph on k vertices in $\log k$ time units. Now, after this second phase all the k vertices of the Knödel graph are informed. Then every vertex of the Knödel graph broadcasts the message optimally within each subtree T' of a binomial tree. It is clear that the first and the third phases of the above broadcast scheme is equivalent to the minimum time broadcast scheme in the optimal tree on $2n/k$ vertices. So, the broadcast time of phases 1 and 3 of the above broadcast scheme will be $1.43 \log \frac{n}{k}$ for large n. Thus, the total broadcast time of H will be $t(H) = 1.43 \log \frac{n}{k} + \log k$ for large values of n.

The number of edges of graph H is $\frac{k \log k}{2} + k(\frac{n}{k} - 1) = \frac{k \log k}{2} + n - k$. Thus, the cost of graph H, $c(n) = (1.43 \log \frac{n}{k} + \log k)(\frac{k \log k}{2} + n - k)$. We pick $k = \frac{n}{\log^2 n}$. So, we obtain $c(n) = (1.43 \log \frac{n}{k} + \log k)(\frac{k \log k}{2} + n - k) \leq (\log n + 0.43 \log(\log^2 n))(n + \frac{n}{2 \log n} + \frac{n}{\log^2 n}) \approx n \log n + 0.86 n \log(\log n) + \Theta(n)$.

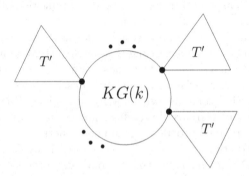

Fig. 7. The structure of the network with near optimal cost

Note that we can make slight improvement by picking $k = \frac{n}{\log^s n}$ for some $1 < s < 2$. Then the coefficient of the second term of the bound from Theorem 1 will be decreased by almost a factor of two. However, the minimum value of function $c(n)$ from above is still $n \log n + \Theta(n \log(\log n))$. The gap between the lower bound from Lemma 6 and the upper bound from Theorem 1 is small. To close the gap one has to construct a graph on n vertices, with at most $n + n/\log n$ edges and the broadcast time $\log n + c$, for some constant c. Another direction, is of course to improve the lower bound.

References

1. Bermond, J.-C., Fraigniaud, P., Peters, J.: Antepenultimate broadcasting. Networks 26, 125–137 (1995)
2. Bermond, J.-C., Harutyunyan, H.A., Liestman, A.L., Perennes, S.: A note on the dimensionality of modified Knödel graphs. Int. J. Found. Comp. Sci. 8, 109–117 (1997)
3. Bermond, J.-C., Hell, P., Liestman, A.L., Peters, J.G.: Sparse broadcast graphs. Discrete Appl. Math. 36, 97–130 (1992)
4. Dinneen, M.J., Fellows, M.R., Faber, V.: lgebraic constructions of efficient broadcast networks. In: Mattson, H.F., Rao, T.R.N., Mora, T. (eds.) AAECC 1991. LNCS, vol. 539, pp. 152–158. Springer, Heidelberg (1991)
5. Elkin, M., Kortsarz, G.: Sublogarithmic approximation for telephone multicast: path out of jungle. In: SODA 2003, Baltimore, pp. 76–85 (2003)
6. Elkin, M., Kortsarz, G.: A combinatorial logarithmic approximation algorithm for the directed telephone broadcast problem. In: Proc. of ACM Symp. on Theory of Computing, pp. 438–447 (2002)
7. Farley, A.M.: Minimal broadcast networks. Networks 9, 313–332 (1979)
8. Farley, A.M., Hedetniemi, S.T., Mitchell, S., Proskurowski, A.: Minimum broadcast graphs. Discrete Math. 25, 189–193 (1979)
9. Fertin, G., Raspaud, A.: Survey on Knödel Graphs. Discrete Appl. Math. 137, 173–195 (2004)
10. Fraigniaud, P., Lazard, E.: Methods and problems of communication in usual networks. Discrete Appl. Math. 53, 79–133 (1994)
11. Gargano, L., Vaccaro, U.: On the construction of minimal broadcast networks. Networks 19, 673–689 (1989)
12. Harutyunyan, H.A.: An Efficient Vertex Addition Method for Broadcast Networks. Internet Mathematics 5(3), 211–225 (2008)
13. Harutyunyan, H.A., Liestman, A.L.: More broadcast graphs. Discrete Applied Math. 98, 81–102 (1999)
14. Harutyunyan, H.A., Liestman, A.L.: On the monotonicity of the broadcast function. Discrete Math. 262(1-3), 149–157 (2003)
15. Harutyunyan, H.A., Liestman, A.L.: Upper bounds on the broadcast function using minimum dominating sets. Discrete Math 312(20), 2992–2996 (2012)
16. Harutyunyan, H.A., Morosan, C.D.: On the minimum path problem in Kndel graphs. Networks 50(1), 86–91 (2007)
17. Harutyunyan, H.A., Morosan, C.D.: The spectra of Knödel graphs. Informatica (Slovenia) 30(3), 295–299 (2006)
18. Hedetniemi, S.M., Hedetniemi, T., Liestman, A.L.: A Survey of Gossiping and Broadcasting in Communication Networks. Networks 18, 319–349 (1988)

19. Johnson, D., Garey, M.: Computers and Intractability: A Guide to the Theory of NP-Completeness. Freeman, San Francisco (1979)
20. Khachatrian, L.H., Haroutunian, H.S.: Construction of new classes of minimal broadcast networks. In: Proceedings 3rd International Colloquium on Coding Theory, Dilijan, Armenia, pp. 69–77 (1990)
21. Khachatrian, L.H., Haroutunian, H.S.: Minimal broadcast trees. In: XIV All-Union School of Computer Networks, Minsk, USSR, pp. 36–40 (1989) (in Russian)
22. Knödel, W.: New gossips and telephones. Discrete Math. 13, 95 (1975)
23. Labahn, R.: A minimum broadcast graph on 63 vertices. Discrete Appl. Math. 53, 247–250 (1994)
24. Labahn, R.: Extremal broadcasting problems. Discrete Applied Mathematics 23(2), 139–155 (1989)
25. Middendorf, M.: Minimum broadcast time is NP-complete for 3-regular planar graphs and deadline 2. Inf. Proc. Lett. 46, 281–287 (1993)
26. Ravi, R.: Rapid rumor ramification: Approximating the minimum broadcast time. In: Proceedings of the IEEE Symposium on Foundations of Computer Science (FOCS 1994), pp. 202–213 (1994)

An Optimal Context for Information Retrieval

Rabeb Mbarek[1], Mohamed Tmar[1], and Hawete Hattab[2]

[1] Multimedia Information systems and Advanced Computing Laboratory,
High Institute of Computer Science and Multimedia, University of Sfax, Sfax, Tunisia
rabeb.hattab@gmail.com, mohamedtmar@yahoo.fr
http://www.miracl.rnu.tn
[2] Umm Al-Qura University
hattab.hawete@yahoo.fr

Abstract. In general, document representation and ranking are dependent on context. In this work, we introduce the notion of *optimal context*, i.e. a context which gives the best ranking. We develop an algorithm to compute this optimal context and we show that it has an effect of query reformulation. Our approach gives substantial improvements in retrieval performance over known models.

Keywords: optimal context, relevance feedback, vector space model.

1 Introduction

Information Retrieval (IR) deals with the retrieval of all and only the documents which contain information relevant to any information need expressed by any user's query. A system matching that definition exists in principle. In practice a system is unable to answer any query with all and only relevant information because of the unsolvable task of understanding both the relevant information enclosed in documents and the information need expressed through any query submitted by any user.

To refine the IR process, it is required to apply the Relevance Feedback (RF) technique. It has been shown that RF is an effective strategy in IR [10].

The RF has been used in several IR models: the vector space model [5,11], the probabilistic model [2,10], the language model [3], and the bayesian network retrieval model [1]. Most of the proposed approaches consist of adding new terms to the initial query and re-weighting original terms [12].

The Latent Semantic Indexing (LSI) [4] foundation is based on the assumption that there are many semantic relations between terms (synonymy, polysemy...), whereas capturing these relations by using semantic resources such as ontologies is complex, an alternative statistical solution could be taken into account by Singular Value Decomposition (SVD). This method results on a new vector space basis with a lower dimension than the original one (all indexing terms), and in which each component is a linear combination of the indexing terms.

Recently, [6] gives a new RF method using a vector space basis change. The authors develop an algorithm to compute a basis in which the relevant document are gathered and the irrelevant ones are kept away from the relevant documents.

Q. Gu, P. Hell, and B. Yang (Eds.): AAIM 2014, LNCS 8546, pp. 323–330, 2014.

Since information needs evolve with many variables like user, place, and time, relevance is context-dependent. Therefore, IR is also context-dependent which represents the high complexity of IR. The mechanism by which contextual information are handled are often rule- based or statistical. Rule-based approaches were common in user-modeling approaches where rules may express contextual variables such as search experience or knowledge. Currently, statistical approaches based on data mining are far more common and there is a growing theoretical strand to modeling context, e.g. [8].

According to [7], a context is modeled by a basis and its evolution is modeled by linear transformations[1] from one base to another. The basic idea is that, first, a vector is generated by a basis just as an information object is generated within a context. Second, every vector can be generated by different bases in our approach and belongs to infinite subspaces; this is consistent with the fact that every information object is generated within different contexts. Finally and as a corollary, the subspace spanned by a basis contains all those vectors that describe information objects in the same context: In this subspace, the vectors are related to each other by a linear combination. In particular, RF is an example of context change [7]. This paper completes the papers [8,7].

In this paper the Vector Space Model (VSM) is adopted as an infrastructure. It is introduced in [14] and [13]. A recent reconsideration of the geometry underlying IR, and indirectly of the VSM, was done in [9]. In VSM, documents and queries are modeled as elements of a vector space. This vector space is generated by a set of basis vectors that correspond to the index terms. Each document can be represented as a linear combination of these term vectors. One can find a nice and short introduction of VSM in [7, Section 3].

The determinant of a triangular matrix is the product of the diagonal elements, its inverse is a triangular matrix and the product of two triangular matrices is a triangular matrix too. These facts assure that the manipulation of triangular matrices is performed within the space of triangular matrices, thus providing advantages at computational level when modeling contexts and describing context changes [7].

In general, document representation and ranking are dependent on context. In fact, if T_1 and T_2 are two context matrices which, respectively, generate documents d_1 and d_2 with the same coefficients a, and if a query is generated by an arbitrary context matrix U with coefficients b, then $a^T.(T_1^T.U).b \neq a^T.(T_2^T.U).b$. From the set of contexts, there exists one that provides the best document ranking. This context is called *the optimal context*. In this paper we give an algorithm to compute this optimal context.

This paper is organized as follows. In Sections 2 we compute an optimal context. In Section 3 some experiments are reported to explore some of the potential of the proposed approach. Section 4 concludes.

---

[1] A base vector models a document or query descriptor. The semantics of a document or query descriptor depends on context. A base can be derived from a context. Therefore, a base of a vector space is the construct to model context. Also, change of context can be modeled by linear transformations from one base to another.

2   Optimal Context

In general, the vector x of the document x authored in its own context is generated by the base T. The latter is in its turn not necessarily equal to the base U that generates, say a query y, or another document. Therefore, x is represented by $x = T.a$ whereas y is represented by $y = U.b$ where a and b are the coefficients used to combine the base vectors of T and U, respectively. If relevance is estimated by the usual inner product, documents are ranked by $x^T.y = (T.a)^T.(U.b) = a^T.(T^T.U).b$.

2.1   Scenario

Let R be the set of relevant documents and S be the set of irrelevant documents. The optimal context C is a context which gives the best ranking that maximizes the inner product between query vector and relevant document vectors and minimizes the inner product between query vector and irrelevant document vectors. Then this optimal context maximizes the sum of inner product between query vector and relevant document vectors and minimizes the sum of inner product between query vector and irrelevant document vectors. Therefore, the optimal context maximises the quotient of the sum of inner product between query vector and relevant document vectors by the sum of inner product between query vector and irrelevant document vectors. So

$$C' = arg \max_{B \in \mathbf{T}_n(\mathbb{R})} \sum_{d \in R} d^T.B^T.q = arg \max_{B \in \mathbf{T}_n(\mathbb{R})} (\sum_{d \in R} d)^T.B^T.q \qquad (1)$$

and

$$C'' = arg \min_{B \in \mathbf{T}_n(\mathbb{R})} \sum_{d \in S} d^T.B^T.q = arg \min_{B \in \mathbf{T}_n(\mathbb{R})} (\sum_{d \in S} d)^T.B^T.q \qquad (2)$$

which implies that, the optimal context satisfies

$$C = arg \max_{B \in \mathbf{T}_n(\mathbb{R})} \frac{(\sum_{d \in R} d)^T.B^T.q}{(\sum_{d \in S} d)^T.B^T.q} \qquad (3)$$

where $\mathbf{T}_n(\mathbb{R})$ is the set of invertible upper triangular matrix of order n.

2.2   Compute of Optimal Context

In this section we attempt to solve equation 3 which leads to the optimal context we look for.

Necessary Condition. If B is a solution of equation 3, then for all $1 \leq i \leq j \leq n$ we have

$$\partial \frac{(\sum_{d \in R} d)^T . B^T . q}{(\sum_{d \in S} d)^T . B^T . q} = 0 \qquad (4)$$

$$\frac{}{\partial b_{ij}}$$

A solution of equation 4 is called a *critical point*. Equation 4 implies:

$$\frac{\partial((\sum_{d \in R} d)^T . B^T . q)}{\partial b_{ij}} . (\sum_{d \in S} d)^T . B^T . q - \frac{\partial((\sum_{d \in S} d)^T . B^T . q)}{\partial b_{ij}} . (\sum_{d \in R} d)^T . B^T . q = 0 \quad (5)$$

that is

$$(\sum_{d \in R} d)^T . (\frac{\partial B}{\partial b_{ij}})^T . q . (\sum_{d \in S} d)^T . B^T . q - \sum_{d \in S} d)^T . (\frac{\partial B}{\partial b_{ij}})^T . q . (\sum_{d \in R} d)^T . B^T . q = 0 \quad (6)$$

where

$$B = \begin{pmatrix} b_{11} & b_{12} & \cdots & b_{1n} \\ 0 & b_{22} & \cdots & b_{2n} \\ \vdots & \ddots & \ddots & \vdots \\ 0 & \cdots & 0 & b_{nn} \end{pmatrix}$$

The matrix $\dfrac{\partial B}{\partial b_{ij}}$ is:

$$\frac{\partial B}{\partial b_{ij}} = \begin{matrix} \\ \\ i \\ \\ \\ \\ \end{matrix} \begin{pmatrix} 0 \ldots 0\,0 \ldots 0 \\ \vdots \ddots \vdots \vdots \ddots \vdots \\ 0 \ldots 0\,1\,0 \ldots 0 \\ 0 \ldots 0\,0 \ldots 0 \\ 0 \ldots 0\,0 \ldots 0 \\ \vdots \ddots \vdots \vdots \ddots \vdots \\ 0 \ldots 0\,0 \ldots 0 \end{pmatrix} \begin{matrix} j \end{matrix}$$

Sufficient Condition (Second Derivative Test). The *Hessian matrix* is a square matrix of second-order partial derivatives of a function. It describes the local curvature of a function of many variables. If the Hessian matrix is negative definite[2] at a critical point x, then it attains a local maximum at x. In this case the function is locally concave.

---

[2] Its eigenvalues are negative.

To apply the second derivative test it suffices to compute the eigenvalues of the Hessian matrix of the function

$$f = \frac{(\sum\limits_{d \in R} d)^T . B^T . q}{(\sum\limits_{d \in S} d)^T . B^T . q} \tag{7}$$

The second-order partial derivatives are: for all $1 \leq i \leq j \leq n$ and $1 \leq k \leq l \leq n$

$$\frac{\partial^2 \frac{(\sum\limits_{d \in R} d)^T . B^T . q}{(\sum\limits_{d \in S} d)^T . B^T . q}}{\partial b_{kl} \partial b_{ij}} = \frac{\partial \frac{(\sum\limits_{d \in R} d)^T . (\frac{\partial B}{\partial b_{ij}})^T . q . (\sum\limits_{d \in S} d)^T . B^T . q - (\sum\limits_{d \in S} d)^T . (\frac{\partial B}{\partial b_{ij}})^T . q . (\sum\limits_{d \in R} d)^T . B^T . q}{((\sum\limits_{d \in S} d)^T . B^T . q)^2}}{\partial b_{kl}} \tag{8}$$

and so the second-order partial derivatives at a critical point are:

$$\partial^2 \frac{\frac{(\sum\limits_{d \in R} d)^T . B^T . q}{(\sum\limits_{d \in S} d)^T . B^T . q}}{\partial b_{ij}^2} = 0 \tag{9}$$

$$\frac{\partial^2 \frac{(\sum\limits_{d \in R} d)^T . B^T . q}{(\sum\limits_{d \in S} d)^T . B^T . q}}{\partial b_{kl} \partial b_{ij}} = \frac{2((\sum\limits_{d \in R} d)^T . (\frac{\partial B}{\partial b_{ij}})^T . q . (\sum\limits_{d \in S} d)^T . (\frac{\partial B}{\partial b_{kl}})^T . q - (\sum\limits_{d \in S} d)^T . (\frac{\partial B}{\partial b_{ij}})^T . q . (\sum\limits_{d \in R} d)^T . (\frac{\partial B}{\partial b_{kl}})^T . q)}{((\sum\limits_{d \in S} d)^T . B^T . q)^2} \tag{10}$$

The Hessian matrix is a symmetric square matrix of order n^2, and so it will have n^2 eigenvalues.

A solution C of equation 3 satisfies the following two conditions:

- The first-order partial derivatives of the function f are equal to 0.
- The eigenvalues of the Hessian matrix of the function f are negative.

2.3   Optimal Query

In the vector space model, the score of a document d vs. a query q is often expressed by the inner product: $RSV(d, q) = d^T . q$

If now the document and the query are generated by the optimal context matrix C, this score becomes:

$$RSV(d, q) = (C.d)^T . C.q = d^T . C^T . C.q$$

This score represents the score of the document d, in the original basis, vs. the query $q' = C^T . C.q$. Hence the optimal context has an effect of query reformulation: q' is the optimal query.

3   Experiments

In this section we give the different experiments and results obtained to evaluate our approach.

3.1   Environnement

The test collection TREC-8 is used in this study. The initial ranking of documents (Baseline Model) were weighted by the $BM25$ formula proposed in [15].
   The experiments consist to re-rank the results of the Baseline Model.
   For our approach the reformulated query is

$$Q_{new} = C^T.C.Q_{int} \tag{11}$$

Where C is the optimal context matrix.
   For the Rocchio model, the reformulated query is

$$Q_{new} = \alpha.Q_{int} + \beta.\frac{1}{|R|}\sum_{d \in R} d + \gamma.\frac{1}{|S|}\sum_{d \in S} d$$

- The initial query Q_{int} is made from the short topic description, and using it the top 1000 documents are retrieved from the collections (weighted $\alpha = 1$).
- R is the set of top ranking 30 documents, assumed to be relevant (weighted $\beta = 0.75$).
- S is the set of retrieved documents $501 - 1000$, assumed to be irrelevant (weighted $\gamma = -0.5$).

For our approach and the Rocchio model, the retrieved documents are ranked by the inner product done by:

$$RSV(Q_{new}, d) = Q_{new}^T.d \tag{12}$$

3.2   Results

Table 1 compares the performance of the Rocchio Model (RM) [11], the IRiX model [8], and our approach : Optimal Context Model (ICM). The performance measures are Precision at 5 ($P@5$), Precision at 10 ($P@10$) and Precision at 15 ($P@15$).

Table 1. Retrieval Performance Comparison

	RM	IRiX	ICM	ΔRM	$\Delta IRiX$
$P@5$	0.41	0.348	0.43	4.9%	23.6%
$P@10$	0.44	0.292	0.46	4.5%	57.5%
$P@15$	0.4	0.276	0.41	2.5%	48.55%

3.3 Analysis

We see that our approach gives improvements in $P@5$, $P@10$ and $P@15$ over the RM model and the IRiX model[3]. At 10 feedback documents, our model and the Rocchio model have the better precision. On a related experiment, we observe that our approach improves Rocchio model by 6% with respect to the Mean Average Precision (MAP). These preliminary results suggest that the optimal context can improve retrieval performance of RF.

4 Conclusion

In this paper we define and compute an optimal context. This context guarantees an optimal representation of documents, that is the relevant documents are gathered and the irrelevant ones are kept away from the relevant documents. On the other hand, the optimal context has an effect of query reformulation. Different Experiments and results show that our approach provides better results than the other ones.

Because the concavity of a function induces the existence and the uniqueness of the maximum, in practice, we use the second derivative test to find the maximum of the function f (equation 7).

This paper proposes an algorithm to separate relevant and non-relevant documents.

References

1. Campos, L.M.D., Huete, J.F., Fernndez-Luna, J.M., Spain, J.: Document Instantiation for Relevance Feedback in the Bayesian Network Retrieval Model (2001)
2. Croft, W.B., Harper, D.: Using Probabilistic Models of Information without Relevance Information. Journal of Documentation 35(4), 285–295 (1979)
3. Croft, W.B., Lavrendo, S.C.T.V.: Relevance Feedback and Personalization: a Language Modelling Perspective. In: CIKM 2006, pp. 49–54 (2006); Proceedings of the Joint DELOS-NSF Workshop on Personalization and Recommender Systems in Digital Libraries (2001)
4. Deerwester, S., Dumais, S.T., Furnas, G.W., Landauer, T., Harshman, R.: Indexing by Latent Semantic Analysis. Journal of the ASIS 41(6), 391–407 (1990)
5. Ide, E.: New Experiments in Relevance Feedback. In: The SMART Retrieval System-Experiments in Automatic Document Processing, pp. 337–354 (1971)
6. Mbarek, R., Tmar, M.: Relevance Feedback Method Based on Vector Space Basis Change. In: Calderón-Benavides, L., González-Caro, C., Chávez, E., Ziviani, N. (eds.) SPIRE 2012. LNCS, vol. 7608, pp. 342–347. Springer, Heidelberg (2012)
7. Melucci, M.: Context modeling and discovery using vector space bases. In: Proceedings of the ACM Conference on Information and Knowledge Management (CIKM), Bremen, Germany, pp. 808–815. ACM Press (2005)

---

[3] Precision using several more feedback documents and Keywords ($n = 2$ and $k = 10$): these parameters give the best results for the IRiX model.

8. Melucci, M.: A Basis for Information Retrieval in Context. ACM Trans. Inf. Syst. 26(3), 1–41 (2008)

9. van Rijsbergen, C.J.: The Geometry of Information Retrieval. Cambridge University Press, UK (2004)

10. Robertson, S., Sparck-Jones, K.: Relevance weighting of search terms. Journal of the American Society of Information Science, 129–146 (1976)

11. Rocchio, J.: Relevance feedback in information retrieval. The SMART Retrieval System-experiments in Automatic Document Processing, 313–323 (1971)

12. Ruthven, I., Lalmas, M., Rijsbergen, K.: Ranking Expansion Terms with Partial and Ostensive Evidence. In: Fourth International Conference on Conceptions of Library and Information Science: Emerging Frameworks and Methods, Seattle, WA, USA, pp. 199–219 (2002)

13. Salton, G., Wong, A., Yang, C.S.: A Vector Space Model for Automatic Indexing. Communications of the ACM 18(11), 613–620 (1975)

14. Salton, G.: Automatic Text Processing: The Transformation, Analysis and Retrieval of Information by Computer. Addison-Wesley (1989)

15. Robertson, S.E., Walker, S., Hancock-Beaulieu, M., Gull, A., Lau, M.: Okapi at TREC. TREC, 21–30 (1992)

An Auction-Bargaining Model for Initial Emission Permits

Lili Ding*, Xiaoling Wang, and Wanglin Kang

College of Economics and Management
Shandong University of Science and Technology
Qingdao, China

Abstract. This paper studies the initial emission permits auction problem from the perspective of government' activities. In the traditional auction models, the basic assumption is that the government, i.e., the auctioneer, only pursues the maximum economic revenue. In this paper, we consider a hybrid auction-bargaining model, which gives new insights on how the government's economic and social goals effect the equilibrium strategies. For this model, we find a symmetric bidding strategy equilibrium for the firms in a sealed bid auction form, which is closely related to the classical results in the auction. Our most important finding is that, compared with the classical auction mechanism, the final trading price is based on not only firm's bidding strategy, but also the application quality of emission permits in the energy consumption market. The results also show that this auction-bargaining mechanism can alleviate distortion by excessive allowance in initial emission permits auction market and promote the social goals in both auction market and consumption market.

Keywords: emission permits, auction-bargaining, equilibrium.

1 Introduction

There is increasingly broad recognition that greenhouse gas emissions are contributing to changes to earth's climate. Emissions trading schemes (ETS) that CO_2 reductions are carried costly, are an important part of the policy response to this problem. The high potential costs of controlling pollutants by emission trading have led to growing interest in economic instruments. One critical issue in designing a tradable emission permit system is how the initial emission permits are distributed.

There are two different approaches existing in the initial emission permits schemes: the grandfathered approach and the auction approach, where the two mainly differ in the costs levied on the producers. Since Montgomery (1972) [1] showed that as long as permits markets were competitive, the initial emission permits allocation schemes might be irrelevant for emission abatement. There

---

* This work is supported by the National Science Foundation of China (No.71373247 and No.71371111).

Q. Gu, P. Hell, and B. Yang (Eds.): AAIM 2014, LNCS 8546, pp. 331–340, 2014.

has been an ongoing debate about these two means. Most studies recognize an auction is preferred to grandfathering[2]. The reason is that auction allows reduced tax distortions, provides more flexibility in distribution of costs, provides greater incentives for innovation, and reduces the need for politically contentious arguments over the allocation of rents. Many important considerations are relevant to the implemented auction design. Firstly, there are a variety of auction formats, i.e., a sealed bid, a descending bid, an ascending bid or an ascending clock auction. Secondly, the pricing rule can be uniform, discriminatory or based on the Ausubel-Vickrey principle[3]. For example, the US Environmental Protection Agency (EPA) auctions for SO_2 permits are in the sealed bid discriminatory price format[4]. Cramton and Kerr [5] explained that when no bidder had significant market power, uniform pricing was nearly as efficient as Vickrey pricing and that among sealed bid auctions, a uniform price auction was probably the best. Betz [6] also proposed an ascending clock auction based on the policy frame work and theoretical as well as experimental findings in the literature, and this auction was later applied by the Australian government. There is broad consensus among economists specializing in auction design that a pay-as-bid auction is not best suited for emissions permits. Even in an idealized perfect competition setting in which all bidders lack market power, the pay-as-bid auction need not lead to an efficient or even an approximately efficient distribution of emission permits[7]. Moreover, many researches study which factors can be used to guide the successful auction design of emission permits[8]. These factors include the ability of the auction to elicit bids that reflect actual valuations by bidders[9], and restricting bidder opportunities for acting strategically in a conclusion way[10].

The above literature is based on the same assumption that the government aims at maximizing the profit to achieve the energy abatement goal. In some situations, a simple emission permits auction's efficiency can be very close to that of the optimal mechanism. Thus, it may be optimal for the government to employ a simple auction to reap most of the efficiency with low implementation costs[11]. However, in many countries there are such situations that grandfathered permits together with auction are more prevalent attributing to its political acceptability. These hybrid forms are perceived to potentially distort inter-firm competitiveness relations in initial emission permits auction market and lead to a lower equilibrium price[12]. There is a significant gap between the governments expected pay-off with a simple emission permits auction and that under the optimal mechanisms. Under such circumstances, the government may search for an intermediate solution to balance between economic goals and environmental goals. This paper introduces a novel hybrid auction mechanism, i.e., auction-bargaining mechanism to bridge the gap. Our proposed approach includes a sealed bid auction, followed by bargaining on payments to ensure that the emission permits transaction will finish.

This paper is also related to the work of [7] and [12], where they consider that the government can reach an environmental goal in an economically efficient way. Their discussion is based on the assumption that the total auctioning quantities of emission permits are exogenous to the auction models. In this paper, this assumption is relaxed and the self implement of energy goal in auction market is not available. Our attention will be focussed on not only the auction revenue in emission permits auction market, but also a further complication, one that will figure extensively in the application of emission permits in the energy consumption market(see Fig.1). This complication is that the winning firms should produce goods by emission permits at an acceptable level of quality, which is also called the quality threshold. This is particularly the case when a government agency is auctioning emission permits on behalf of the public. In the second phase, the government bargains with the chosen firm over the final price of the emission permits. The outcome of bargaining results in the final trading price that is based on the government's preference to economic revenue or social reward. It concludes that one of quality threshold, i.e., mandatory green standard in energy consumption market, impacts on the final trading price. Our chief finding is that the hybrid auction-bargaining generates alleviated impact on price violations in the emission permits auction market and promote the environmental goals.

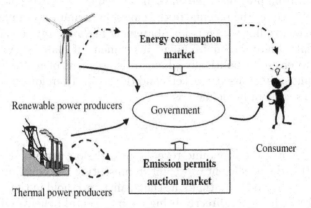

Fig. 1. The two markets included in the government's policy discussion

2   The Model

We next discuss our model and then proceed with the analysis. For the initial emission permits auction problem, we establish a two-stage dynamic game model, in which there are two main participants, i.e., government and firms. The time line of the game is as follows. In the auction stage, the government announces the mandatory green standard ε, and determines the temporary winner(s) based on the auction rules in the first stage. After receiving the invitation, the firms submit their bids according to their own abilities.

We will make a number of simplifications in order to compare various factors with equilibrium strategies in the hybrid auction-bargaining mechanism. The major ones, which we will hope to relax in later work, are that the maximum number of bidders is two and that these begin from the same position: thus we restrict ourselves to a symmetric case. It is supposed that the risk neutral firm

i (i=1,2) wants to buy some units k of emission permits for its production as in [13]. Since emission permits are homogeneous goods, the auctioning number of permits always does not influence the bidding strategies. For simplicity, we assume that $k = 1$. In order to allow for analytical solutions, the firm's quadratic abatement costs are assumed to be $\frac{c_i}{2}\varepsilon^2$, in which c_i is the cost coefficient of each firm i(see details in [14]). Here, c_i is the private information only known by each firm and is uniformly distributed over $[0, \overline{c}]$. Thus, firm i's net profit is given by

$$u_i = g_i - \frac{c_i}{2}\varepsilon^2 - b_i = v_i - b_i \tag{1}$$

where g_i is the profit received by firm i from production based on purchased emission permits and b_i ($b_i \in [0, \overline{b}]$) is firm i's bid. To keep the model simple, it also assumes that g_i is uniformly distributed over $[0, \overline{g}]$. In this paper, we take the same assumption as [9] that each firm is truthfully to present his bid. Thus firm i's valuation v_i of auctioning emission permits satisfies $v_i = g_i - \frac{c_i}{2}\varepsilon^2$.

Different from previous literature, in this paper, the government not only considers the aspect of the revenue in the emission permits auction market, but also focuses on controlling the emission abatement in the energy consumption market. For example, there are environment instrument of mandatory green standard, which is to ensure a politically planned deployment of renewable energy technologies under liberalised market conditions[15]. Therefore, the benefit of the government U_G can be defined as

$$U_G = \alpha R + (1 - \alpha)\varepsilon. \tag{2}$$

where R denotes the economic revenue in emission permits auction market, α ($\alpha \in [0, 1]$) is the coefficient of weight, representing the government's preference to these two objects. Note that ε as the quality threshold in the energy consumption market, which can directly bring environmental benefit. Of course, quality threshold has many dimensions. For the environmental governance problem, the government knows exactly what it wants, and can obtain perfect information on the quality achieved, then including quality considerations within an auction is relatively straightforward. Thus, we denote the quality threshold by the mandatory green standard denoted by ε, one kind of climate policies, which can reach lower emissions and develop future renewable industries. Thus, the specific function of ε is not necessary for our two-stage model.

3   Auction Stage

The two-stage game can be solved by backward induction. We look for a subgame perfect Nash equilibrium (SPNE), defined by a set of strategies for the firms and the government. Firstly, we analyze the bidding strategies b_i of firm i ($i = 1, 2, \cdots, n$) in the auction under the condition of announced mandatory green standard ε by the government. According to the assumptions about c_i and g_i, we can compute the distribution function $F_i(v_i)$ of firm i's valuation as follows.

$$F_i(v_i) = \begin{cases} \frac{v_i}{\bar{g}} + \frac{\varepsilon^2 \bar{c}}{4\bar{g}}, \ 0 \le v_i \le \bar{g} - \frac{\varepsilon^2 \bar{c}}{2} \\ 1 - \frac{(v_i - \bar{g})^2}{\varepsilon^2 \bar{g}\bar{c}}, \ \bar{g} - \frac{\varepsilon^2 \bar{c}}{2} < v_i \le \bar{g} \\ 1, \ \bar{g} < v_i \end{cases} \tag{3}$$

Notice that each firms' valuation about the auctioning emission permits is relative to mandatory green standard ε. With the increase of ε, the valuation of emission permits of each firm decreases.

Let the bidding strategy of firm i be β_i, where $\beta_i(\cdot)$ is the function of firm i's valuation v_i. If the cost information is the firm's private information and all the firms are all symmetric, the firms have the symmetric behaviors. Thus this assumption is available and do not influence our results. We achieve that $\beta_i(0) = 0$ and $\beta_i(\bar{g}) = \bar{b}$. Thus, the firm i's valuation function is β_i^{-1}, i.e., the inverse functions of β_i. Set $\beta_i^{-1} = \phi_i(b)$. When firm i chooses b as its the equilibrium bidding strategy, $\phi_i(b)$ is the valuation of the auctioning emission permits.

Lemma 1. *When firm i chooses b as its the equilibrium bidding strategy, the probability of its success is*

$$Prob(firm\ i\ wins) = \prod F_{-i}(\phi_{-i}(b)) \tag{4}$$

Proof. These are n firms strictly compete for the auction stage. If firm i wins the game, the other firms denoted by $-i$ lose it. Thus, when the bidding strategy of firm i is b , the probability of its success can be denoted by

$$\begin{aligned} Prob(firm\ i\ wins) &= Prob(\beta_1(v_1) < b) \cdot Prob(\beta_2(v_2) < b) \cdots Prob(\beta_n(v_n) < b) \\ &= F_1(\phi_1(b)) \cdots F_{i-1}(\phi_{i-1}(b)) \cdot F_{i+1}(\phi_{i+1}(b)) \cdots F_n(\phi_n(b)) \\ &= \prod F_{-i}(\phi_{-i}(b)) \end{aligned} \tag{5}$$

Theorem 1. *The equilibrium bidding strategies b_1^* and b_2^* for firm 1 and firm 2 respectively are solutions of the following inverse bidding functions.*

$$\begin{cases} b_1^* = \{b | b + \frac{H_2(b)}{h_2(b)} - \phi_1(b) = 0\} \\ b_2^* = \{b | b + \frac{H_1(b)}{h_1(b)} - \phi_2(b) = 0\} \end{cases} \tag{6}$$

Proof. The first firm's expect profit π_1 is denoted by

$$\begin{aligned} \pi_1(v_1, b) &= Prob(firm\ 1\ wins)[v_1 - b] + Prob(firm\ 1\ loses) * 0 \\ &= Prob(firm\ 1\ wins)[v_1 - b] \\ &= F_2(\phi_2(b))[v_1 - b] \end{aligned} \tag{7}$$

In this paper we assume that the collusion behavior which will damage the benefit of the government is not available. Thus, the optimal bidding strategy b^* must satisfy the condition as follows.

$$b^* \in argmax\{F_2(\phi_2(b))[v_1 - b]\} \tag{8}$$

Setting $\partial\pi_1(v_1, b)/\partial b = 0$, we find that

$$\frac{\partial\pi_1(v_1, b)}{\partial b} = F_2'(\phi_2(b))[v_1 - b] - F_2(\phi_2(b)) = 0 \tag{9}$$

For simplicity, it assumes that $H_i(b) \equiv F_i(\phi_i(b))$ and $h_i(b) \equiv H_i'(b)$. Solving the Eq.(9), we can attain the first firm's valuation function, i.e., inverse bidding function, as follows.

$$\phi_1(b) = b + \frac{H_2(b)}{h_2(b)} \tag{10}$$

As the same discuss as the first firm, the second firm's inverse bidding function is that

$$\phi_2(b) = b + \frac{H_1(b)}{h_1(b)} \tag{11}$$

In Fig.2, notice that the firm i's valuation will decrease with ε or c_i. If the firm estimates the emission permits will bring it more benefit, it would like to pay more for the emission. Otherwise, if the firm costs a lot to reach the announced quality threshold ε, it would like to pay less for the emission permits. Furthermore, we analyze the relation between g_i and valuation. We have that $\frac{\partial F}{\partial g_i} < 0$. Therefore, the bidding strategy of the firms for the emission permits will increase with the firm's except benefit and decrease with the mandatory green standard.

4   Bargaining Stage

In this stage, we discuss the final trading price based on the government's equilibrium strategy, i.e., the equilibrium mandatory green standard ε^*. It assumes that the equilibrium trading price of the emission permits is P. Suppose that the distribution function of auctioning price p is $G(p)$ and its density function is $g(p)$. In the emission permits auction, if there are only two firms, then $P = max(b_1, b_2)$, where b_1 and b_2 are independent from each other. Thus, $G(p)$ can be expressed by $G(p) = Prob(P \leq p) = Prob\{max(b_1, b_2) \leq p\} = F_1(\phi_1(p))F_2(\phi_2(p))$.

Lemma 2. In the bargaining stage, the final trading price p^* is

$$p^* = argmax\{\alpha(\bar{b} - \int_0^{\bar{b}} G(p)dp) + (1 - \alpha)(1 - \frac{\bar{c}^2\varepsilon^4}{16\bar{g}^2})\varepsilon\} \tag{12}$$

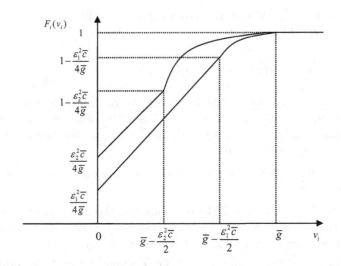

Fig. 2. Firm's valuation distribution as a function of $\bar{c}$ and ε when g_i is set as $\bar{g}$

Proof. When one of two firms wins the auction game, the government can achieve the expected revenue ER from the auctioning activity in the emission permits auction market. ER is given by

$$ER = \int_0^{\bar{b}} pg(p)dp = pG(p)|_0^{\bar{b}} - \int_0^{\bar{b}} G(p)dp = \bar{b} - \int_0^{\bar{b}} G(p)dp \qquad (13)$$

Except for the revenue from the auctioning activity, the government also pay attention to the reward from the mandatory emission standard, i.e., the reward of climate policies.

According to Eq.(3), if there is no firms participating into game, then $Prob(v_i \leq 0) = \frac{\varepsilon^2 \bar{c}}{16\bar{g}^2}$. Moreover, if there is at least one participant in the auction, then the probability of this case is $1 - Prob(v_1 \leq 0)Prob(v_2 \leq 0) = 1 - \frac{\bar{c}^2 \varepsilon^4}{16\bar{g}^2}$. Therefore, the mandatory emission standard ε can bring the expected reward denoted by MR for the government as follows.

$$MR = (1 - \frac{\bar{c}^2 \varepsilon^4}{16\bar{g}^2})\varepsilon \qquad (14)$$

Furthermore, the expected benefit of the government can be achieved as follows.

$$U_G = \alpha(\bar{b} - \int_0^{\bar{b}} G(p)dp) + (1 - \alpha)(1 - \frac{\bar{c}^2 \varepsilon^4}{16\bar{g}^2})\varepsilon \qquad (15)$$

Theorem 2. *When* $\alpha = 0$, *the government's equilibrium strategy about green standard is* $\varepsilon^* = \sqrt[4]{\frac{16\bar{g}^2}{5\bar{c}^2}}$

Proof. In order to maximize the government's benefit function, we take the derivative of Eq.(15) as follows.

$$\frac{\partial U_G}{\partial \varepsilon} = \alpha(\bar{b}'(\varepsilon) - G(\bar{b}(\varepsilon))\bar{b}'(\varepsilon)) - (1-\alpha)(1 - \frac{5\bar{c}^2\varepsilon^4}{16\bar{g}^2}) \qquad (16)$$

Set $\alpha = 0$. It means that the key concern of the government is to set mandatary green standard rate for social goals, and has little idea of the revenue of the auction. Thus, the government's equilibrium strategy about green standard is the optimal solution of Eq.(16).

Theorem 3. *When* $\alpha = 1$, *the government's equilibrium strategy about green standard is* $\varepsilon^* = 0$.

Proof. In the situation of $\alpha = 1$, the key concern of the government is the auction revenue of the emission permits auction. Since the higher green standard ε, the lower bidding price the firm will submit. Based on Eq.(16), for $b'(\varepsilon) \le 0$, the derivative $\frac{\partial U_G}{\partial \varepsilon} = \bar{b}'(\varepsilon)[1 - G(\bar{b}'(\varepsilon))] \le 0$. Thus, when $\varepsilon^* = 0$, the government can get the maximum of expect revenue. The constraint of green standard becomes the incredible threat, so the valuation price of the firms will fully depend on the expect revenue of the emission permit auction, i.e., $v_i = g_i$. The same result also can be found in [7], when only considering the government's economic goals.

Theorem 4. *When* $0 < \alpha < 1$, *the government's equilibrium strategy about green standard is* ε^*, *which is the solution of*

$$\varepsilon^* = \{\varepsilon | \bar{b}(\varepsilon) - \int_0^{\bar{b}} G(p)dp + \varepsilon - \frac{\bar{c}^2\varepsilon^5}{16\bar{g}^2} = 0\} \qquad (17)$$

Proof. Based on Eq.(15), we can get the optimal mandatory green standard ε^*. This means that the increasing of the government's total benefit is generated by economic revenue and social reward. The optimal point is that the increase of economic revenue is equal to the reduction of social reward. After designing the weight of the economic revenue and that of social reward about mandatory green standard, the government achieves its equilibrium strategy as follows.

$$\frac{\partial U_G^*}{\partial \alpha} = (\bar{b}(\varepsilon) - \int_0^{\bar{b}} G(p)dp) + (\varepsilon - \frac{\bar{c}^2\varepsilon^5}{16\bar{g}^2}) = 0 \qquad (18)$$

The derivative $\partial U_G^*/\partial \alpha$ determines the optimal green standard ε^*. We need to choose the optimal α^* in Eq.(18) and then achieve ε^*. Notice that in the process of α increase from 0 to 1, the goal of choosing the weight is to achieve the target that the government both focus on maximizing the revenue and setting the green standard.

Corollary 1. If there is no bargaining stage, the final trading price of emission permits is only based on the private information of c_i. A lower equilibrium price always exists (see detail in [11]).

Corollary 2. If there are two stages, the optimal bidding strategy b_i^* in the auction stage and the optimal mandatory green standard ε^* in the bargaining stage constitute a subgame perfect Nash equilibrium.

5 Conclusion

For initial emission permits auction problem, this paper introduces a hybrid form, auction-bargaining mechanism, into the traditional emission permits auction market. We highlight the strategic interaction between the government's economic actions and social actions in the auction-bargaining model, which can explain how the government's economic and social goals effect the equilibrium strategies. We also find that mandatory green standard will be the equilibrium strategy in some circumstances. It concludes that this hybrid mechanism can alleviate distortion by excessive allowance in initial emission permits auction market. This result can be applied to a more general question regarding the choice between economic goals and social goals. We make a number of simplifications, e.g., the number of bidders and the distribution form, and future studies can relax these assumptions.

References

1. Montgomery, W.D.: Markets in Licenses and Efficient Pollution Control Programs. Journal of Economic Theory 5(3), 395–418 (1972)
2. Rao, C.J., Zhao, Y., Li, C.F.: Asymmetric Nash Equilibrium in Emission Rights Auctions. Technological Forecasting & Social Change 79, 429–435 (2012)
3. Benz, E., Loschel, A., Sturm, B.: Auctioning of CO_2 Emission Allowances in Phase 3 of the EU Emissions Trading Scheme. Climate Policy 10(6), 705–718 (2010)
4. Porter, D., Rassenti, S., Shobe, W.: The Design, Testing and Implementation of Virginias NOx Allowance Auction. Journal of Economic Behavior & Organization 69(2), 190–200 (2009)
5. Genc, T.S.: Discriminatory Versus Uniform-price Electricity Auctions with Supply Function Equilibrium. J. Optim. Theory Appl. 140, 9–31 (2009)
6. Cramton, P., Kerr, S.: Tradeable Carbon Permit Auctions: How and Why to Auction not Grandfather. Energy Policy 30(4), 333–345 (2002)
7. Betz, R., Seifert, S., Cramton, P., Kerr, S.: Auctioning Greenhous Gas Emissions Permits in Australia. Aust. J. Agric. Resour. Econ. 54(2), 219–238 (2010)
8. Lai, Y.B.: Auctions or Grandfathering: the Political Economy of Tradable Emission Permits. Public Choice 136(1-2), 181–200 (2008)
9. Ausubel, L.M.: An Efficient Ascending-bid Auction for Multiple Objects. The American Economic Review 94(5), 1452–1475 (2004)
10. Mougeot, M., Naegelen, F., Pelloux, B., et al.: Breaking Collusion in Auctions through Speculation: An Experiment on CO2 Emission Permit Markets. Journal of Public Economic Theory 13(5), 829–856 (2011)
11. Yu, Y.: An Optimal Ad Valorem Tax/Subsidy with an Output-Based Refunded Emission Payment for Permits Auction in an Oligopoly Market. Environmental and Resource Economics 52(2), 235–248 (2012)

12. Sunnevag, K.J.: Auction Design for the Allocation of Emission Permits in the Presence of Market Power. Environmental and Resource Economics 26(3), 385–400 (2003)
13. Goeree, J.K., Palmer, K., Holt, C.A.: An Experimental Study of Auctions versus Grandfathering to Assign Pollution Permits. Journal of the European Economic Association 8(2-3), 514–525 (2010)
14. Jensen, S.G., Skytte, K.: Simultaneous Attainment of Energy Goals by Means of Green Certificates and Emission Permits. Energy Policy 31(1), 63–71 (2003)
15. Harstad, B., Eskeland, G.S.: Trading for the Future: Signaling in Permit Markets. Energy Policy 94(9), 749–760 (2010)

Author Index